D0072212

Cellulose Derivatives

ACS SYMPOSIUM SERIES **688**

Cellulose Derivatives

Modification, Characterization, and Nanostructures

Thomas J. Heinze, EDITOR
Friedrich-Schiller-Universität Jena

Wolfgang G. Glasser, EDITOR
Virginia Polytechnic Institute and State University

Developed from a symposium sponsored by the Division
of Cellulose, Paper, and Textiles at the 212th National Meeting
of the American Chemical Society,
Orlando, Florida,
August 25–29, 1996

American Chemical Society, Washington, DC

Library of Congress Cataloging-in-Publication Data

Cellulose derivatives : modification, characterization, and nanostructures / Thomas Heinze, editor, Wolfgang Glasser, editor.

p.   cm.—(ACS symposium series, ISSN 0097–6156; 688)

"Developed from a symposium sponsored by the Division of Cellulose, Paper, and Textiles at the 212th National Meeting of the American Chemical Society, Orlando, Florida, August 25–29, 1996."

Includes bibliographical references and indexes.

ISBN 0–8412–3548–1

1. Cellulose—Congresses. 2. Cellulose—Derivatives—Congresses.

I. Heinze, Thomas, 1958–   . Glasser, Wolfgang G., 1941–   . III. American Chemical Society. Cellulose, Paper, and Textiles Division. IV. American Chemical Society. Meeting (212th : 1996 : Orlando, Fla.) V. Series.

QD323.C39   1998
661'.802—dc21
                                                                98–14355
                                                                CIP

PRINTED IN THE UNITED STATES OF AMERICA

Foreword

THE ACS SYMPOSIUM SERIES was first published in 1974 to provide a mechanism for publishing symposia quickly in book form. The purpose of the series is to publish timely, comprehensive books developed from ACS sponsored symposia based on current scientific research. Occasionally, books are developed from symposia sponsored by other organizations when the topic is of keen interest to the chemistry audience.

Before agreeing to publish a book, the proposed table of contents is reviewed for appropriate and comprehensive coverage and for interest to the audience. Some papers may be excluded in order to better focus the book; others may be added to provide comprehensiveness. When appropriate, overview or introductory chapters are added. Drafts of chapters are peer-reviewed prior to final acceptance or rejection, and manuscripts are prepared in camera-ready format.

As a rule, only original research papers and original review papers are included in the volumes. Verbatim reproductions of previously published papers are not accepted.

ACS BOOKS DEPARTMENT

Contents

CHEMICAL AND MOLECULAR STRUCTURE

SUPRAMOLECULAR STRUCTURE

INDEXES

Preface

IMPRESSIVE ADVANCES in the ability to modify, regenerate, and reshape cellulose and polysaccharide derivatives with unique chemical, physical, and physiological properties have raised the interest in this most important biological macromolecule over the past decade. Cellulose derivatives have received much attention from authors with diverse research, clinical, and business interests; and this interest has created an opportunity for a broad display of topical discussions with varying degrees of technical depth. The launching of a new, nationwide research program in Germany focusing on the design of molecular and supramolecular structures based on cellulose and cellulose derivatives, by the Deutsche Forschungsgemeinschaft (DFG, the equivalent of the National Science Foundation in the United States), has attracted scientists from a variety of disciplines.

This book was developed from a symposium titled "Recent Advances in Cellulose Modification", held at the 212th National Meeting of the American Chemical Society, in Orlando, Florida, August 25–29, 1996. The symposium provided a forum for organizing an integrated discussion of the current state of the art. It was organized with the intent of bringing together scientists from academia and industry in the expectation that the new insights gained would be useful for the development of novel, value-added materials from this polymer which is basic to all plants.

Impulses for the new focus on cellulose derivatives originated from several sources:

- availability of new cellulose sources, especially bacterial cellulose
- new cellulose solvents and their corresponding regenerated fibers
- new regioselective modification methods
- new insights into the anisotropic solution states of (especially lyotropic liquid-crystalline) cellulose derivatives
- a new understanding of the enzyme systems involved in cellulose degradation
- new, chirally active cellulose-based separation materials
- new, highly ordered thin film architectures of cellulose derivatives prepared by the Langmuir–Blodgett technique

This book highlights advances in (1) both homogeneous and heterogeneous phase modification of cellulose to create unusual derivatives, often with regioselective substitution patterns, (2) analysis of selectively and specifically modified derivatives, (3) issues such as the self-assembly of cellulosic macromolecules in dilute and concentrated solutions as well as in solids, and (4) supramolecular architectures potentially useful in novel sensors, immunoassays, membranes, and biocompatibilized surfaces.

The first section of this book, "Modification Chemistry", shows that recent research on chemical conversion of cellulose is mainly directed toward the synthesis of functionalized derivatives with well-defined primary structures, both within the anhydroglucose repeat unit and along the polymer chain. Moreover, nonconventional functional groups with special properties are playing a dominant role in advanced cellulosics. In order to design new polymers based on cellulose, special synthesis concepts are introduced to create reactive microstructures by, for example, induced phase separation. Nucleophilic displacement reactions and selective cellulose oxidations are discussed as well. Examples are given for the advanced manufacture of conventional esters and ethers.

A fast and reliable supply of comprehensive analytical data is an indispensable prerequisite for considering and pursuing new routes of synthesis and for controlling chemical processes in cellulose functionalization. In the section "Chemical and Molecular Structure", recent results in adapted instrumental techniques are presented. In particular, NMR, Fourier-transform infrared, and chromatographic techniques are providing new insights into molecular structures and intermolecular interactions. The determination of molecular weights and molecular-weight distributions, and their changes during chemical, physical, and enzymatic modifications, is discussed as well.

Based on the synthesis and structure characterization of functionalized cellulose derivatives, the processing of materials with defined supramolecular architectures is one of the most important areas of cellulose research. The present state of the art in the design of supramolecular structures in the liquid and solid state, and the search for promising applications of these structures, is summarized in the section "Supramolecular Structures". The successful engineering of nanostructures and defined colloids yields insight into self-organization principles. The study of mesophase formation of cellulose derivatives with a defined primary structure provides a better understanding of liquid-crystalline systems. Tailored cellulosic compounds are employed to design ordered supramolecular structures that can find application as sensors, light-wave conductors, and selective membranes.

We express our appreciation to the American Chemical Society's Cellulose, Paper, and Textile Division for sponsoring the symposium. The editors are indebted to their respective institutions, the Friedrich Schiller University of Jena, Germany and the Virginia Polytechnic Institute and State University, Blacksburg, VA, for financial and logistic support of this endeavor. We also thank Mark Fitzgerald and David Orloff of the American Chemical Society Books Department, and Mary Holliman of Pocahontas Press, Blacksburg, VA for their conscientious efforts to ensure timely review and completion of the book.

THOMAS HEINZE
Institut für Organische Chemie und
  Makromolekulare Chemie der
Friedrich Schiller-Universität
Humboldtstrasse 10
D–07743 Jena, Germany

WOLFGANG G. GLASSER
Department of Wood Science and Forest Products
Virginia Polytechnic Institute and State University
Blacksburg, VA 24061–0324

MODIFICATION CHEMISTRY

Chapter 1

The Role of Novel Solvents and Solution Complexes for the Preparation of Highly Engineered Cellulose Derivatives

Thomas Heinze[1] and Wolfgang G. Glasser[2]

[1]Institute of Organic Chemistry and Macromolecular Chemistry, University of Jena, Humboldtstrasse 10, D-07743 Jena, Germany
[2]Biobased Materials/Recycling Center, and Department of Wood Science and Forest Products, Virginia Polytechnic Institute and State University, Blacksburg, VA 24061

Novel solvent systems for cellulose are reviewed in terms of their potential for supporting chemical modification reactions that lead to highly engineered derivatives. Systems capable of sustaining homogeneous phase conditions include (1) dissolution with non-derivatizing solvents; (2) dissolution with partial derivatization with reactive solvents; (3) dissolution by derivatization with protective substituents; and (4) dissolution by derivatization with solubilizing substituents that provide access to subsequent replacement reactions (ie., leaving groups). Numerous examples and applications are cited for each type of homogeneous phase reaction system.

Chemical modification continues to provide a dominant route towards cellulose utilization in polymeric materials. The discovery of novel solvents and solution complexes for cellulose in the past three decades has created opportunities for the application of significantly more diverse synthesis pathways and derivative types. Although functionalization opportunities with cellulose are limited to OH-groups, significant distinctions exist in different solvent systems between the reactivities of primary and secondary OH groups, and between two different secondary OH-groups located on the anhydroglucose repeat unit (AGU). The control of substituent distribution within the AGU and along the polymer chain presents a relatively recent challenge to the polysaccharide chemist. Homogeneous phase chemistry has opened doors for the derivatization with more than one functional group as well as for both partial and complete functionalization of OH-groups with such functionalities as chromophores and fluorophores, redox-active groups as well as substituents with special magnetic, optical, and biological activity that are indispensable prerequisites for the design of highly-engineered materials as well as "smart materials." Access to novel reaction methodologies requires the support from analytical tools qualified to describe quantitatively chemical and molecular structures in a cost effective and timely manner. This support has become available through product-adapted ^{1}H-, ^{13}C NMR- and two-dimensional NMR-spectroscopy, either applied to the parent polymer (1-5) or to the partially or completely depolymerized molecule (6), and through FAB and MALDI-TOF-mass spectroscopy (7,8), as well as through HPLC following hydrolytic chain degradation (9,10).

Highly-engineered cellulose derivatives, with carefully selected sites of functionalization within the AGU and with controlled pattern of derivatization, offer the potential for materials with specific solubility, enzyme degradability (13-15) (carboxymethyl and methyl cellulose), and blend compatibility (cellulose/poly(ethylene oxide)) (16). Physiological activities have recently been found to be significantly related to the distribution of sulfuric half ester functions within the AGU, which seem to influence the interaction with human blood (17). In addition, the potential to self-organized in liquid crystalline and monomolecular ultra thin film structures also appears to be influenced by the specific design of cellulose derivatives (18-21).

Highly engineered cellulose derivatives by homogeneous phase modification can, in general, be accessed via the following four strategies:

- Dissolution with non-derivatizing solvents;
- Dissolution with partial functionalization, i.e., by the application of derivatizing solvents;
- Dissolution by derivatization, with the introduction of specific solubilizing substituents that protect specific OH functional groups during subsequent reactions; and
- Dissolution by chemical modification with solubilizing substituents that provide access to subsequent replacement reactions (i.e., leaving groups).

The intent of this review is to highlight examples of recent research which illustrate the variety of the different synthesis chemistries, and to focus on the opportunities and limitations they present. Comprehensive reviews of the field have recently appeared elsewhere (22-25).

1. Dissolution With Non-derivatizing Solvents

Cellulose dissolves in a wide variety of non-aqueous solvents and solvent complexes (Table I). These may involve single or multiple components. Although the list of potential solvent systems for cellulose is impressive, indeed (Table I), only a few systems have demonstrated the capability for supporting controlled and homogenous chemical modification. Among the limitations are high toxicity, high

Table I. Typical Non-Aqueous Cellulose Solvents.*

Number of components	Solvent group	Examples
Single (solvent)	N-alkylpyridinium halogenide	Ethylpyridinium chloride
	Oxide of tertiary amine	Triethylamine-N-oxide N-methylpiperidine-N-oxide N-methylmorpholine-N-oxide N,N-dimethylcyclo-hexylamine-N-oxide
Multiple (solvent complex)	Dimethyl sulfoxide (DMSO)-containing solvents	DMSO/methyl amine DMSO/KSCN DMSO/CaCl$_2$
	Liquid ammonia/sodium or ammonium salts	NH$_3$/NaI (NH$_4$I) NH$_3$/NaSCN (NH$_4$SCN) NH$_3$/NaNO$_3$ NH$_3$/N(C$_3$H$_5$)$_4$Br
	Dipolar aprotic solvents/LiCl	N,N-Dimethyl acetamide (DMA)/LiCl N-Methylpyrrolidone/LiCl N-Methylcaprolactam/LiCl N-Hexamethyl posphoric acid triamide/LiCl Dimethyl urea/LiCl
	Pyridine or quinoline containing systems	Pyridine/resorcinol Quinoline/Ca(SCN)$_2$
	Liquid SO$_3$/secondary or tertiary amines	SO$_3$/triethyl amine
	NH$_3$ or amine/salt/polar solvent	NH$_3$/NaCl/morpholine NH$_3$/NaCl/DMSO Ethylene diamine/NaI/N,N-dimethyl formamide
	NH$_3$ or amine/SO$_2$ or SOCl$_2$/polar solvent	diethylamine/SO$_2$/DMSO

*A preactivation of the cellulose is required in most cases.

reactivity of the solvent leading to undesired side reactions, and loss of solubility resulting in non-homogeneous reaction conditions. Two solvent systems have found considerable interest for cellulose modification; these are N,N-dimethylacetamide (DMAc) in combination with lithium chloride (LiCl) and dimethyl sulfoxide (DMSO) in combination with SO_2 and diethylamine.

The synthesis of cellulose esters, carbamates, and lactones in the DMAc/LiCl solvent system yields derivatives with high purity and high uniformity, in high yields, at moderate temperatures and with modest reagent concentrations (26,28). Prominent known side reactions involve derivatives which are susceptible to nucleophilic attack in the reaction system. Examples (Scheme 1) include the formation of chlorodeoxy celluloses during the reaction with N-chlorosuccinimide-triphenylphosphine (29), and the homogeneous bromination of cellulose with N-bromosuccinimide-triphenylphosphine in DMAc/LiBr (30). However, relatively pure cellulose-p-toluenesulfonates (**1a**) have been prepared with a DS as high as 2.3 with low chlorine content (less than 0.5%) (31). Tosylation proved more effective at O-6 than at 0-2,3. Cellulosemethylsulfonates have also been reported (32). The DMAc/LiCl solvent system also has supported the derivatization of cellulose with the fluorophore 5-dimethylamino-1-naphthalenesulfonate (**1b**) which allows studying cellulose solution properties by means of absorption and fluorescence spectroscopy (33).

The synthesis of uniform unconventional cellulose esters, having fatty and waxy ester substituents as large as C-20 (eicosanic acid, **1c**), has been supported by the DMAc/LiCl solvent system in combination with acid anhydrides in the presence of N,N-dicyclohexylcarbodiimide and 4-pyrrolidinopyridine and/or p-toluenesulfonic acid (34-36). The homogeneous phase carbanilation of cellulose in DMAc/LiCl with phenylisocyanate produces a completely substituted derivative which has been widely adopted for the determination of molecular weights in nonaqueous solvents by means of gel permeation chromatography (GPC) (37).

The synthesis of cellulose ethers in DMAc/LiCl has been accomplished using triethylamine and pyridine as catalysts (38). Films of cellulose 4,4'-bis(dimethylamino) diphenylmethyl ether (**1d**) cast from DMF-solution showed photo conducting behavior (39). Although this example provides testimony to the ability of DMAc/LiCl to support cellulose etherification reactions, the lack of solubility of strong bases in this solvent system presents a significant limitation to most typical ether-forming reactions. There has not appeared to exist a particular advantage of this homogeneous phase reaction system over conventional heterogeneous processes for cellulose ether production (40). However, comparing conventionally prepared carboxymethyl cellulose derivatives (CMC) with corresponding derivatives synthesized in a suspension of solid NaOH in DMAc/LiCl, it was revealed (by HPLC analysis), that the substituent pattern may vary dramatically in relation to reaction conditions (10,41). The CMC prepared in the presence of DMAc/LiCl contained a significantly higher amount of both

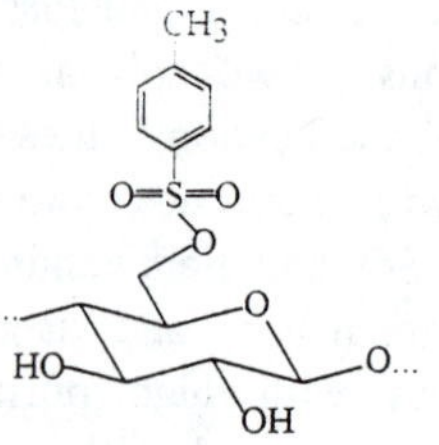

1a

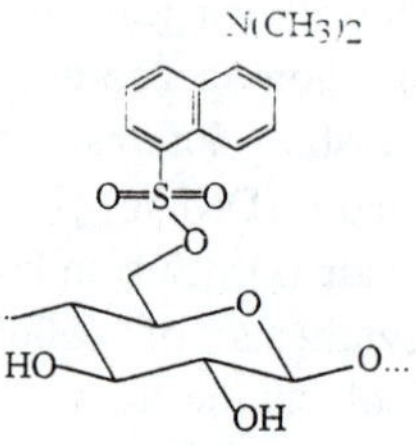

1b

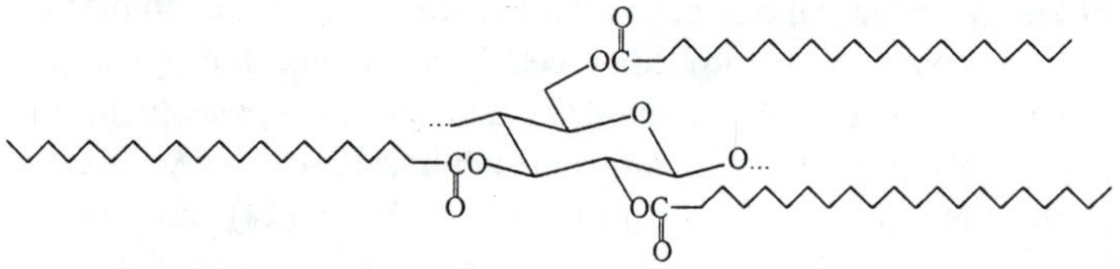

1c

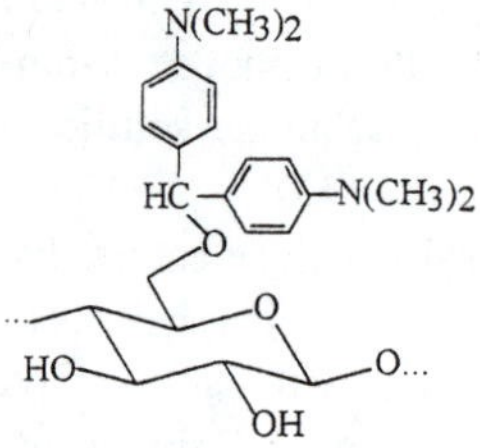

1d

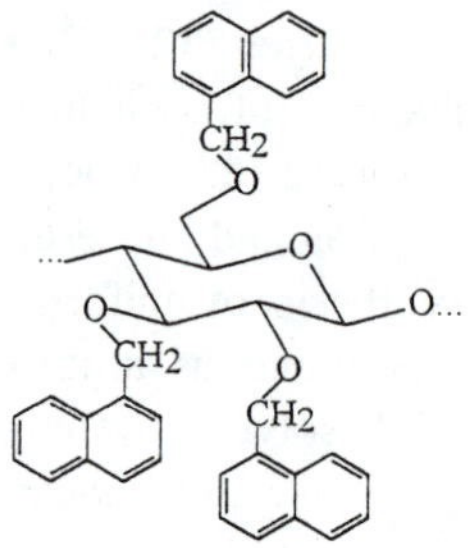

1e

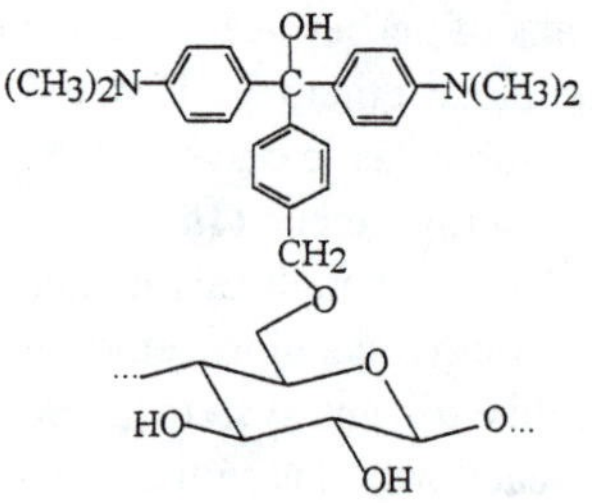

1f

Scheme I Representative monomeric units of: cellulose-*p*-toluenesulfonate **1a**, cellulose-5-dimethylamino-l-naphthalenesulfonate **1b**, cellulose eicosate **1c**, 4,4'-bis(dimethylamino)diphenylmethyl cellulose **1d**, tri-*O*-naphthylmethyl cellulose **13**, triphenylcarbinol containing cellulose **1f**.

tricarboxymethylated and unsubstituted units than those derivatives obtained by the conventional slurry method of cellulose in isopropanol/water. The DMAc/LiCl solvent system seems to give rise to a non-statistical distribution of monomeric units, and this is responsible for several unconventional properties. It is suggested that reactions using the induced phase separation method are not limited to CMC synthesis, and that this may represent a new synthesis strategy for cellulose esters with unconventional distribution of functional groups.

A more universally adaptable homogeneous phase cellulose etherification system involves the DMSO/SO$_2$/diethylamine solvent. The conversion of cellulose dissolved in this system with suspended sodium hydroxide and benzyl chloride has been studied extensively by Isogai et al (42). This system revealed superiority by comparison to the solvents N$_2$O$_4$/DMF and DMAc/LiCl in terms of reaction rates and yields. Among the derivatives reported were tri-O-arylmethylethers containing double bonds (43) and tri-O-α-naphthylmethyl ether (**1e**) which exhibits liquid crystalline behavior (44). A cellulose ether derivative containing a triphenyl moiety (**1f**) assumed conformations which were subject to photoregulation (45).

2. Dissolution With Derivatizing Solvents (Reactive Intermediates)

Effective cellulose dissolution requires the complete disruption of all cellulose to cellulose hydrogen bonds during solvation. Few solvent systems exist which accomplish this task without at least partial polymer derivatization. Examples of these derivatizing solvents, their reactive intermediates and typical reactions for which these intermediates are used, are given in Table II. Although cellulose dissolution by the formation of reactive (soluble) intermediates by partial derivatization is an acceptable method for cellulose regeneration, the use of these solvent systems for chemical modification has remained limited. Unreproducibility has been attributed to the undefined structure of the reactive intermediates, to the limited stability of the intermediates during rapidly-changing reaction conditions, and to the general dynamics of the complexation with the solubilizing derivatizing group during the process of modification. Among the reactions listed in Table II, the N,N-dimethylformamide (DMF)/N$_2$O$_4$ solvent system yielding cellulose nitrite as intermediate has found considerable interest in the synthesis of inorganic cellulose esters despite its highly toxic nature. Various cellulose sulfates and phosphates have been prepared by this method (46-48).

The derivatization of cellulose via the formation of reactive intermediates proceeds with greater predictability if the respective intermediate is isolated (and characterized) following formation and prior to conversion into a final cellulose derivative product. Reactive intermediates are typically soluble in common organic solvents, and they can be the starting materials for a wide variety of highly engineered cellulose derivatives. Among popular reactive intermediates are the

Table II. Examples of Cellulose Dissolution with Partial Derivatization; their Reactive Intermediates Formed; and Typical Subsequent Conversions *in situ.*

Dissolving agent	Reactive cellulose intermediate	Subsequent reactions *in situ*	Ref.
Formic acid	Formate	--	--
Trifluoroacetic acid	Trifluoroacetate	Acylation	50
Trichloroacetic acid/polar solvent	Trichloroacetate	--	--
N,N-dimethyl formamide (DMF)/N_2O_4	Nitrite	Sulfation, Phosphation	94
Paraformaldehyde/DMSO	Methylol	Etherification	95
		Silylation	96
Trimethylchlorsilane/DMF	Trimethylsilyl	--	--
Chloral/DMSO	Trichloroacetal	Chlorination	97

formate, the trifluoroacetate, and the trimethylsilyl ether (Table III). These reactive cellulose intermediates vary widely in DS as well as in derivative stability.

Trifluoroacetylation takes place preferably at C-6 (49,50). The derivatives are conveniently characterized regarding their structures following permethylation, saponification and degradation by means of ^{13}C-NMR and HPLC analysis (51). The dissolution involves a mixture of trifluoroacetic acid and trifluoroacetic anhydride at room temperature for 4 h, and this results in DS-values of circa 1.5 with complete substitution at C-6. Higher DS-values are obtained by the addition of chlorinated hydrocarbons as co-solvents during trifluoroacetylation. Owing to its high C-6-selectivity, the DS 1.5-trifluoroacetate of cellulose has been used as starting material for a wide variety of esters involving subsequent esterification with acid chlorides, mixed anhydrides of p-toluenesulfonic and carboxylic acids, or carboxylic acids in combination with N-N-carbonyldiimidazole (52) (Table III). (Trifluoroacetates do not transesterify under these conditions as was determined by the reaction of pure 2,3-di-O-methyl-6-O-trifluoroacetyl cellulose (**2a**) in aprotic medium with acetic acid/acetic anhydride in the presence of pyridine at 80°C with subsequent analysis by ^{1}H-^{1}H-COSY-NMR spectroscopy (Figure 1)). A remarkable O-3 selectivity was established for the sulfation of cellulose trifluoroacetate with pyridine/SO_3 (53). The primary trifluoroacetate substituent is easily removed in an aqueous work-up procedure. In addition to the use of cellulose formates for cellulose regeneration (54), these DMF-soluble derivatives were isolated with DS-values of up to 1.2 (55). ^{13}C-NMR spectroscopy revealed that the ease of formylation is in the order of O-6 > than O-2 > than O-3 (56,57). Sulfation of the reactive formate intermediate was found to occur with partial substituent removal (transesterification) as well as at the free OH-groups (58). Since transesterification can be avoided by choice of reaction conditions, (52) an inverse pattern of substituention can be obtained after subsequent removal of the formate substituents by treatment with water. This route, however, is limited by the high DS-value of the starting formate. Both trifluoroacetylation and formylation are highly degradative reactions which often produce cellulose derivatives of low DP. Depolymerization can be avoided by silylation.

Trimethylsilyl (TMS) celluloses were first described by Schuyten (59), and they were extensively studied for their regeneration potential by simple treatment with acids (60). A wide variety of silylated cellulose derivatives with a broad range of DS-values has been reported (61-63). TMS cellulose has shown to react either with its free OH-groups or with its trimethylsiloxy groups during treatment with acid chlorides, depending on reaction conditions (64,65). Whereas tertiary amines were found to catalyze the reaction of free OH-groups with acid chlorides, nitrobenzene at 160°C caused reaction only at the TMS functionalities (63). This, however, is not universally true since a partially silylated TMS-polyvinylalcohol has recently been shown to produce esters at both the TMS and OH-functional groups by acylation in nitrobenzene at 160°C (66).

10

Table III. Examples of Subsequent Reactions on Isolated Cellulose Intermediates.

Reactive cellulose intermediate[a]	Reagents	Cellulose derivative produced	Ref.
Trifluoroacetate (1.5)	Pyridine/SO₃	Cellulose sulfate[b]	53
	N,N-carbonyldiimidazole/4-nitrobenzoic acid	Cellulose-4-nitrobenzoate[c]	52
	4-Nitrobenzoic acid/tosyl chloride/	Cellulose-4-nitrobenzoate[d]	52
	Palmitoyl chloride/tosyl chloride	Cellulose palmitate[d]	52
	Phenyl isocyanate	Cellulose phenylcarbamate[d]	52
	Na-monochloroacetate/NaOH	Carboxymethyl cellulose[c]	67
Formate (2.2)	Pyridine/SO₃	Cellulose sulfate[d]	52
	N,N-dimethyl formamide/SO₃	Cellulose sulfate[c]	58
	Phenyl isocyanate	Cellulose phenylcarbamate[d]	52
	Na-monochloroacetate/NaOH	Carboxymethyl cellulose[c]	67
Trimethylsilyl cellulose (1.6)	3,4-Dinitrobenzoyl chloride (in triethylamine)	Cellulose-3,4-dinitrobenzoate[d]	62
Trimethylsilyl cellulose (2.0)	4-Bromobenzoyl chloride/4-dimethylamino pyridine (in benzene/triethylamine)	Cellulose-4-bromobenzoate[d]	62
Trimethylsilyl cellulose (2.5)	4-Nitrobenzoyl chloride	Cellulose-4-nitrobenzoate	63
Trimethylsilyl cellulose (1.1)	Na-monochloroacetate/NaOH	Carboxymethyl cellulose[c]	68

[a]Degree of substitution given in ( ).
[b]0-3 substitution
[c]Partial or total removal of functional groups of the reactive intermediate during the reaction.
[d]Inverse pattern of functionalization.

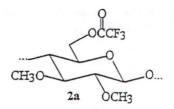

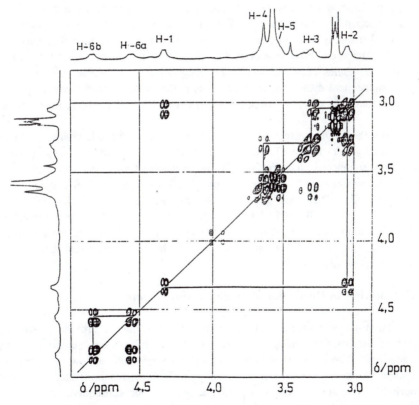

Fig. 1 ^{1}H-^{1}H-COSY-NMR spectrum of 2,3-di-*O*-methyl-6-*O*-trifluoroacetyl cellulose **2a** after treatment with acetic acid/acetic anhydride in the presence of pyridine for 8 h at 80°C proving that neither transesterification nor split off of trifluoroacetyl functions occurred.

The soluble reactive intermediates of Table II, and even cellulose acetate, can be converted into cellulose ethers with high DS by removing the solubilizing ester groups in a one step synthesis in DMSO with suspended solid NaOH powder as base (67). This pseudo-homogeneous conversion proceeds to cellulose ethers that have a substitution pattern that is different from that of a heterogeneously prepared products (68).

3. Dissolution by Derivatization with Protective Groups

Two major groups of protective agents have been of interest to cellulose chemists. The triphenylmethyl (trityl) group is a widely used protective group for the C-6 position which provides for subsequent selective functionalization of the remaining secondary OH-groups (69-71). The reaction of cellulose with methoxy-substituted triphenyl chlorides in homogeneous DMAc/LiCl solution proceeds with little degradation and high reproducibility (72). The monomethoxy-trityl (**3a**), in particular, combines a fast and specific blocking step of well-soluble polymers with an easy subsequent deblocking reaction. This pathway has recently been employed for the preparation of 2,3-di-O-carboxymethyl cellulose (73), among many others (74-77).

Another OH-protective agent with great regio-selectivity is 6-O-thexyldimethylchlorosilane (TDMS). 6-O-TDMS-celluloses (**3b**) were synthesized using a heterogeneous phase reaction in the presence of ammonia-saturated polar-aprotic solvents at -15°C. The conversion of cellulose with TDMS chloride in N-methylpyrrolidone/NH_3 was found not to proceed past a total DS_{Si} of 1.0; it yielded a derivative with 96% silylation at O-6 (78). This well-protected cellulose derivative is a useful starting material for the subsequent synthesis of various regio-specifically modified cellulose materials (79). This field has been subjected to several recent reviews (12,25).

4. Dissolution by Derivatization with Leaving Groups

Homogeneous phase reaction conditions with cellulose provide convenient access to derivatives which may be subject to subsequent substitution by nucleophiles. The chemical activation of cellulose proceeds at low temperatures and short reaction times, and it is virtually free of side-reactions and associated impurities. The active groups of particular interest for subsequent nucleophilic replacement are sulfonates (especially tosylate) and halodeoxy cellulose derivatives.

In contrast to heterogeneous sulfonation, which is subject to numerous potential side reactions, homogeneous mesylation (32) and tosylation (80,81) in DMAc/LiCl produces uniform and well-defined products with DS ranging from 0.4 to 2.3 (31). The derivatives are typically soluble in a variety of organic solvents and contain only traces of halogen and no nitrogen. The use of cellulose sulfonates

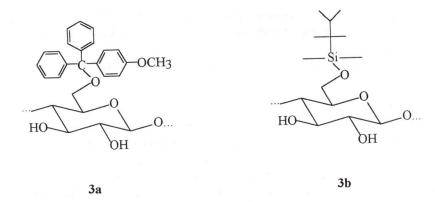

3a   **3b**

Scheme 2. Representative monomeric units of monomethoxytrityl cellulose (3a) and 6-O-texyl dimethylsilyl cellulose (36).

for a variety of nucleophylic substitutions has been summarized by Belyakova (82) and Hon (83). In contrast to cellulose triflates, which are highly reactive and sensitive to even mildly nucleophylic species, cellulose tosylates are rather stable. Various novel cellulose derivatives were synthesized from cellulose tosylates of varying DS-levels and molecular weights by both acylation (84) and sulfation of the unsubstituted OH-groups. The amphiphilic cellulose sulfate tosylates are soluble in both water and DMSO at an appropriate DS-balance, and they are promising starting materials for self-organizing polymer systems (85).

Chlorodeoxy cellulose can be prepared heterogeneously (in an inert medium such DMF or pyridine/chloroform with thinoyl chloride, phosphorus oxychloride, sulfuryl chloride or methanesulfonyl chloride), or homogeneously in DMF/chloral (29,86). More elegantly, cellulose chlorination may also proceed in DMAc/LiCl with sulfuryl chloride (87) or tosyl chloride via the in-situ formed cellulose sulfonates (88). A convenient and relatively selective chlorination of cellulose in DMAc/LiCl involves N-chlorosuccinimide-triphenylphosphine. While the reaction initially derivatizes only the C-6 position, DS-levels as high as 1.86 have been reported (29). The corresponding N-bromosuccinimide-triphenylphosphine gave rise to bromodeoxy cellulose in three different solvents (DMF, N-methylpyrrolidone, and DMAc), in combination with LiBr. The derivatives had a maximum DS of 0.9 and were selectively brominated at C-6 (30). An alternative bromination reagent consists of tribromoimidazole, triphenylphosphine and imidazole (90). Homogeneous reactions in DMAc/LiBr produced polymers with DS-values of up to 1.6. The halodeoxy celluloses are also accessible via nucleophilic substitution of tosylate functions by halides (91). A large number of follow-up replacement reactions using halodeoxy celluloses have been described, and these include conversions with thio-urea, with iminodiacetic acid (92) or with thiols (93).

Replacement reactions based on cellulose silylethers have been discussed in a previous section.

CONCLUSION

With the advancement of solvents and solution complexes for cellulose, modification reactions have become feasible that provide the tools necessary for tailoring highly engineered cellulose derivatives to specific end-uses. The pathways to selectively and specifically modified cellulose derivatives involve dissolution in single component solvents; dissolution via reactive intermediate by partial derivatization (i.e., derivatizing solvents); dissolution as protected cellulose derivatives; and dissolution by chemical activation for subsequent substituent replacement by nucleophilic attack. These systems provide a wide range of opportunities for the preparation of highly-engineered cellulose derivatives, and they make it possible that cellulose derivatives be used in such molecular recognition concepts as self-organizing supra-molecular systems, in nano-structures, in environmentally responsive (smart) materials, and in materials useful in sensors as well as in chiral templates. Much future work is needed to explore and exploit the full benefits of homogeneous phase reactions with polysaccharides.

Literature Cited

1. Nehls, I.; Wagenknecht, W.; Philipp, B.; Stscherbina, D. *Progr. Polym. Sci.* **1994**, *19*, 29.
2. Buchanan, C.M.; Hyatt, J.A.; Lowman, D.W. *J. Am. Chem. Soc.* **1989**, *111*, 7312.
3. Stein, A.; Klemm, D. *Papier (Darmstadt)* **1995**, *49*, 732.
4. Tezuka, Y.; Tsuchiya, Y. *Carbohydr. Res.* **1995**, *273*, 83.
5. Tezuka, Y.; Tsuchiya, Y.; Shiomi, T. *Carbohydr. Res.* **1996**, *291*, 83.
6. Baar, A.; Kulicke, W.M.; Szablikowski, K.; Kiesewetter, R. *Macromol. Chem. Phys.*, **1994**, *195*, 1483.
7. Arisz, P.W.F. *Ph. D. Thesis*, University of Amsterdam, The Netherlands, **1995**.
8. Mischnick, P.; Kühn, G. *Carbohydr. Res.* **1996**, *290*, 199.
9. Erler, U.; Mischnick, P.; Stein, A.; Klemm, D. *Polym. Bull.* **1992**, *29*, 349.
10. Heinze, Th.; Erler, U.; Nehls, I.; Klemm, D. *Angew. Makromol. Chem.* **1994**, *215*, 93.
11. Kamide, K.; Saito, M. *Macromol. Symp.* **1994**, *83*, 233.

12. Klemm, D.; Stein, A.; Heinze, Th.; Philipp, B.; Wagenknecht, W. *Polymeric Materials Encyclopedia: Synthesis, Properties and Applications,* Salamone, J.C. Ed.; CRC Press, Inc., Boca Raton, USA, vol. 2, **1996**, p. 1043-1053.

13. Gelman, R.A. *J. Appl. Polym. Sci.* **1982**, *27*, 2957.

14. Takahaishi, S.-I.; Fujimoto, T.; Miyamoto, T.; Inagaki, H. *J. Polym. Sci., Part A: Polym. Chem.* **1987**, *25*, 987.

15. Nojiri, M.; Kondo, T. *Macromolecules* **1996**, *29*, 2392.

16. Kondo, T.; Sawatari, C. *Polymer* **1994**, *35*, 4423.

17. Klemm, D.; Heinze, Th.; Wagenknecht, W. *Ber. Bunsenges. Phys. Chem.* **1996**, *100*, 730.

18. Guo, J.X.; Gray, D.G. Lyotropic Cellulosic Liquid Crystals, In: *Cellulosic Polymers, Blends and Composites,* Gilbert, R.D. Ed.; Hanser Publ., Munich, Vienna, New York, **1994**, p. 25.

19. Zugenmaier, P. Polymer Sovent Interactions in Lyotropic Liquid Crystalline Cellulose Derivative Systems, In: *Cellulosic Polymers, Blends and Composites,* Gilbert, R.D. Ed.; Hanser Publ., Munich, Vienna, New York, **1994**, p. 71.

20. Schaub, M.; Fakirov, C.; Schmidt, A.; Lieser, G.; Wenz, G.; Wegner, G.; Albony, P.A.; Wu, H.; Foster, M.D.; Majrkzak, M.; Satija, S. *Macromolecules* **1995**, *28*, 1221.

21. Wegner, G.; Schaub, M.; Wenz, G.; Stein, A.; Klemm, D. *Adv. Mater.* **1993**, *5*, 919.

22. Philipp, B. *J.M.S.-Pure Appl. Chem.* **1993**, *A30*, 703.

23. Johnson, D.C. Solvents for Cellulose, In. *Cellulose Chemistry and its Applications,* Nevell, T.P.; Zeronian, S.H. Eds.; E. Horwood Ltd., Chichester, **1985**, p.181.

24. Herlinger, H.; Hengstberger, M. *Lenzinger Ber.* **1985**, *59*, 96.

25. Philipp, B.; Wagenknecht, W.; Nehls, I.; Klemm, D.; Stein, A.; Heinze, Th. *Polymer News* **1996**, **21**, 155 .

26. McCormick, C.L.; Lichatowich, D.K. *J. Polym. Sci., Polym. Lett. Ed.* **1979**, *17*, 479.

27. Dawsey, T.R. Applications and Limitations of LiCl/N,N-Dimethylacetamide in the Homogeneous Derivatization of Cellulose, In: *Cellulosic Polymers, Blends and Composites,* Gilbert, R.D. Ed.; Hanser Publ., Munich, Vienna, New York, **1994**, p. 157.

28. Morgenstern, B.; Kammer, H.-W. *TRIP* **1996**, *4*, 87.

29. Furuhata, K.-I.; Chang, H.-S.; Aoki, N.; Sakamoto, M. *Carbohydr. Res.* **1992**, *230*, 151.

30. Furuhata, K.-I; Koganai, K.; Chang, H.-U.; Aoki, N.; Sakamoto, M. *Carbohydr. Res.,* **1992**, *230*, 165.

31. Rahn, K.; Diamantoglou, M.; Klemm, D.; Berghmans, H.; Heinze, Th. *Angew. Makromol. Chem.* **1996**, *238*, 143.

32. Frazier, C.E.; Glasser, W.G. *Polymer Preprints* **1990**, *31*, 634.

33. Heinze, Th.; Camacho Gomez, J.A.; Haucke, G. *Polym. Bull.* **1996**, *37*, 743.

34. Samaranayake, G.; Glasser, W.G. *Carbohydr. Polym.* **1993**, *22*, 1.

35. Glasser, W.G.; Samaranayake, G.; Dumay, M.; Dave, V. *J. Polym. Sci., Part B: Polym. Phys.* **1995**, *33*, 2045.

36. Sealey, J. E.; Samaranayake, G.; Todd, J.G.; Glasser, W.G. *J. Polym. Sci., Part B: Polym. Phys.* **1996**, *34*, 1613.

37. Terbojevich, M.; Cosani, A.; Camilat, M.; Focher, B. *J. Appl. Polym. Sci.* **1995**, *55*, 1663.

38. Erler, U.; Klemm, D.; Nehls, I. *Makromol. Chem., Rapid Commun.* **1992**, *13*, 195.

39. Heinze, Th.; Erler, U.; Heinze, U.; Camacho, J.; Grummt, U.-W.; Klemm, D. *Macromol. Chem. Phys.* **1995**, *196*, 1937.

40. Dawsey, T.R. *Polym. Fiber Sci.: Recent Adv.* **1992**, Forners, R.E.; Gilbert R.D. Eds.; VCH New York, p. 157.

41. Heinze, Th.; Heinze, U.; Klemm, D. *Angew. Makromol. Chem.* **1994**, *220*, 123.

42. Isogai, A.; Ishizu, A.; Nakano, J. *J. Appl. Polym. Sci* **1984**, *29*, 2097, 3873.

43. Isogai, A.; Ishizu, A.; Nakano, J. *J. Appl. Polym. Sci* **1986**, *31*, 341.

44. Dave, V.; Frazier, C.L.; Glasser, W.G. *J. Appl. Polym. Sci.* **1993**, *49*, 1671.

45. Arai, K.; Kwabata, Y. *Macromol. Chem. Phys.* **1995**, *196*, 2139.

46. Akelah, A.; Sherrington, D.C. *J. Appl. Polym. Sci.* **1981**, *26*, 3377.

47. Wagenknecht, W.; Nehls, I.; Philipp, B. *Carbohydr. Res.* **1992**, *237*, 211.

48. Wagenknecht, W.; Nehls, I.; Philipp, B. *Carbohydr. Res.* **1993**, *240*, 245.

49. Hasegawa, M.; Isogai, A.; Onabe, F.; Usuda M., *J. Appl. Polym. Sci* **1992**, *45*, 1857.

50. Salin, B.N.; Cemeris, M.; Mironov, D.P.; Zatsepin, A.G. *Khim. Drev.* **1991**, *3*, 65, CA 116(8): 61812b.

51. Liebert, T.; Schnabelrauch, M.; Klemm, D.; Erler, U. *Cellulose* **1994**, *1*, 249.

52. Liebert, T. *Ph. D. Thesis*, University of Jena, Germany, **1995**.

53. Klemm, D.; Heinze, Th.; Stein, A.; Liebert, T. *Macromol. Symp.* **1995**, *99*, 129.

54. Rudy, H. *Cellulosechemie* **1931**, *13*, 49.

55. Schnabelrauch, M.; Vogt, S.; Klemm, D.; Nehls, I.; Philipp, B. *Angew. Makromol. Chem.* **1992**, *198*, 155.

56.   Takahashi, S.I.; Fujimoto, T.; Barua, B.M.; Miyamoto, T.; Inagaki, H. *J. Polym. Sci., Polym. Chem. Ed.* **1986**, *24*, 2981.

57.   Fujimoto, T.; Takahashi, S.I.; Tsuji, M.; Miyamoto, T.; Inagaki, H. *J. Polym. Sci., Polym. Lett.* **1986**, *24*, 495.

58.   Philipp, B.; Wagenknecht, W.; Nehls, I.; Ludwig, J.; Schnabelrauch, M.; Rim, K.H.; Klemm, D. *Cellul. Chem. Technol.* **1990**, *24*, 667.

59.   Schuyten, H.A.; Weaver, J.W.; Reid, J.D.; Jürgens, J.F. *J. Am. Chem. Soc.* **1948**, *70*, 1919.

60.   Weigel, P.; Gensrich, J.; Wagenknecht, W.; Klemm, D.; Erler, U.; Philipp, B. *Papier (Darmstadt)* **1996**, *50*, 483.

61.   Schemp, W.; Krause, Th.; Seifried, U.; Koura, A. *Papier (Darmstadt)* **1984**, *38*, 607.

62.   Klemm, D.; Schnabelrauch, M.; Stein, A.; Philipp, B.; Wagenknecht, W.; Nehls, I. *Papier (Darmstadt)* **1990**, *44*, 624.

63.   Stein, A.; Klemm, D. *Makromol. Chem., Rapid Commun.* **1988**, *9*, 569.

64.   Wagenknecht, W.; Nehls, I.; Stein, A.; Klemm, D.; Philipp, B. *Acta Polymerica* **1992**, *43*, 266.

65.   Klemm, D.; Stein, A.; Erler, U.; Wagenknecht, W.; Nehls, I.; Philipp, B. New Precedures for regioselective synthesis and modification of trialkylsilylcelluloses, In: *Cellulosics: Materials for Selective Separation and Other Technologies*, Kennedy, J.F.; Phillips, G.O.; Williams P.A. Eds.; E. Horwood Ltd., New York, London, Toronto, Sydney, Tokyo, Singapore **1993,** p. 221.

66.   Mormann, W.; Wagner, Th. *Macromol. Chem. Phys.* **1996**, *197*, 3463.

67.   Liebert, T.; Klemm, D.; Heinze, Th. *J.M.S.-Pure Appl. Chem.* **1996**, *A33*, 613.

68.   Heinze, Th.; Liebert, T. *ACS Symp. Ser.* **1997**, in press.

69.   Green, J.W. Triphenylmethyl Ethers, In: *Methods in Carbohydr. Chem.*, Whistler, R.L.; Green, J.W.; BeMiller, J.N. Eds.; vol 3, **1963**, p. 327.

70.   Harkness, B.R.; Gray, D.G. *Macromolecules* **1991**, *24*, 1800.

71.   Kondo, T.; Gray, D.G. *Carbohydr. Res.* **1991**, *220*, 173.

72.   Camacho Gomez, J.A.; Klemm, D.; Erler, U. *Macromol. Chem. Phys.* **1996**, *197*, 953.

73.   Heinze, Th.; Röttig, K.; Nehls, I. *Macromol. Rapid Commun.* **1994**, *15*, 311.

74.   Iwata, T.; Azuma, J.-I.; Okamura, K.; Muramoto, M. B. Chun, *Carbohydr. Res.* **1992**, *224*, 277.

75.   Kondo, T. *Carbohydr. Res.* **1993**, *238*, 231.

76.   Itagaki, H.; Takahashi, I.; Natsume, M.; Kondo, T. *Polym. Bull.* **1994**, *32*, 77.

77.   Kasuya, N.; Iiyama, K.; Meshituska, G.; Ishizu, A. *Carbohydr. Res.* **1994**, *260*, 251.

18

78.   Klemm, D.; Stein, A. *J.M.S.-Pure Appl. Chem.* **1995**, *A32*, 899.
79.   Koschella, A.; Klemm, D. *Macromol. Symp.* **1997**, in press.
80.   Dawsey, T.R.; Newman, J.K.; McCormick, C.L. *Polym. Prepr. (Am. Chem. Soc., Div. Polym. Chem.)* **1989**, *30*, 191.
81.   McCormick, C.L.; Dawsey, T.R.; Newman, J.K. *Carbohydr. Res.* **1990**, *208*, 183.
82.   Belyakova, M.K.; Gal`braikh, L.S.; Rogovin, Z.A. *Cellul. Chem. Technol.* **1971**, *5*, 405.
83.   Hon, D. N.-S. Chemical Modification of Cellulose, In: *Chemical Modification of Lignocellulosic Materials,* Hon D. N.-S. Ed.; Marcel Dekker, New York, Basel, Hong Kong, **1996**, p. 114.
84.   Heinze, Th.; Rahn, K.; Jaspers, M.; Berghmans, H. *Macromol. Chem. Phys.* **1996**, *197*, 4207.
85.   Heinze, Th.; Rahn, K. *Macromol. Rapid Commun.* **1996**, *17*, 675.
86.   Schnabelrauch, M.; Heinze, T.; Klemm, D. *Acta Polymerica* **1990**, *41*, 113.
87.   Furubeppu, S.; Kondo, T.; Ishizu, A. *Sen`i Gakkaishi* **1991**, *47*, 592.
88.   McCormick, C. L.; Callais, P. *Polymer* **1987**, *28*, 2317.
89.   Furuhata, K.-I.; Aoki, N.; Suzuki, S.; Arai, N.; Sakamoto, M.; Saegusa, Y.; Nakamura, S. *Carbohydr. Res.* **1994**, *258*, 169.
90.   Furuhata, K.-I.; Aoki, N.; Suzuki, S.; Sakamoto, M.; Saegusa, Y.; Nakamura, S. *Carbohydr. Polym.* **1995**, *26*, 25.
91.   Heinze, Th.; Rahn, K. *Papier (Darmstadt)* **1996**, *50*, 721.
92.   Mentasti, E.; Sarzanini, C.; Gennora, M.C.; Porta, V. *Polyhedron* **1987**, *6*, 1197.
93.   Aoki, N.; Koganei, K.; Chang, H-S.; Furuhata, K.; Sakamoto, M. *Carbohydr. Polym.* **1995**, *27*, 13.
94.   Philipp, B. *Polymer News* **1990**, *15*, 170.
95.   Nicholson, M.D.; Johnson, D.C. *Cellul. Chem. Technol.* **1977**, *11*, 349.
96.   Shiraishi, N.; Miyagi, Y. *Sen-i Gakkaishi* **1979**, *35*, 466.
97.   Ishii, T.; Ishizu, A.; Nakano, J. *Carbohydr. Res.* **1977**, *59*, 155.

Chapter 2

Regiocontrol in Cellulose Chemistry: Principles and Examples of Etherification and Esterification

D. O. Klemm

**Institute of Organic and Macromolecular Chemistry,
Friedrich Schiller University, D-07743 Jena, Germany**

Regiocontrol means site-selective reactions of cellulose and related polysaccharides that form functionalization patterns essential for new properties. Experimental results are presented in synthesis, analysis, and subsequent reactions of silylethers, cellulose p-toluenesulfonates, and deoxy-thiosulfates (Bunte salts). Regiocontrol succeeds by selective loosening supramolecular structures, as well as by selective protection, activation, and migration of functional groups.

The topics of this paper relate to the essential interaction of cellulose functionalization with organic synthesis and structure analysis, as well as with product properties and applications. After some remarks on principles of regioselectivity in cellulose modification, the selective synthesis and subsequent reactions of cellulose ethers, the functionalization via ester intermediates of controlled stability and reactivity, and examples of structure properties relationships will be described.

General Principles of Regioselective Cellulose Chemistry

Up to now there have been known four principal synthesis pathways to obtain functionalized celluloses (Figure 1): The first is the well-known polymeranalogous reaction of cellulose after isolation from plants or bacteria culture media. The second is the biosynthesis of functionalized celluloses using, for example, the copolymerization of ß-ᴅ-glucose with N-acetylglucoseamine by *Acetobacter xylinum* (*1*). The third way is enzymatic in vitro synthesis starting, for example, from 6-O-methyl-ß-cellobiosyl fluoride and purified cellulases in a stereo- and regioselective polymerization (*2*). The fourth is the chemical synthesis starting from glucose. Results from last year demonstrated the first chemosyntheses of functionalized celluloses by ring-opening polymerization of 3,6-di-O-benzyl-α-ᴅ-

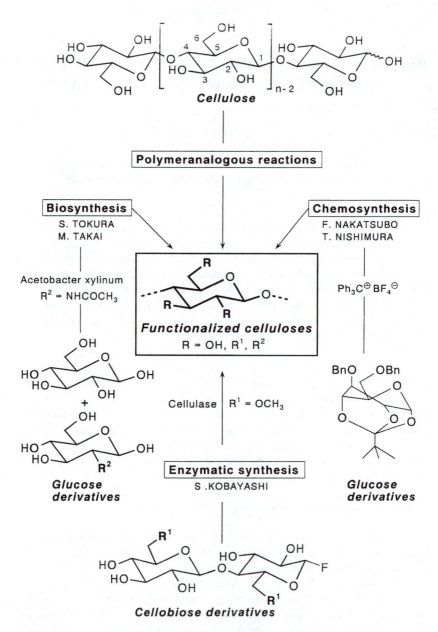

Figure 1. Principal synthesis pathways to obtain functionalized celluloses.

glucose-1,2,4-orthopivalate (*3*), and by stepwise reaction of selectively protected allylethers of ß-ᴅ-glucose (*4*).

In the case of polymeranalogous functionalization of cellulose, the chemical reaction proceeds in a statistic or regioselective way. The term "regioselectivity" means an exclusively or significantly preferential reaction at one or two of the three sites 2,3, and 6 of the anhydroglucose unit (AGU) as well as along the polymer chains. Typically simple examples are selectively C-6 or C-2,3 modified celluloses (A) and copolymers with block-like structures (B) as shown schematically in Figure 2. The symbols used (●,✱) denote different types of functional groups.

The energy profile of a typical regioselective reaction demonstrates that steric hindrance (by bulky groups), entropic acceleration (caused, for example, by HO----H bonds), or electronic promotion (e.g., electronically withdrawn substituents) are essential to differentiate among the free enthalpies of activation as well as of rate constants and concentrations of the isomeric products formed. From the point of view of such differences, the varying reactivity of the OH groups in the AGU of cellulose - primary 6-OH, more acidic 2-OH - may be used for selective functionalization of cellulose. Moreover, a selective loosening of the supramolecular structure may be of importance in regiocontrol.

In regard to the question: "Why regiocontrol in cellulose chemistry?" three important fields should be pointed out:
- Basic investigations, e.g., on structure and interaction in solution, as well as on formation of well-defined supramolecular structures using cellulose derivatives with known patterns of functionalization;
- Design of advanced materials and nanoscale architectures in interdisciplinary research at the interface of organic and macromolecular chemistry. Potential fields of application are liquid crystalline polymers, selective membranes, multilayered assemblies, sensor matrices, recognition devices, and bioactive materials;
- Better knowledge of reactions, mechanisms, and product structures and properties of present as well as of new commercial types of cellulosics of industrial large-scale processes. Further knowledge of the control of processing and end-use properties by the pattern of functionalization.

For example, regioselectively modified celluloses are of importance in investigations into solution properties as well as into enzymatic degradation of cellulose derivatives. This work deals with fringed micelles at different aggregation stages in solution as demonstrated by Burchard (*5*), with the characterization of the reaction behaviour of cellulases using 6-O- and 2,3-di-O-methyl celluloses (*6*), and with the investigation of HO----H bond systems and gelation of cellulose derivatives (*7, 8*) as published by Kondo.

Our investigations on regiocontrol in polysaccharide chemistry have been based up to now on cellulose and starch including ß-cyclodextrine as a model compound. Regioselectivity in the AGU and along the polymer chains may be low, high, or 100 % and may lead to copolymers (degree of substitution, DS, < 1) or homopolymers (DS = 1) by reaction at one preferred site of the AGU. Regiocontrol of the primary and of subsequent reactions succeeds by the reagent (e.g., steric

22

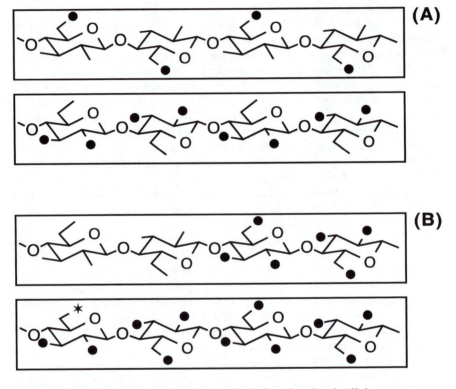

Figure 2. Examples of regioselectively functionalized celluloses.

hindrance) and by substituent effects (protection, activation, migration) and by the supramolecular structure in the reaction media (stepwise loosening of HO---H bonds, phase separation during reaction, structure in solution). Preferentially used cellulose types are cellulose powder obtained by acid hydrolysis and mechanical disintegration (degree of polymerization, DP, 150-300), spruce sulfite dissolving pulp, pine sulfate dissolving pulp (DP 600-900), scoured and bleached cotton linters (DP 800-2000), and cellulose prepared by *Acetobacter xylinum* (DP 1000-2000).

Important synthesis routes to obtain regioselectively functionalized celluloses have been developed: Firstly, the complete functionalization of cellulose followed by a site selective reaction of the introduced groups and by the elimination of the primary groups. Secondly, the regioselective partial functionalization of cellulose and subsequent reactions of the free or functionalized OH groups followed by the elimination of the primary groups. Typical results of these investigations are summarized in Klemm et al. (*9*). Silylcelluloses (soluble in organic solvents, DS 0.4-3.0, deblocking by HCl/H$_2$O resp. fluoride ions), tritylethers (soluble, detritylation by HCl), and cellulose p-toluenesulfonates (DS up to 2.3, soluble in organic solvents, reactive in nucleophilic substitutions), have been developed as suitable cellulose intermediates for regioselective functionalization.

Regioselective Synthesis and Subsequent Reactions of Cellulose Ethers
The silylation of cellulosic OH groups leads to a preferred 6-O- or 2,6-di-O-functionalization. These silylether intermediates open up a wide field of subsequent reactions. The chemical modification of starch is included in the investigations.

The trimethylsilylation (R = Me, cf. Figure 3) of cellulose dissolved in N-methylpyrrolidone (NMP)/LiCl leads to DS values up to 3.0. At DS values of about 1.0, 60 % of the silylether groups are located at C-6. In a cellulose suspension in N-methylpyrrolidone/NH$_3$ at -33 to -15 °C, the trimethylsilyl-cellulose dissolves at DS 1.3 after a heterogeneous reaction, and the silylation takes place up to DS values of 3.0. At DS values of about 1.0, 80 % of the silylether groups are located at C-6.

In the case of thexyldimethylsilylation (R = CMe$_2$CHMe$_2$ = Thx), this higher 6-O-regioselectivity leads to a very high selectivity at the primary OH groups and to the formation of pure 6-O-thexyldimethylsilyl cellulose (*10-12*). This thexyldimethylsilylation in NMP/NH$_3$ stops at the DS of 1.0. No further silylation takes place, even with additional amounts of chlorosilane, either at higher temperatures or after isolation and dissolution of the polymer in pyridine. We assume a low accessibility of the OH groups at positions 2 and 3 after swelling and decrystallization with ammonia, in contrast to OH groups in the cellulose dissolved in NMP/LiCl or N,N-dimethylacetamide (DMA)/LiCl (*13*).

The low selectivity in case of dissolved cellulose is suitable to prepare 2,6-di-O-thexyldimethylsilyl cellulose even with the bulky reagent thexyldimethyl-chlorosilane.

As summarized in Figure 3, starting from primary 6-O-protected celluloses we

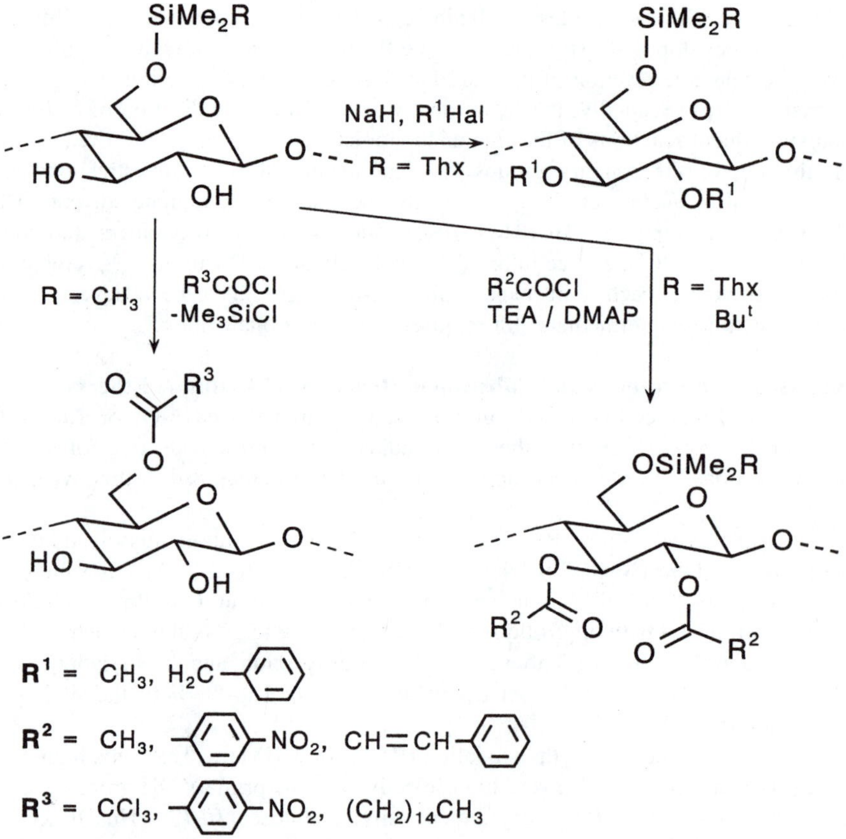

Figure 3. Subsequent reactions of 6-0-silylated celluloses.

developed effective ways to obtain different types of 2,3-di-O-esters and -ethers by reactions of the free OH groups as well as of esters by reaction of the silylether group (*14*).

As in the case of the cellulose products, the subsequent acylation of 6-O- and 2,6-di-O-thexyldimethylsilyl starch leads to the 2,3-di-O- resp. 3-O-starch esters. Whereas the alkylation of 2,6-di-O-silyl celluloses is a suitable way to prepare the 3-O-ethers, the related starch derivatives form the 2-O-ethers with migration of the silyl groups to position 3. The same results could be observed in the benzylation of ß-cyclodextrine thexyldimethylsilylated resp. t-butyldimethyl-silylated in positions 2 and 6. The formation of the 2-O-alkyl-3,6-di-O-silyl derivatives was mainly concluded from ^{1}H NMR spectra by the typical high field shift of the proton at C-2 after alkylation (*15*).

In cooperation with P. Mischnick (*16*), we investigated the silyl group migration in alkylation of oligo- and polysaccharides with α–glycosidic linkages as a suitable synthesis way to obtain the corresponding 2-O-ethers. Figure 4 shows 2-O-methylation of 2,6-di-O-thexyldimethylsilyl starch as a typical example. The cyclic intermediates possessing two trans-diequatorial OH groups are formed under the reaction conditions because of interaction with the neighboring 1α–oxygen. After desilylation of the 2-O-ethers, a subsequent acylation resp. alkylation proceeds to form regioselectively trifunctionalized starches (cf. Figure 12).

A further and well-known 6-O-blocking group in cellulose chemistry represents the trityl group suitable for the preparation of many of 2,3-di-O-derivatives (*17-20*). With the aim of a complete detritylation under mild conditions, we investigated the formation and properties of methoxy substituted tritylethers with this result: The higher the methoxy substitution, the higher is the rate of tritylation and detritylation (*21, 22*). With respect to product stability, the p-mono-methoxytrityl group is very suitable for 6-O-protection, e.g., for synthesis of 2,3-di-O-carboxymethyl cellulose, including a complete deprotection with ethanolic HCl (*23*) as demonstrated in Figure 5.

Functionalization via Ester Intermediates of Controlled Stability and Reactivity

The introduction of the p-toluenesulfonyl (tosyl) group into different types of celluloses in the DP range of 150-2000 is a suitable way to prepare 6-functionalized derivatives under "Umpolung" of the reactivity at position 6. Donor agents such as sodium thiosulfate lead to cellulose-deoxy-thiosulfates (Bunte salts) with high 6-regioselectivity (*22*).

The cellulose p-toluenesulfonates were synthesized in an N,N-dimethylacet-amide/LiCl solution of cellulose in the presence of triethylamine (TEA). In the DS range of about 1 - controlled by the mol equivalents of the reagent tosyl chloride - 80 % of the sulfonate groups are located at position 6 (*24, 25*). The subsequent reaction (DS values of cellulose p-toluenesulfonates up to 2.3) with sodium thio-sulfate in a mixture of dimethylsulfoxide and water resulted in nucleophilic substitution of the primary tosylate groups forming water-soluble polymers (Figure 6). This Bunte salt formation takes place in the same way after subsequent functio-nalization of free OH groups in the cellulose p-toluenesulfonates.

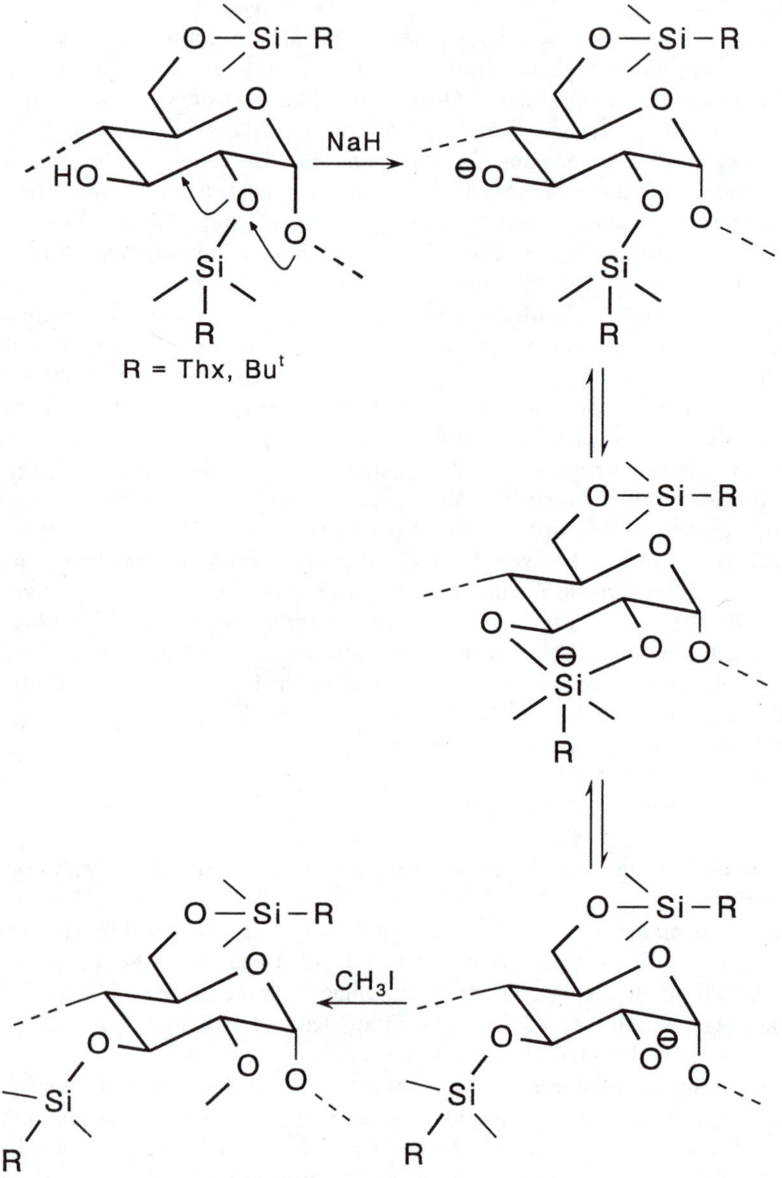

Figure 4. Silyl group migration in alkylation of 2,6-di-O-thexyldimethylsilyl starch.

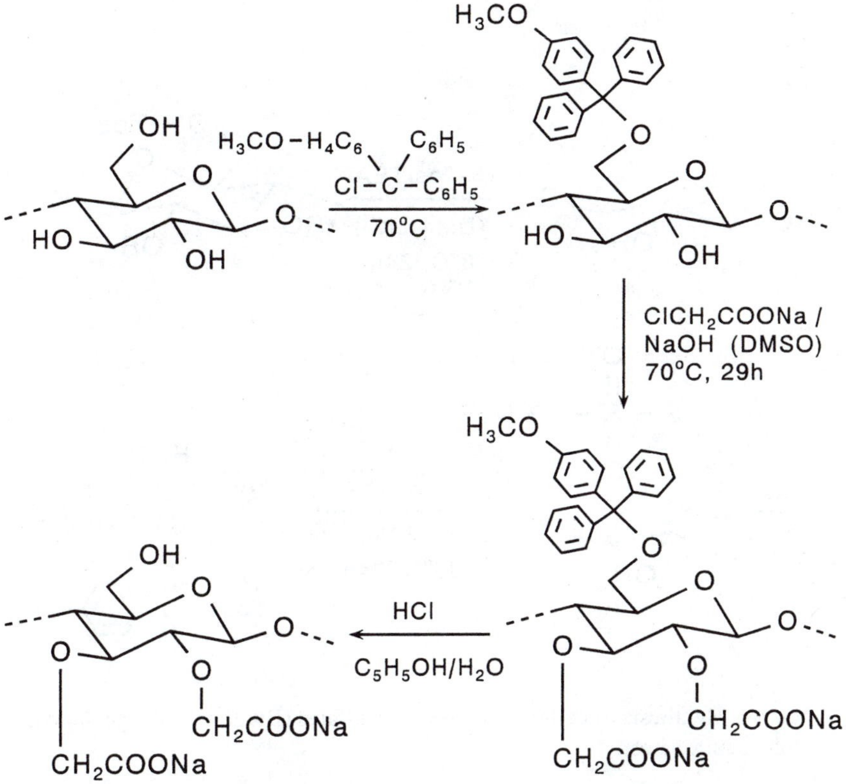

Figure 5. Methoxytrityl ethers as very suitable 6-O-protective group in cellulose chemistry.

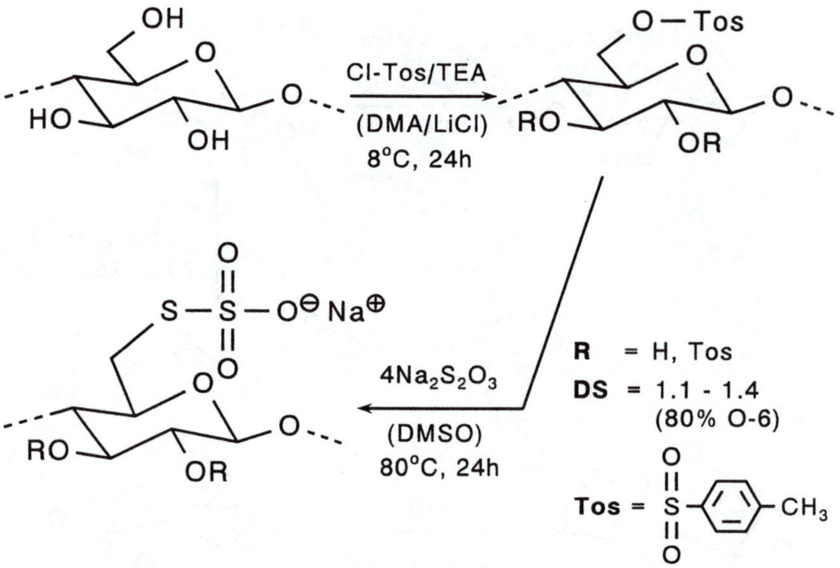

Figure 6. Synthesis of cellulose-deoxy-thiosulfates (Bunte salts) via cellulose p-toluensulfonates.

Typical examples are the formation of the corresponding esters, cellulose-O-(4-chloro-phenylcarbamate)-deoxy-thiosulfate and cellulose acetate-deoxy-thiosulfate, soluble in organic solvents such as tetrahydrofurane and acetone (cf. R^1 in Figure 8).

Further synthesis pathways to obtain Bunte salts of cellulose and cellulose derivatives consist in the reaction of different types of chlorinated and unsaturated cellulose esters described in Camacho Gómez (22). Examples are substitution reactions of cellulose-O-(3-chloropropionate), cellulose-O-(4-chloro-methylbenzoate), cellulose-O-(2-chloroethylcarbonate), and cellulose-O-(N-2-chloroethylcarbamate) with sodium thiosulfate in a mixture of dimethylsulfoxide and water at 80 °C for 24 h (cf. R^2 in Figure 8), as well as addition reactions of cellulose-O-monomaleate with this reagent.

Three important types of subsequent reactions of low-molecular Bunte salts are known in relation to the conditions (26) as shown in Figure 7.

For the polymeric and watersoluble cellulose-deoxy-thiosulfates, the introduced anionic S-S(O_2)O-groups are useful as protected alkene and thiol functions, suitable for subsequent crosslinking reactions in solution, as well as in films, ultrathin layers, recognition matrices, and spherical supports. These investigations are part of our ongoing research on regioselective functionalized celluloses and starches, including the preparation of artifical supramolecular architectures (27). First results demonstrate effective crosslinking of Bunte salts of cellulose and cellulose derivatives in the presence of oxidizing reagents such as iodine or hydrogen peroxide (cf. Figure 8). As a result of the crosslinking, the polymers precipitated from solutions and layers formed from water or organic solvents were insoluble after the described treatment.

Analysis of the Molecular Structure

An important part of an investigation on regiocontrol in polysaccharide chemistry is the analysis of the functionalization patterns prepared. As suitable methods we used ^{13}C and ^1H NMR spectroscopy of the polymers and HPLC after chain degradation. In the case of HPLC a complete methylation of the free OH groups takes place (methyl triflate, ditert. butyl pyridine, 4 h 60 °C, 16 h 25 °C) before acidic degradation (0.2 N trifluoroacetic acid in water) occurs. Typical examples are described in Erler et al. (28).

A suitable way to take $^1H^1H$ COSY spectra of high molecular cellulose and starch derivatives is based on the stepwise introduction of methyl and acetyl groups under complete functionalization of free OH groups. Starting from the regioselectively functionalized silylethers, the subsequent chemical modification leads, e.g., to 6-O-acetyl-2,3-di-O-methyl cellulose (Figure 9).

The ^1H NMR spectra of the prepared polymers show a high resolution. The very intensive signals of the original silyl groups are eliminated, the methylation causes a signal shift of the proton at the functionalized C atom to higher field, and the acetylation of the former silylated OH groups a signal shift to lower field. Typical examples are presented in Figure 10 as well as in (12, 13, 29).

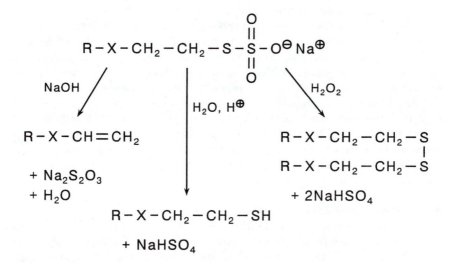

Figure 7. Subsequent reactions of Bunte salts.

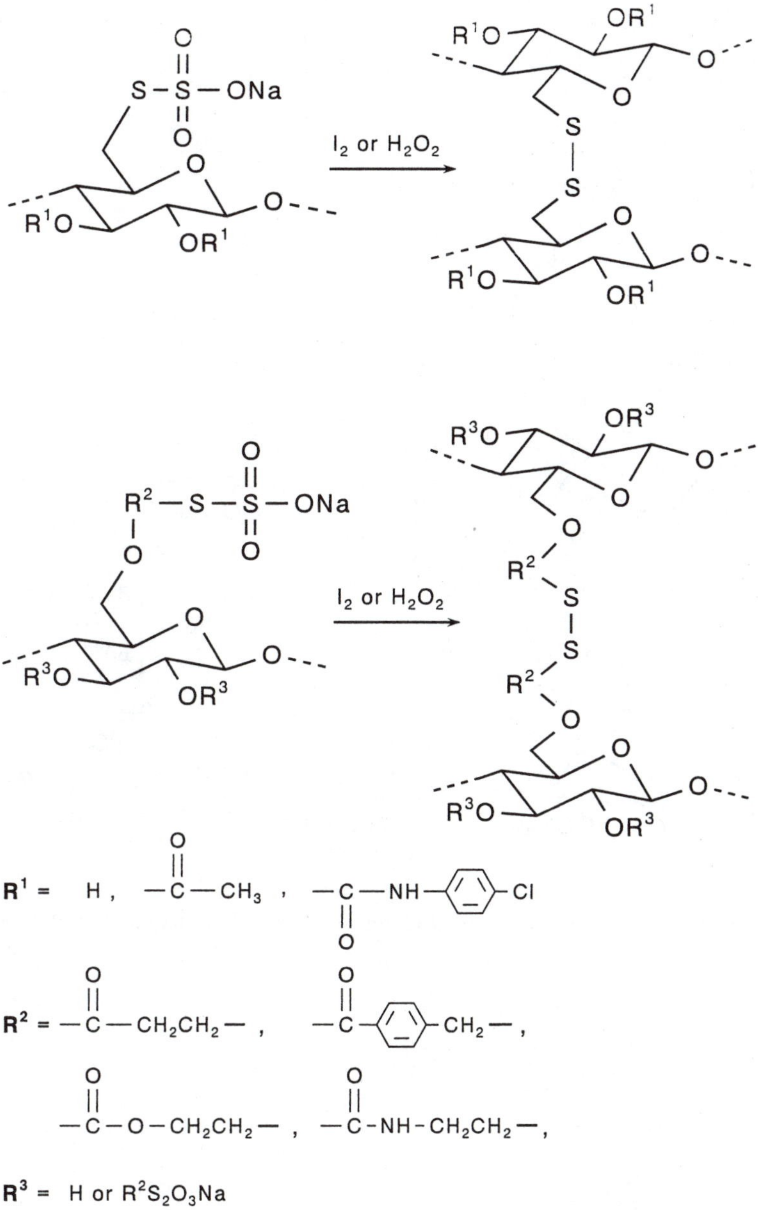

Figure 8. Crosslinking of Bunte salts of cellulose in the presence of oxidizing reagents.

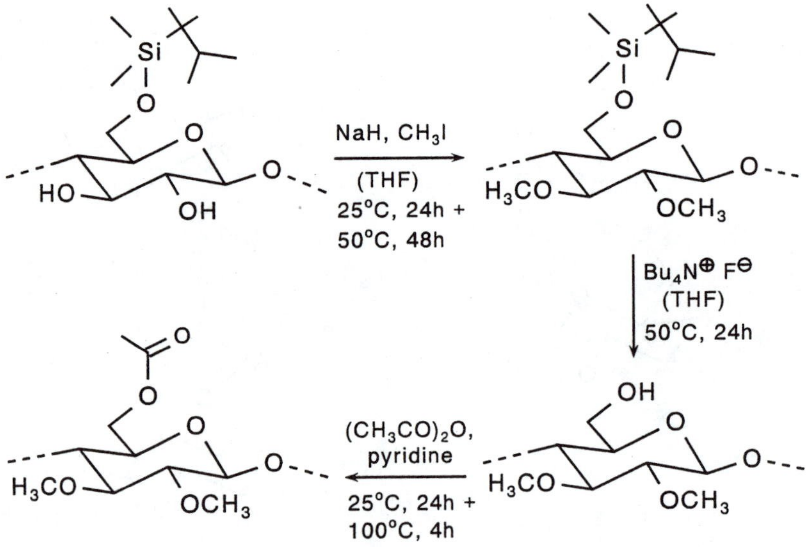

Figure 9. Synthesis of acetyl-methyl celluloses for ^{1}H^{1}H COSY NMR spectroscopy.

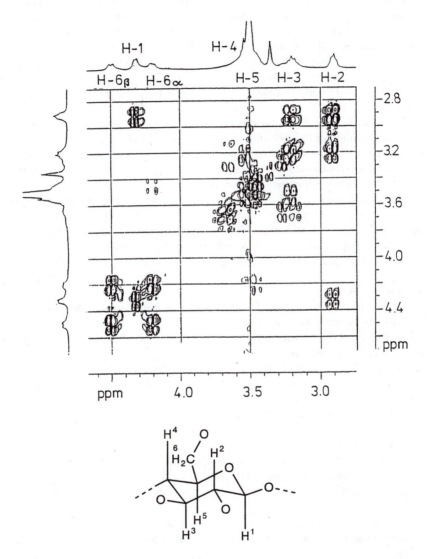

Figure 10. ^{1}H^{1}H COSY NMR spectrum of 6-O-acetyl-2,3-di-O-methylcellulose prepared from 6-O-thexyldimethylsilyl cellulose (DS = 0.78).

Conclusion

The silylation of cellulose and starch with the bulky thexyldimethylchlorsilane represents an important example of a regioselective polysaccharide functionalization. This etherification is controlled by a partial (6-O-silylation) or complete (2,3-di-O selectivity) loosening of the supramolecular structure caused by decrystallization and solvation of the polymers.

In the case of cellulose these silylethers are suitable intermediates to prepare 2,3-di-O and 3-O ethers and esters (protection group technique) as well as to synthesize 6-O and 2,6-di-O esters (activation by silylation of the OH groups) as summarized in Figure 11.

The corresponding starch functionalization additionally leads to 2-O ethers as a result of migration of the 2-O-silyl groups to position 3 (Figure 12).

In all cases the subsequent reactions may be used to prepare different types of completely functionalized polymers suitable for ^{1}H NMR analysis and investigation into structure properties relationships.

As an important example of ester intermediates in regioselective cellulose functionalization, water-soluble and film-forming Bunte salts (deoxy-thiosulfates) represent crosslinkable polymers with a wide range of additional functional groups. Prepared from cellulose p-toluenesulfonates, Bunte salts react like anionic protected alkene and thiol groups suitable for controlled polymer modification.

Acknowledgements

The author is indebted to Dr. Th. Heinze, Dr. Katrin Petzold, Dr. Armin Stein, Dipl.-Chem. Andreas Koschella, Dipl.-Chem. Juan Chamacho Gómez, and Dipl.-Chem. Kerstin Rahn for their creative work in the field of the described ethers and esters of cellulose and starch.

The financial support of the work on regioselective functionalization by the Deutsche Forschungsgemeinschaft, by the Bundesministerium für Ernährung, Landwirtschaft und Forsten, as well as by the Fonds der Chemischen Industrie of the FRG, is gratefully acknowledged.

Literature Cited

(1) Shirai, A.; Takahashi, M.; Kaneko, H.; Nishimura, S.; Ogawa, M.; Nishi, N.; Tokura, S. *Int. J. Biol. Macromol.* **1994**, *16*, 297.

(2) Kobayashi, S.; Kashiwa, K.; Kawasaki, T.; Shoda, S. *J. Am. Chem. Soc.* **1991**, *113*, 3079.

(3) Nakatsubo, F.; Kamitakahara, H.; Hori, M. *J. Am. Chem. Soc*.**1996**, *118*, 1677.

(4) Nishimura, T.; Takano, T.; Nakatsubo, F.; Murakami, K. *Mokuzai Gakkaishi* **1993**, *39*, 40.

(5) Burchard, W. *Adv. In Colloid and Interface Sci.* **1996**, *64*, 45.

(6) Nojiri, M.; Kondo, T. *Macromolecules* **1996**, *29*, 2392.

(7) Kondo, T. *J. Polym. Sci., Polym. Phys.* **1994**, *B32*, 1229.

(8) Hagaki, H.; Takahashi, I.; Natsume, M.; Kondo, T. *Polymer Bull.* **1994**, *32*, 77.

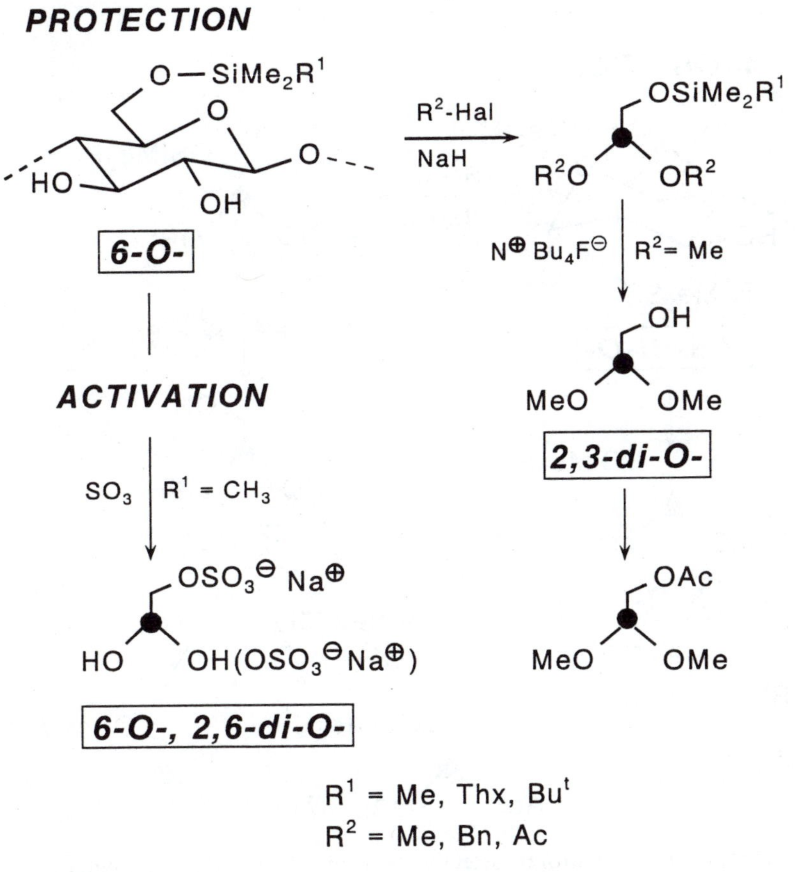

Figure 11. Regiocontrol by selective protection and activation.

MIGRATION

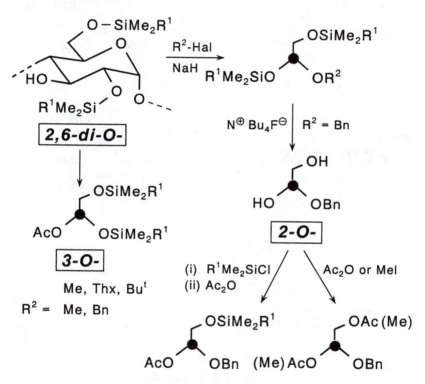

Figure 12. Regiocontrol by selective protection and silyl group migration.

(*9*) Klemm, D.; Stein, A.; Heinze, Th.; Philipp, B.; Wagenknecht, W. In *Polymeric Materials Encyclopedia*; Salamone, J.C., Ed.; CRC Press, Inc.: Boca Raton, FL, USA, **1996**, Vol. 2c; pp 1032-1054.

(*10*) Stein, A. *PhD Thesis*; University of Jena, Germany, **1991**.

(*11*) Henze-Wetkamp, H.; Zugenmaier, P.; Stein, A.; Klemm, D. *Macromol. Symp.* **1995**, *99*, 245.

(*12*) Klemm, D.; Stein, A. *J. Macromol. Sci., Pure Appl. Chem.* **1995**, *A32*, 899.

(*13*) Klemm, D.; Heinze, Th.; Stein, A.; Liebert, T. *Macromol. Symp.* **1995**, *99*, 129.

(*14*) Stein, A.; Klemm, D. *Macromol. Chem., Rapid Commun.* **1988**, *9*, 569.

(*15*) Petzold, K. *PhD Thesis*; University of Jena, Germany, **1996**.

(*16*) Mischnick, P.; Lange, M.; Gohdes, M.; Stein, A.; Petzold, K. *Carbohydr. Res.* **1995**, *277*, 179.

(*17*) Harkness, B.R.; Gray, D.G. *Macromolecules* **1990**, *23*, 1452.

(*18*) Harkness, B.R.; Gray, D.G. *Macromolecules* **1991**, *24*, 1800.

(19) Kondo, T.; Gray, D.G. *Carbohydr. Res.* **1991**, *220*, 173.

(*20*) Kondo, T. *Carbohydr. Res.* **1993**, *238*, 231.

(*21*) Camacho, Gómez, J.A.; Erler, U.W.; Klemm, D. *Macromol. Chem. Phys.* **1996**, *197*, 953.

(*22*) Camacho Gómez, J. *PhD Thesis*; University of Jena, Germany, **1997**

(*23*) Heinze, Th.; Röttig, K.; Nehls, I. *Macromol. Rapid Commun.* **1994**, *15*, 311.

(*24*) Rahn, K.; Diamantoglou, M.; Klemm, D.; Berghmans, H.; Heinze, Th. *Angew. Makromol. Chem* **1996**, *238,* 143.

(*25*) Heinze, Th.; Rahn, K.; Jaspers, M.; Berghmans, H. *J. Appl. Polym. Sci.* **1996**, *60*, 1891.

(*26*) Milligan, B.; Swan, J.M. *Rev. pure appl. Chem.* **1962**, *12*, 72.

(*27*) Schaub, M.; Wenz, G.; Wegner, G.; Stein, A.; Klemm, D. *Adv. Mater.* **1993**, *5*, 919.

(*28*) Erler, U.; Mischnick, P.; Stein, A.; Klemm, D. *Polymer Bulletin* **1992**, *29*, 349.

(*29*) Stein, A.; Klemm, D. *Das Papier* **1995**, *49*, 732.

Chapter 3

Long-Chain Cellulose Esters: Preparation, Properties, and Perspective

Kevin J. Edgar[1], Thomas J. Pecorini[1], and Wolfgang G. Glasser[2]

[1]**Eastman Chemical Company, P.O. Box 511, Kingsport, TN**
[2]**Biobased Materials/Recycling Center, and Department of Wood Science and Forest Products, Virginia Polytechnic Institute and State University, Blacksburg, VA 24061**

Obstacles to the heterogeneous esterification of cellulose with long-chain alkanoic acids were overcome (a) by reacting cellulose in homogeneous phase solution (in DMAc/LiCl) in the presence of dicyclohexyldicarbodiimide (DCC) and 4-pyrrolidinopyridine (PP) or tosyl chloride; or (b) by reaching cellulose heterogeneously, in DMAc-suspension, with anhydrides in the presence of Ti (IV) isopropoxide. Long-chain cellulose esters (LCCEs) were prepared with constant and uniform degree of substitution (DS), and with stoichiometric control. Cellulose ester structure was found to influence solubility, flexural modulus, melt viscosity, thermal transitions, and enzyme recognition. LCCEs showed melt-rheology behaviors that are normally found only in plasticized esters. Melt viscosity and flexural modulus of these internally plasticized cellulose esters were found to be related to the solubility parameter over a broad range of ester-substituents and DS. Practical synthesis of LCCEs is possible leading to melt-processable (uncompounded) thermoplastic polymers on cellulose basis.

Classically, cellulose esters are prepared using mineral acid catalysts and carboxylic anhydrides in the corresponding carboxylic acid diluent with acetic, propionic, butyric and valeric acids, or mixtures thereof. Due to slow reaction rates and competitive cellulose chain cleavage, it is difficult to prepare esters with longer chain acids by this method (1). Cellulose esters with short chain carboxylic acids (C-2 and C-3) have high glass transition and melting temperatures and relatively low decomposition temperatures. In some cases (low-DS-acetates, for example), the decomposition temperature is actually lower than the T_g and T_m, which is similar to parent cellulose. Cellulose esters are therefore normally melt-processed in the presence of plasticizers that enlarge the processing window. Attachment of longer chain ester groups, if it could be accomplished efficiently, should provide an effective internal plasticizer which potentially eliminates the need for external plasticizers which may be prone to extraction or volatilization. This would also substantially reduce the solubility parameter of the cellulose ester, enhancing (a) the solubility in less polar solvents, and (b) the blending with less polar polymers than are miscible with currently available materials.

Previous attempts to prepare long-chain cellulose esters have been severely limited in scope and in practicality. The first route was the so-called "impeller" method by Clarke and Malm in 1932 (2). They attempted to compensate for the low reactivity of the sterically encumbered long-chain acyls by converting them to mixed anhydrides with reactive acyls such as chloroacetyl or methoxyacetyl. Thus, reaction of cellulose with acetic acid, stearic acid, and chloroacetic anhydride in chloroacetic acid, with magnesium perchlorate catalysis, gave a cellulose acetate stearate. The "impeller" acids efficiently catalyze chain cleavage and make the degree of polymerization (DP) difficult to control. Residues of the impeller acid may also be found on the product. A modern version of the impeller method, which uses trifluoroacetic anhydride as the impeller (3), effectively conserves DP but provides access only to fully substituted esters.

A second approach to the preparation of long-chain cellulose esters was described by Malm and coworkers in 1951 (1). They found that acylation with the appropriate acid chloride in 1,4-dioxane with pyridine as an acid acceptor and catalyst afforded cellulose triesters ranging from acetate to hexanoate to palmitate. Surprisingly, only a moderate amount of DP-reduction occurred in this reaction. Although this is a convenient method to form triesters, this method is limited to high-DS cellulose esters, and furthermore it requires amorphous, and thus highly reactive, regenerated cellulose as starting material. This introduces the inefficiency of having to acylate cellulose twice, with an intervening deacetylation step. However, the availability of this method allowed Malm et al. (4) to prepare the entire series of cellulose triesters and investigate their properties. A profound impact on such properties as melting point, solubility, moisture regain, density, and tensile strength was firmly established. This research indicated an area of promising new materials which was limited only by the missing dimension of DS-modification.

Recently Tao et al. have described a new and more direct method to synthesize cellulose esters with long-chain acids (5). The extremely simple method involves the reaction of mercerized cellulose with an acid chloride at elevated temperature under vacuum to facilitate the removal of the by-product HCl. Partially substituted cellulose esters with a long chain (palmitate) were prepared in this manner for the first time, with DS ranging from zero to 2.5. Although simple, the method is limited by the tendency of the co-produced HCl to cause chain cleavage and loss of DP. Furthermore, esterification does not proceed past DS 1.5

unless at least 10 equivalents of the acid chloride are used, even if no solvent is used. Most seriously, the product esters are not homogeneously substituted and are not soluble in organic solvents (6).

Long-chain Cellulose Esters (LCCEs) by Solution Acylation

In the 1980s, Turbak (7) and McCormick et al. (8) showed that cellulose dissolves in a solution of lithium chloride (LiCl) in N,N-dimethylacetamide (DMAc) under well-defined conditions. Of equal importance was the subsequent work by several investigators, with major contributions from the McCormick laboratory, which showed that reactions of cellulose could be performed in this solvent system (9). With respect to ester synthesis, McCormick, Diamantoglou (10), Samaranayake and Glasser (11,12), and others (13-15) showed that cellulose could be induced to react with carboxylic anhydrides, acid chlorides, and other electrophilic acyl derivatives using mineral acid or alkaline catalysts to afford partially substituted cellulose esters directly, without the need for a hydrolysis step. This was significant because conventional cellulose esterification is a heterogeneous reaction until the cellulose is nearly fully reacted; therefore, the only previous way to obtain processable (solvent-soluble or melt-flowable) cellulose esters was to esterify the cellulose fully and then back-hydrolyze the derivative to the desired DS. By reacting cellulose in DMAc/LiCl solution with 3.8 eq of propionic anhydride and 0.1 eq of acetic anhydride at 100°C, and without any catalyst, we now confirmed that cellulose may react with short-chain anhydrides to form a mixed cellulose acetate propionate with DS_{pr} 2.5 and DS_{ac} 0.10.

Application of the DMAc/LiCl solution process to long-chain cellulose esters is complicated by two factors that are related (a) to solubility of reagents and reaction products, and (b) to the susceptibility of cellulose to depolymerization. As chain-length gets higher than hexanoic anhydride, the requisite anhydrides either are poorly soluble or insoluble in the reaction mixture, or they contribute to a substantial inefficiency of esterification due to the loss of the by-product carboxylic acid as a consequence of esterification. Loss of solubility results in heterogeneous acetylation and a heterogeneous and insoluble product (see Entry 1, Table I). We were able to address this problem by adapting the impeller method of Clarke and Malm. Mixed anhydrides of reactive acyl groups were formed with either 4-pyrrolidinopyridine (PP) or with tosic acid; or by using the corresponding acid chlorides with either pyridine or triethylamine as acid acceptor. The use of the impeller method or the acyl chloride improved efficiency by preventing the loss of the by-product carboxylic acid during esterification with anhydrides as well as contributing to solubility in general.

The use of the mixed anhydride with PP was pioneered by Samaranayake and Glasser (11). The reaction is mediated by the presence of dicyclohexyl-

Table I. LCCEs by Solution Esterification in DMAc/LiCl.

Entry	Acylating Agent	Equivalents per AHG	Acid Scavenger	Reaction Time (h)	Reaction Temp. (C)	DS	IV (DMSO)	M_n^d (10^3)	Solubility
1	Stearic Anhydride	1.00	None	1	110	0.95[a]	--	--	Insoluble
2	Hexanoyl Chloride	1.00	Pyridine	0.5	60	0.89[b]	1.10	41	DMSO, NMP, Pyridine
3	Hexanoyl Chloride	2.00	Pyridine	0.5	60	1.70[b]	0.76	90	Acetone, MEK, CHCl$_3$, HOAc, THF, DMSO, NMP, Pyridine
4	Lauroyl Chloride	2.00	Pyridine	0.5	60	1.83[b]	1.00[c]	67	Pyridine
5	Stearoyl Chloride	1.00	Pyridine	1	105	0.79[a]	0.11[e]	27	Acetone, MEK, CHCl$_3$, HOAc, THF, DMSO, NMP, Pyridine
	Acetic Anhydride	3.00				1.93[b]			

[a]DS by alcoholysis/GC. [b]DS by ^{1}H NMR. [c]Phenol/Tetrachloroethane. [d]M_n by GPC in THF except Entry 2 (NMP).
[e]Partly insoluble. AHG = Anhydroglucose

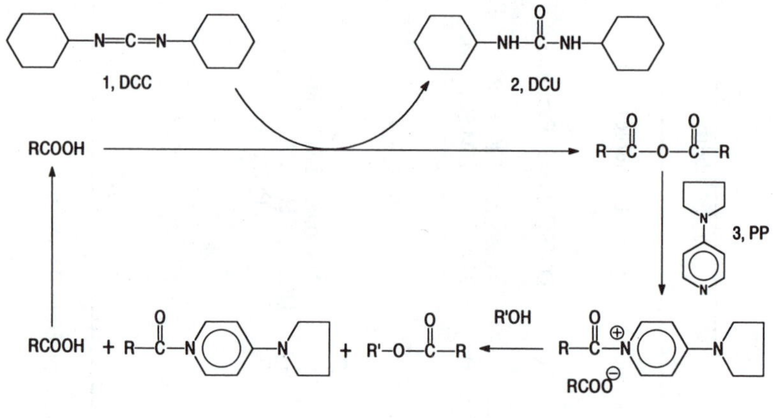

Scheme 1. Mechanism of esterification with DCC/PP reagent.

carbodiimide (DCC), a powerful condensation agent which is well-known to couple amines and carboxylic acids in peptide and protein chemistry (16), but it is rarely used in polymer modification reactions. When DCC is applied in combination with anhydrides, the by-product carboxylic acid is recycled by formatting a mixed anhydride with PP (Scheme 1). Mixed anhydrides are soluble in the reaction mixture up to hexanoic acid; beyond C-6 carboxylic acids, the reaction mixtures become heterogeneous (11).

The second application of the impeller method has involved the use of tosyl chloride (TsCl) (15). When we added a solution of long-chain alkanoic acids (C-12 to C-20) in DMAc to a cellulose solution in DMAc/LiCl, the non-polar aliphatic carboxylic acid precipitated (14). This process was subsequently reversed by adding TsOH in DMAc. The homogenization of the reaction mixture indicates the formation of a mixed anhydride of the alkanoic acid with TsOH. By heating the reaction mixture to 50 to 70°C in the presence of sufficient acid acceptor (pyridine or equivalent), a long-chain cellulose ester (LCCE) is formed that is (a) soluble in the reaction mixture; (b) that has a well-preserved DP; and (c) that is completely devoid of tosyl or Cl-substitution. C-12 to C-20 derivatives with DS between 2.8 and 2.9 were obtained using a stoichiometric ratio of two equivalents of acid per cellulose hydroxyl (14).

Another method of overcoming the solubility problems routinely encountered during cellulose esterification in DMAc/LiCl involves the use of acid chlorides combined with pyridine (or equivalent) as acid acceptor. Although acid chlorides are soluble in the DMAc/LiCl solvent mixture, they quickly become insoluble when the necessary acid acceptor is added. Triethylamine is less useful for this purpose than is pyridine. Reaction mixture homogeneity could also be triggered by product insolubility. We found that this difficulty could be circumvented by adding the acid chloride in solution in a small amount of a co-solvent which was a better solvent for the product; tetrahydrofuran (THF) proved useful for this purpose. Soluble and melt-processable products were obtained that are represented by Entries 2 to 5 of Table I. We were able to make long-chain esters and mixed esters of cellulose with hexanoate, laurate, and stearate groups, with DS ranging from 0.9 to 2.7, using this procedure.

These procedures provide access to a broad range of materials, with one or more long-chain ester substituents, having any DS and any chain length. This flexibility has not previously been available to cellulose chemists.

Heterogeneous Acylation Using Ti Catalyst in Amide Solvent

The methods we have so far described provide access to a class of cellulose esters which was previously inaccessible, and so expand the armament of the materials scientist. These methods have the advantage that they make available the entire range of ester-DS and chain length by simply choosing the acylating agent and stoichiometry; however, the DMAc/LiCl system is less than ideal from the perspective of efficiency. Because of the viscosity of cellulose solutions in this system, cellulose concentrations above about 8% are impractical. The procedure for dissolving cellulose in this system is complex, all of the components are quite hygroscopic, and the expense of lithium chloride is a practical issue. It would be desirable to find a more efficient route to these long-chain esters.

The use of titanium (IV) alkoxides as catalysts for cellulose esterification in carboxylic acid diluents has been known since the work of Tamblyn and Touey in the 1960s (17). These Lewis acid catalysts require much higher temperature (120 - 180°C) than conventional mineral acid catalysts. We have found that titanium (IV) isopropoxide is an excellent catalyst for the reaction of cellulose in the appropriate carboxylic acid solvent with short-chain anhydrides for preparing triesters. We have not found this heterogeneous system to be a practical direct route to partially substituted esters. We wondered if a more powerful solvent for partially substituted esters, such as a polar aprotic solvent like a dialkylcarboxamide, would facilitate their preparation as homogeneous materials. Reaction of cellulose with a mixture of hexanoic and acetic anhydrides (2 eq each), in DMAc diluent, with titanium (IV) isopropoxide catalyst, afforded a homogeneous solution. Isolation of the product by precipitation gave a cellulose acetate hexanoate, with DS_{ac} 1.88, DS_{hex} 0.91 (Entry 1, Table II). Several features of this reaction are of interest; the product is partially substituted and apparently homogeneous, as judged by its solubility in a wide range of solvents. Despite the high reaction temperature, a high molecular weight product was obtained, which is perhaps attributable to the moderating effect of the mildly alkaline amide solvent on the acidic catalyst. While the cellulose clearly reacted preferentially with the shorter-chain anhydride, a substantial amount of reaction with the hexanoic anhydride did occur.

The generality of this method is illustrated in Table II. In each case (except Entry 7), DMAc was the diluent and titanium (IV) isopropoxide (3% by weight based on cellulose) was the catalyst. These seven examples illustrate the fact that it is possible by this method to esterify cellulose with acyl groups as large as palmitate (C-16). High temperatures are required, but even so, the products are of relatively high molecular weight. The total DS of these products ranges from 2.5-2.7. It should be noted here that the esterification with this catalyst in the amide solvent is rather slow, with reaction times of 6-12 h common. This slow esterification, however, is what permits easily repeatable isolation of partially substituted esters. The esterification rate, when the cellulose fully dissolves, is sufficiently slow that dissolution may be used as an endpoint, with highly repeatable results in terms of product DS. The product T_g varies in a predictable way depending on the relative content of acetyl and long-chain acyl. All of these products are soluble in a wide range of solvents, including acetone, acetic acid, NMP, $CHCl_3$, and THF. Even where only long-chain anhydride is used (Entry 3), a partially substituted, homogeneous long-chain ester is smoothly obtained. It is especially interesting that in this case the product contains a DS 0.12 of acetyl, even though no acetic anhydride was used. We believe that the acetyl in the product must have come from reaction with the N,N-dimethylacetamide solvent. The results of Entry 4, Table III, support this contention. The hypothesis is further confirmed by Entry 7 of Table II in which a solution is identified that allows exclusive attachment of a long-chain acyl. In this reaction, we substituted a tetraalkylurea, N,N-dimethyl-2-imidazolidinone (DMI), for DMAc as diluent. Because there is no solvent acetyl to react with cellulose, a cellulose hexanoate without acetyl groups is expected and that is what is obtained. There does appear to be slightly more chain cleavage when the DMI diluent is used; the reaction rate and efficiency with respect to the anhydride (DS_{hex} 2.73 obtained with only 3 eq hexanoic anhydride used) are also substantially higher than with DMAc. It is clear from the results in Table II that the cellulose acylation rate with a given anhydride declines predictably as the anhydride chain length increases. Comparing the results of reacting cellulose with two equivalents each of acetic anhydride and a long-chain anhydride, proceeding up the series from hexanoic to nonanoic to lauric to palmitic

Table II. LCCEs by Titanium (IV) Isopropoxide-Catalyzed Esterification in DMAc.

Entry	Anhydride(s)	Equivalents per AHG	Reaction Time (h)	Reaction Temp. (C)	DS (¹H NMR)	IV (NMP)	M_n/1000[a]	M_w/1000[a]	T_g (DSC)
1	Acetic	2.00	9	155	1.91	1.39	40	164	149°C
	Hexanoic	2.00			0.75				
2	Acetic	1.00	9	155	1.38	0.90	35	113	122°C
	Hexanoic	3.00			1.36				
3	Acetic	0.00	6	155	0.12	0.94	33	245	119°C
	Hexanoic	4.50			2.39				
4	Acetic	2.00	11	145	2.03	1.18	44	177	129°C
	Nonanoic	2.00			0.70				
5	Acetic	3.50	12	140	2.40	2.12	96	295	165°C
	Lauric	1.00			0.20				
6	Acetic	2.00	12	145	2.06	0.29[b]	33	125	156°C
	Palmitic	2.00			0.42				
7[c]	Acetic	0.00	7	140	0.00	0.44	23	61	104°C
	Hexanoic	3.00			2.73				

[a]By GPC in NMP. [b]Partly insoluble. [c]N,N-Dimethylimidazolidinone (DMI) used as solvent instead of DMAc. AHG = Anhydroglucose.

Table III. Cellulose Acetate Nonanoates by Titanium (IV) Isopropoxide-Catalyzed Esterification in DMAc.

Entry	Anhydride	Equivalents per AHG	Reaction Time (h)	Reaction Temp. (C)	DS (¹H NMR)	IV (NMP)	M_n/1000	M_w/1000	T_g (DSC)	T_m (DSC)
1	Acetic	2.00	11	145	2.03	1.18	44	177	129	183[a]
	Nonanoic	2.00			0.70					
2	Acetic	3.00	8	145	2.44	1.71	43	220	161	180[a]
	Nonanoic	1.00			0.26					
3	Acetic	1.00	13	155	1.59	1.16	44	182	118	174[b]
	Nonanoic	3.00			1.11					
4	Acetic	0.00	13	160	1.11	0.89	31	200	110	167[b]
	Nonanoic	4.00			1.35					

[a]First scan only. [b]By modulated DSC. AHG = Anhydroglucose. M_n, M_w by GPC in NMP.

(Entries 1, 4, 5, and 6, Table II) shows that the DS of long-chain acyls in the product declines with increasing chain length. It remained to be seen what chain length was necessary to obtain sufficient processability for a self-plasticized polymer.

We performed a series of experiments (Table III) reacting cellulose with acetic and nonanoic anhydrides with titanium (IV) isopropoxide in DMAc diluent to learn more about the nature and properties of materials available via this chemistry. To maintain practical reaction rates, the reaction temperatures ranged from 145°C for the more acetic anhydride-rich runs, to 160°C for the run with only nonanoic anhydride. The cellulose acetate nonanoates obtained were all of reasonably high molecular weight, as indicated by IV and GPC. The T_g decreases with increasing nonanoyl content, as expected; nonanoyl content can in turn be rationally controlled by the ratio of anhydrides used. The maximum nonanoyl content obtainable is DS 1.35, with a DS_{ac} of 1.11 (Entry 4); clearly, nonanoic anhydride has sufficient bulk (hydrophobicity may also be a factor in early stages of the reaction with the hydrophilic cellulose) that the reaction of cellulose with DMAc becomes competitive in rate. The melting temperature also seemed to decline with increasing nonanoyl substitution; modulated DSC was an invaluable aid in identifying the T_m in the high-nonanoyl esters.

Mechanical and Rheological Properties

In several respects, esterification of cellulose with long-chain substituents can provide superior properties over their conventional short side-chain counterparts. To their disadvantage, currently commercial cellulose esters (CA, CAP, CAB) possess melt viscosities that are too high to permit melt processing at reasonable temperatures without adding external plasticizer. The compounding step associated with that addition not only increases the cost of the final product, but external plasticizers may also leach out of the compounded plastic, producing undesired smell, taste or feel. Furthermore, selection of an effective plasticizer is limited by solubility. Therefore, the only good plasticizer for a particular mixed ester may possess undesirable properties.

In contrast, increasing the amount of long-chain ester essentially increases the amount of covalently bonded internal plasticizer, and does not produce the problems cited above. The chemical structure of a typical LCCE with covalently bonded internal plasticizer is illustrated in Scheme 2. The waxy substituents that surround the cellulose backbone provide significant plasticization at temperatures at which most cellulose esters are glassy. We found that LCCEs with fatty acid substituents larger than C-12 had distinctly dual morphology, with both the cellulosic and the waxy phase undergoing separate glass-to-rubber and rubber-to-melt transitions (14). The T_g-, T_c- and T_m-transitions of the waxy (plasticizing) phase of LCCEs having substituents in the range of C-12 to C-20 reveal significant

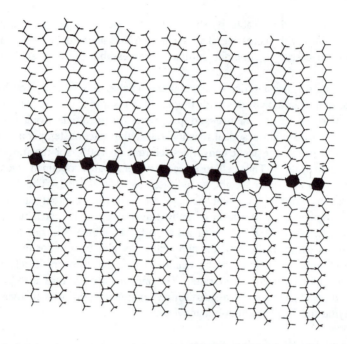

Scheme 2. Molecular model of cellulose eicosanoate (C-20) with ester substituents aligned regularly as they might be in a crystal lattice (From ref. 14).

molecular mobility in the temperature range between -60 to +60°C (Fig. 1) (14). As expected, the mechanical and rheological properties of several of the LCCEs, listed in Tables IV and V, vary with changes in ester side-chain length and degree of substitution similar to the effect of adding greater amounts of external plasticizer. Indeed, with sufficient side-chain length and degree of substitution, the properties (particularly modulus and melt viscosity) equal those of plasticized CA and CAP, affording LCCEs that can be easily processed without external plasticizer. These properties all reflect the ability of a plasticizer, both external and internal, to increase free volume. For example, T_g decreases with increasing external plasticizer content as well as with increasing DS and chain length (see Figure 2). Resins with lower T_g should exhibit lower melt viscosities when measured at 220°C. However, as shown in Figure 3, at a given T_g, cellulose esters with longer side-chain lengths and higher DS have lower melt viscosities than externally plasticized resins of shorter chain length and lower DS. This suggests that long side-chains are more

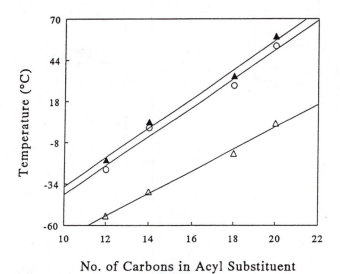

Figure 1. Thermal transitions of the waxy phase of LCCEs in the temperature range of -60 to +70°C. T_m by DSC: -▲-; T_g by DMTA: -Δ-; T_c by DSC: -O-. (In accordance with ref. 14).

Table IV. Rheological and Mechanical Data for CA, CAP, CAB, CAH and CAN.

Material	DS of long chain	Plasticizer Content (%)	Absolute Mw (g/mole)	Viscosity at 1/s (Pa·s)	Viscosity at 1/100 s (Pa·s)	Viscosity @ 100 rad/s normalized to Mw = 100,000 (Pa·s)	T_g by DMTA (C)	Modulus (mPa)
CA	0	33	61000	2783	329	1766	103	1650
CA	0	28	61000	4009	317	1702	107	2000
CA	0	18	61000	10320	1243	6673	132	2910[a]
CA	0	0	61000			47500[a]	203	4410[a]
CAP	0.73	25	79000	919	261	582	120	1520
CAP	0.73	20	79000	2126	419	934	125	1930
CAP	0.73	0	79000			8500[a]	188	3580[a]
CAP	2.60	14	103000	600	306	277	114	1430
CAP	2.60	10	103000	1350	450	407	122	1660
CAP	2.60	6	103000	3000	830	751	131	1960
CAP	2.60	0	103000	10600	1258	1138	156	2410[a]
CAB	1.67	0	90000				141[b]	2070
CAH	0.75	10	164000	2971	702	131	129	1670
CAH	0.75	5	164000	7032	1107	206	137	2120
CAH	0.75	0	164000	855		340[a]	149[b]	2410[a]
CAH	1.36	0	113000	855	353	233	128/122[b]	1310
CAH	2.39	0	245000	1175	104	5	111/119[b]	620
CAN	0.70	0	177000	10300	952	137	129[b]	
CAN	1.11	0	182000	1929	371	48	118[b]	690
CAN	1.35	0	200000	547	73	7	110[b]	

[a] Extrapolated; [b] by DSC.

Table V. Physical Property Data for Compounded CAH and CAN.

Cellulose Ester		CAH	CAH	CAH	CAH	CAN	CAN
DS		2.39	1.36	0.75	0.75	1.35	1.35
Plasticizer	%DOA	0.0	0.0	5	10	0.0	5
PROPERTY	UNITS						
Density	g/cm^3	0.877	1.17	1.21	1.198	1.11	1.1
IV after molding			0.998	1.194	1.194	0.77	0.73
Flex Strength	MPa	20.1	40.7	57.7	45.7	19.0	13.4
Flex Modulus	MPa	668	1322	2115	1674	606	434
Creep Modulus[a]	MPa	179	627	792	578	158	124
0.45 MPa HDT	C	61	82	100	87	65	57
1.82 MPa HDT	C	43	62	68	58	51	47
Hardness	R,L scale	R28.3	R87.3	L36.6	L23.5	R23.6	R8.0
23C Notched Izod	J/m	194	184	231	230	48	123
OC Notched Izod	J/m	10	136	134	127	8	58
-40C Notched Izod	J/m	23	24	67	102	10	9
23C UnNotched Izod	J/m	NB	NB	2016	NB	NB	NB
OC UnNotched Izod	J/m	NB	NB	1663	NB	NB	NB
-40C UnNotched Izod	J/m	267	706	1512	1260	121	207

[a] 6.89 MPa, 23 C, 200 h; NB = No Break.

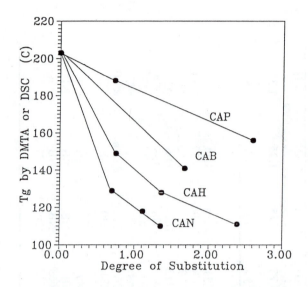

Figure 2. Relationship between T_g -values of cellulose esters (without plasticizer)
and DS.

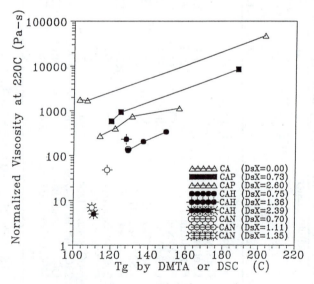

Figure 3. Normalized viscosities of plasticized and non-plasticized cellulose
esters plotted as a function of T_g.

effective as flow aids than are external plasticizers, which may be related to reduced hydrogen bonding or the creation of free space between the cellulose backbones due to the addition of long non-polar side chains. In addition to the plasticizing effect of bulky substituents, it can not be ruled out that liquid crystallinity factors may not also contribute to the facilitation of melt processing. Although we did not detect any direct, overt evidence of liquid crystallinity, we did not specifically and systematically look for it.

Because the side chains are comprised of CH_2-units having a low solubility parameter, one would expect the overall solubility parameter of the LCCEs to decrease with increasing DS and chain length. This expectation is borne out when solubility parameters calculated by the method of Coleman (18) are related to the DS (Table VI). Long-chain esterification, therefore, could allow the use of alternative external plasticizers that are insoluble in currently commercial cellulosic materials and may also offer the opportunity to create miscible blends of cellulosics with more hydrophobic polymers than are compatible with conventional cellulose esters such as CA, CAP and CAB. Furthermore, the solubility parameter as it is calculated for these materials, in essence measures how much (internal) plasticizer is present. Material properties that are controlled by plasticization, such as T_g, normalized melt viscosity, and modulus, all strongly correlate with the solubility parameter (Figure 4). It is possible, therefore, to design a cellulosic material with any desired properties (as far as they are related to solubility parameter) simply by identifying the matching solubility parameter from Figure 4, and esterifying with the associated combination of DS and chain length. The principal advantage of selecting a low-DS long-chain cellulose derivative over a high-DS short-chain ester with similar effective plasticization for melt processing rests with (a) the long side chains effectiveness as processing aids; (b) with their non-fugitive (permanent) character; and (c) with their likely better compatibility with more hydrophobic polymers in blending. While the latter is supported by solubility parameter data, the experimental evidence for it still needs to be developed.

Other Properties

Mixed cellulose esters show crystallinity across the entire spectrum of substituent mixing. T_m declines sharply as DS of the bulky substituent rises from 0 to 1, followed by a more moderate decline as DS increases from 1 to 3 (Figure 5) (13). When the number of carbons in the long-chain ester substituent increases beyond 12 (laurate), a series of thermal transitions is reported which represents motion by both ester substituents and the cellulosic main chain (14). Broad crystallization and melting transitions attributed to side-chain crystallinity range between -19 to +55°C, and these side-chain T_m- and T_c- transition temperatures increased by 10°C per carbon atom of the ester substituent (13, 14).

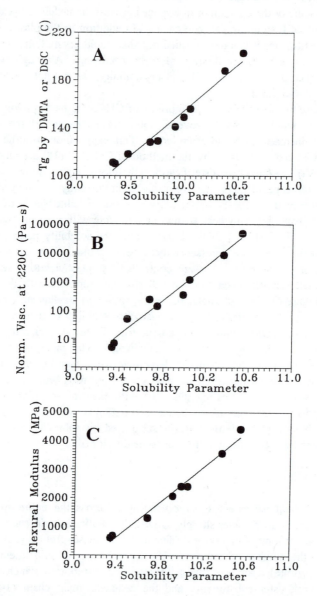

Figure 4. Properties for various cellulose esters without plasticizer plotted as a function of calculated solubility parameter: A) T_g-values, B) normalized viscosities, and C) flexural modulus.

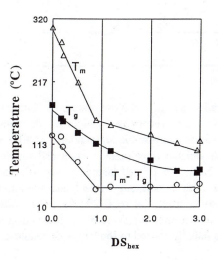

Figure 5. Relationship between thermal transitions (T_m, T_g and T_m - T_g) and DS with hexanoyl groups, of CAHs (In accordance with ref. 13).

Table VI. Solubility Parameters of Selected Cellulose Esters.

Material	DS (H,N,P,B) Total DS = 2.7	Solubility Parameter $(cal/cm^3)^{1/2}$
CA	0.00	10.53
CAP	0.73	10.36
CAP	2.60	10.04
CAB	1.70	9.90
CAH	0.75	9.98
CAH	1.36	9.67
CAH	2.39	9.33
CAN	0.70	9.74
CAN	1.11	9.47
CAN	1.35	9.35
Cellulose		14.02
Polyethylene		8.00

The biodegradability of cellulose esters is limited by the ability of cellulolytic enzymes to recognize cellulose derivatives as degradable polysaccharides. Cellulase enzyme-biodegradability (CEB) was probed using a series of LCCEs with variable substituent content and substituent size (19). Our results suggest that the ability of cellulase enzymes to recognize cellulose esters declines rapidly with both substituent content and substituent size (Fig. 6). As substituent size increases from C-4 to C-20, the DS at which a significant (i.e., greater than 20% glucose formation) degradation occurs decreases from 1.8 to 0.5. This rate of decline is linearly related to the length of cellulose ester side chain (Fig. 6) (19).

Conclusions:

1. The synthesis of unusual cellulose esters is possible by homogeneous-phase reaction chemistry using DMAc/LiCl as a solvent system, or by heterogeneous reaction of cellulose with titanium (IV) isopropoxide as catalyst and DMAc as diluent.

2. Homogeneous-phase reactions may involve carboxylic anhydrides as well as acyl chlorides and a series of mixed anhydrides. These may involve the traditional (impellers) of Clarke and Malm (i.e., chloroacetyl or methoxyacetyl) or 4-pyrrolidino pyridine (in combination with DCC) or tosyl chloride.

3. The new reaction systems afford greater reagent efficiency, greater uniformity, and greater product control via stoichiometric reagent mixing.

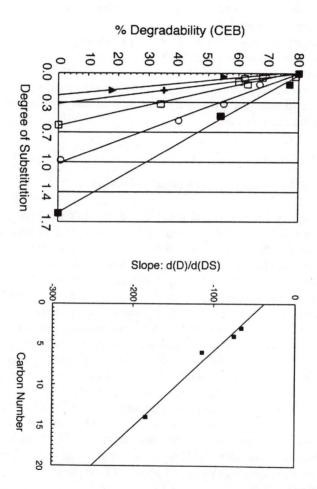

Figure 6. Cellulolytic enzyme biodegradability (CEB) of cellulose esters in relation to DS (A); and the relationship between the slope of CEB/DS vs. substituent size in number of carbons per acyl substituent (B) (From ref. 19).

4. Cellulose ester structure dramatically influences solubility, flexural modulus, melt viscosity, thermal transitions, and enzyme recognition.

5. Cellulose esters with variable substituent content and substituent type, and cellulose mixed esters, represent a novel class of engineered biobased materials with potential use in biodegradable and biocompatible thermoplastic materials.

6. Relationships of certain physical properties (e.g. flexural modulus, melt viscosity) with easily calculated solubility parameters have been established for cellulose esters over a broad range of ester chain-length and DS. The materials scientist can now *design* a cellulose ester whose properties (those related to solubility parameter) may be predicted with confidence *before* preparation in the lab.

APPENDIX

Experimental Evaluation of Mechanical Properties

To identify the mechanical properties of the LCCEs, larger quantities of several of the cellulose ester derivatives were pelletized on a two-roll mill. After the compounded pellets were dried at 80°C for 16 h, test specimens were molded on a Newbury molding machine using a melt temperature of 220°C and a mold temperature of 35°C. The CAH (DS_{hex} 0.75) was compounded with 5% or 10% di(2-ethylhexyl)adipate (DOA) plasticizer. The other two CAH materials (DS_{hex} 2.39 and DS_{hex} 1.36) were compounded without adding plasticizer. The CAN (DS_{non} 1.35) was split into two halves for compounding with either 0% or 5% DOA plasticizer. In addition, data are also shown for several commercial cellulose acetate (CA), cellulose acetate propionate (CAP) and cellulose acetate butyrate (CAB) materials. The CA material was plasticized with DEP and the CAP materials (DS_{pr} 0.73 and DS_{pr} 2.60) were plasticized with DBP and DOA, respectively. The CAB was also plasticized with DOA. (The total DS of all these materials was 2.7.) Physical properties were measured by standard ASTM methods using standard conditions. Dynamic Mechanical Thermal Analysis (DMTA) was run on a Polymer Laboratory DMTA in flexure at 1 Hz with a heating rate of 4°C/min. Creep modulus was measured after 200 h of flexural creep at 1000 psi applied load and at 23°C.

Melt rheology was performed on a Rheometrics Dynamic Analyzer (RDA) in a nitrogen blanket. Frequency scans were taken between 1/s and 1/100 s. The viscosities were all measured at 220°C, which is a reasonable processing temperature for cellulose esters, sufficiently below the temperature at which significant degradation occurs (~260-280°C). Table IV lists melt viscosity at 1/s and 1/100 s, as well as melt viscosities at 1/100 s normalized to a molecular weight of 100,000. Because melt viscosity is influenced by molecular weight in addition to ester type, an attempt to filter out the effect of molecular weight was performed by normalizing all the data to a molecular weight of 100,000 using the relation η C * $Mw^{3.4}$ (20). Prior investigations using a variety of cellulose esters (CAP, CAB and CA) had demonstrated a strong linear relation between plasticizer content and log(viscosity); this relation was then applied to the plasticized CA, CAP and CAB materials to obtain values of melt viscosity for these materials at 0% plasticizer content. These extrapolations are shown in Figure 7, and also listed in Table IV. The flexural modulus values for unplasticized CA, CAP and CAB shown in Table IV were also generated by extrapolating known values back to the 0% plasticizer condition, as shown in Figure 8.

Literature Cited

1.   Malm, C. J.; Mench, J. W.; Kendall, D. L.; Hiatt, G. D. *Ind. Eng. Chem.* 1951, 43, 684-688.

2.   Clarke, H. T.; Malm, C. J. U. S. 1,880-808, 1932.

3.   Morooka, T.; Norimoto, M.; Yamada, T.; Shiraishi, N. *J. Appl. Poly. Sci.*. 1984, 29, 3981-3990.

4.   Malm, C. J.; Mench, J. W.; Kendall, D. L.; Hiatt, G. D. *Ind. Eng. Chem.* 1951, 43, 688-6921.

5.   Kwatro, H. S.; Caruthers, J. M.; Tao, B. Y. *Ind. Eng. Chem., Res.* 1992, 31, 2647-2651.

6.   Tao, B. Y., private communication.

7.   Turbak, A. F.; El-Kafrawy, A.; Snyder, F. W.; Auerbach, A. B. U. S. Patent 4,302,252, 1981.

8.   McCormick, C. L. U. S. Patent 4,278,790, 1981.

9.   Dawsey, T. R.; McCormick, C. L. *Rev. Macromol. Chem. Phys.* 1990, C30, 405-420.

10.   Diamantoglou, M.; Kuhne, H. *Das Papier* 1988, 42, 690-696.

11.   Samaranayake, G., W. G. Glasser. Carbohydrate Polymers, 1993, 22, 1-7.

12.   Samaranayake, G., W. G. Glasser. Carbohydrate Polymers, 1993, 22, 79-86.

13.   Glasser, W. G., G. Samaranayake, M. Dumay, V. Dave, J. Polym. Sci.: Pt. B, 1995, 33, 2045-2054.

14.   Sealey, J. E., G. Samaranayake, J. G. Todd, W. G. Glasser. J. Polym. Sci.: Pt. B, 1996, 34, 1613-1620.

15.   Shimizu, Y., J. Hayashi, Cell. Chem. Technol., 1989, 23, 661.

16.   Haslam, E., Tetrahedron, 1980, 30, 2409-33.

17.   Touey, G. P.; Tamblyn, J. W., 1961, U. S. Patent 2,976,277.

18.   M. M. Coleman, C. J., Serman, D. E. Bhagwager, and P. C. Painter Polymer, 1990, 31, 1187.

19.   Glasser, W. G., B. McCartney, G. Samaranayake. Biotech. Progr. 1994, 10, 214-219.

20.   U. W. Gedde, "Polymer Physics," 1995, Chapman & Hall, London.

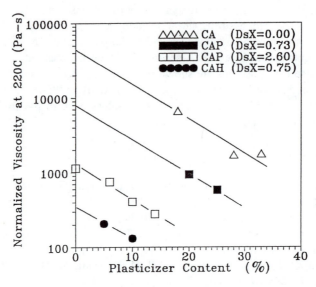

Figure 7. Normalized viscosities of plasticized cellulose esters extrapolated back to the zero plasticizer condition.

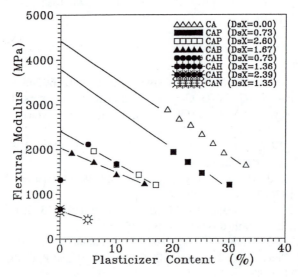

Figure 8. Flexural moduli of plasticized cellulose esters extrapolated back to the zero plasticizer condition.

Chapter 4

Induced Phase Separation: A New Synthesis Concept in Cellulose Chemistry

T. Liebert and Thomas Heinze[1]

Institute of Organic Chemistry and Macromolecular Chemistry, University of Jena, Humboldtstrasse 10, D-07743 Jena, Germany

A new synthesis concept in cellulose chemistry is described. Cellulose dissolved in the solvent system N,N-dimethylacetamide/LiCl as well as organo-soluble cellulose derivatives of different hydrolytic stability (dissolved in dimethyl sulfoxide) were converted into cellulose ethers of unusual substituent patterns after an induced phase separation with solid NaOH particles. As confirmed by means of HPLC analysis, these cellulose ethers contain a significantly higher amount of both 2,3,6-tri-O-functionalized and unsubstituted units in the polymer chain than those obtained by conventional etherification reactions in cellulose slurries. Moreover, the cellulose ethers show a preferred functionalization of O-6 within the modified anhydroglucose units as revealed by means of ^{1}H-NMR-spectroscopy.

Cellulose is a very uniform homopolymer composed of D-anhydroglucopyranose units (AGU) which are linked together by ß-1→4 glycosidic bonds. Based on this unique naturally occurring structure it is challenging to develop synthesis paths for the design of advanced cellulosic materials usable for self-organizing and controlled interactions in solutions.

Recent research in cellulose chemistry is in particular directed to the development of synthesis capabilities for a regiocontrolled introduction of functional groups in the polymer (*1*). The majority of methods studied so far are based on the difference in reactivity of the primary and secondary OH groups within the AGU. Consequently, protective group techniques have found considerable interest. Especially the bulky triphenylmethyl moiety (*2,3*) was established to protect the primary positions. Recently, trialkylsilyl derivatives of cellulose have been synthesized and it was found that the activated polymer (ammonia-saturated dipolar aprotic solvent) converted under heterogeneous conditions yields a 6-O-thexyldimethylsilyl cellulose of remarkable high uniformity (see chapter 1). The free OH-groups of these protected polymers could be modified to regioselectively substituted ethers and esters

---

[1]Corresponding author.

and the obtained deblocked products show differences in properties in comparison with conventionally prepared samples (4,5).

Besides the controlled distribution of functional groups on the level of the anhydroglucose unit (AGU) a more complex problem is the regioselective conversion of segments of the polymer chains as well as the analytical determination of such structures. Recently, a model consisting of oligoacetylcellulose- and oligodihexanoyl-chitin blocks was synthesized using non-glycosidic urethane linkages (6). Poly-saccharide block copolymers containing glycosidic linkages only are expected to possess unique properties.

In general, cellulose reactions are carried out with dissolved reagents and highly swollen or dissolved (so far in lab-scale synthesis only) cellulose polymers to avoid both concentration and accessibility gradients. Consequently, no differences in reactivity of the chemically equivalent glucose repeating units exist. Thus, it is necessary to establish new synthesis concepts employing segments of different reactivity within the polymer chains of the polysaccharide.

We introduce a new and general approach for the preparation of cellulosics with an unconventional distribution of substituents along the polymer chains. This approach is based on an induced high concentration gradient (phase separation), i.e. the formation of specific areas of reactivity (regioselectivity) within the polymer chains. The results of structure determination studies are compared with those of conventional products.

Results and Discussion

Today large-scale production of cellulose ethers like the most important ionic one, carboxymethyl cellulose (CMC), is exclusively carried out by slurry processes, i.e. by conversions of alkali cellulose swollen in an organic liquid and aqueous NaOH with an appropriate etherifying agent. In the case of CMC it was revealed by means of ^{13}C- and ^{1}H-NMR spectroscopy of hydrolytically degraded samples that the carboxymethyl functions are distributed in the order O-2 $\geq$ O-6 > O-3 within the AGU (7,8). Although a time-consuming mathematical processing of the ^{13}C-NMR spectra reveals the monomer composition, i.e. the molar ratio of all differently modified units (anhydroglucose unit, AGU; 2-, 3-, and 6-mono-O-AGU; 2,3-, 2,6-, and 3,6-di-O-AGU as well as 2,3,6-tri-O-AGU), more appropriate methods have to be developed to gain these information. Using a convenient HPLC method we confirmed for a broad variety of CMC that by the conventional slurry process no significant deviation from a statistical pattern of functionalization is achievable (9). This rapid and convenient method represents an inalienable prerequisite for studying the influence of synthesis conditions on the amount of the different monomer units. It was useful to determine the un-, mono-, di- and tricarboxymethylated units only, i.e. to neglect the different positions of both the mono- and dicarboxymethylated units. Figure 1 shows a typical HPL chromatogram as well as the assignment of the peaks.

It seems appropriate to control the functionalization patterns within the cellulose chains by using differences in reactivity due to their accessibility gradients caused by the well known supramoleculare structure. To activate the polymer and to initiate the etherification reaction it is common to use aqueous sodium hydroxide solutions. Dependent on the concentration of alkali, the cellulose swells to various extents and yields a more or less decrystallized polymer. That means the degree of activation (accessibility) may be controlled by the alkali concentration and thereby a control of the content of the different monomeric units might be possible. In a first series of experiments, cellulose **1** (spruce sulfite pulp) slurried in isopropanol was converted with monochloroacetic acid after activating the polymer with aqueous NaOH in the concentration range from 5 to 30 % (m/v). Regarding the results (Table I), it has to be underlined that the total degree of substitution (DS_{CMC}) reached a maximum value of 1.24 at a concentration of 15 % (m/v) aqueous NaOH. In order to analyze the mole fractions of the monomeric units, the CMCs **6a-f** were degraded using $HClO_4$. Subsequent removal of most of the perchloric acid as $KClO_4$ after precipitation with KOH gives solutes which were analyzed directly by HPLC. A separation on a polystyrene-based strong cation-exchange resin resulted in the desired mole fractions of unsubstituted-, mono-, di-, and tricarboxymethylated glucose units (see Figure 1). The results obtained are graphically displayed as a function of the total DS_{CMC} in Figure 2. The curves (shown in Figure 2 as well) were calculated on the basis of a statistical model (bionominal distribution) for the arrangement of substituents in cellulose derivatives (*9, 10*). Obviously, the determined mole fractions are in good agreement with the statistical model independence of the aqueous NaOH concentration used (see Experimental). That means that the totally heterogeneous carboxymethylation is mainly determined by statistics even at a low activation.

Table I. Degree of substitution (DS_{HPLC}) of carboxymethyl cellulose dependent on concentration of aqueous NaOH (reaction of cellulose 1 in isopropanol and aqueous NaOH with monochloroacetic acid for 5 h at 55 °C)

NaOH concentration[a] (%, w/v)	Carboxymethyl cellulose No.	DS_{HPLC}
5	**6a**	0.59
8	**6b**	0.93
10	**6c**	1.00
15	**6d**	1.24
20	**6e**	1.03
30	**6f**	0.95

[a] see Experimental

Thus, new synthesis concepts had to be established employing segments of different reactivity within the polymer chain.

In a recent research project on carboxymethylation of cellulose it was observed that a treatment of cellulose dissolved in the solvent system *N,N*-dimethylacetamide (DMA)/LiCl with solid NaOH-particles (size about 1 mm) suspended in DMA induces a phase separation (*11*). The gel particles formed can be isolated by solidification with diethyl ether. From deconvoluted FTIR-analysis of the gel particles it was concluded that cellulose II is regenerated. The phase separated systems were converted with monochloroacetic acid resulting in CMC with unconventional properties. While CMC samples, prepared via the conventional totally heterogeneous path in

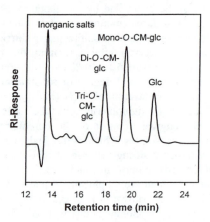

Figure 1. HPLC analysis of the carboxymethyl cellulose sample **6d** after hydrolysis with dilute HClO$_4$.

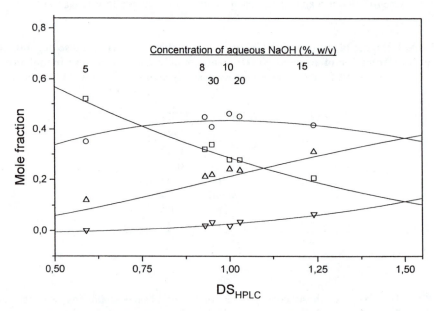

Figure 2. The mole fractions of repeating units (□, glucose, ○, mono-*O*-carboxymethyl-, Δ, di-*O*-carboxymethyl-, and ∇, 2,3,6-tri-*O*-carboxymethylated glucose) in hydrolyzed CMC samples (CMCs were synthesized via the conventional slurry process) plotted as function of DS$_{HPLC}$·dependent on the aqueous NaOH concentration. The curves are calculated as described in the text, see ref. (*9*) as well.

isopropanol/aqueous NaOH, dissolve already at a DS_{CMC} of about 0.5 in water, the samples synthesized in DMA/LiCl are water-soluble not until 1.5 (*12*). Representative results are listed in Table II. An analysis of the CMC polymers by means of HPLC after degradation with $HClO_4$ reveals a significant deviation of the amounts of the monomeric units with values calculated (see Figure 3). These CMCs contain a significantly higher amount of both tricarboxymethylated and unsubstituted units than those obtained in a slurry of cellulose in isopropanol/aqueous NaOH at comparable DS_{CMC} values, i.e. a non statistic distribution of the monomeric units occurs.

To confirm that these unconventional patterns of functionalization are due to the induced phase separation before the chemical conversion, various cellulose intermediates resp. derivatives of different hydrolytic stability were reacted under comparable water-free conditions. Solutions of cellulose trifluoroacetate **2**, prepared with trifluoroacetic acid/trifluoroacetic anhydride, DS_{CTFA} 1.5, DP 460 (*13*), cellulose formate **3**, prepared with formic acid/$POCl_{1.5}(OH)_{1.5}$; DS_{CF} 2.2, DP 260 (*11*), commercial cellulose acetate **4** (DS_{CA} 1.8, DP 220) as well as trimethylsilyl (TMS) cellulose **5**, prepared with TMS chloride after activation with NH_3, DS 1.1, DP 220 (*14*), in DMSO (5.7 %, w/v, polymer) were treated with solid NaOH particles suspended in dimethyl sulfoxide (DMSO).

In any case, phase separation and gel formation upon addition of the NaOH particles was observed. An interesting way to study the phase behavior during this first stage of reaction was found to be the application of polarized-light microscopy. While the solutions of the cellulosics mentioned represent a homogeneous system, after addition of the NaOH/DMSO suspension a growth of crystals (sodium salts of trifluoroacetic-, formic- and acetic acid, respectively) was observed. The regenerated cellulose II is not detectable as a polymeric particle. Consequently, the polymer has to be fixed mainly on the solid NaOH. The phase separated systems formed were allowed to react with monochloroacetic acid. Typical experimental data and analytical results are listed in Table II. It can be seen that a maximum DS_{CMC} of 2.2 was reached even in an one step synthesis. The samples obtained become water soluble starting from 1.5. A number of representative data of the mole fractions determined by means of HPLC are graphically displayed as a function of the total DS_{HPLC} (Figure 3). As can be concluded from the comparision with the values calculated according to the binominal distribution, the polymers consist of higher amounts of glucose and tricarboxymethylated anhydroglucoses and lower amounts of mono- and dicarboxymethylated anhydroglucoses. The deviation from the statistics is exact in the same range as found for CMCs prepared in DMA/LiCl. Consequently, the induced phase separation and subsequent carboxymethylation yield samples with a non-statistical distribution of substituents within the polymer chains.

An important question was, how the system behaves if the alkali becomes mobile in the reaction mixture by addition of an appropriate sovent like e.g. water. For this purpose a cellulose acetate (DS 0.8) was converted in the above mentioned manner with solid NaOH (20 mol/mol modified AGU) and monochloroacetic acid (10 mol/mol modified AGU) but in the presence of 1 % (v/v) of water. Results of the HPLC analysis are shown in Figure 4 as absolute deviation from the statistic values. The CMCs obtained exhibit a much higher DS (2.1) compared to a synthesis under water-free conditions as well as a strictly statistical pattern of substitution. It is believed that this observation is due to the fact that the water in the system is able to partially dissolve the solid NaOH and does thereby level out the phase separation

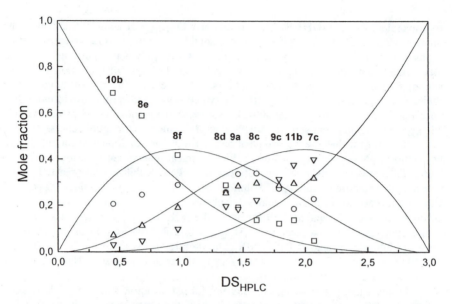

Figure 3. The mole fractions of repeating units (□, glucose, O, mono-*O*-carboxymethyl-, Δ, di-*O*-carboxymethyl-, and ∇, 2,3,6-tri-*O*-carboxymethylated glucose) in hydrolyzed CMC samples (CMCs were synthesized via induced phase separation) plotted as function of DS_{HPLC}. The curves are calculated as described in the text, see ref. (*9*) as well.

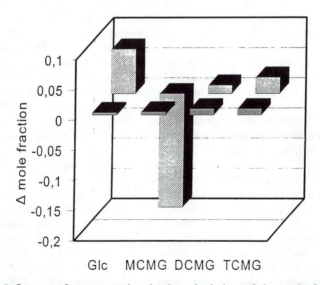

Figure 4. Influence of water on the absolute deviation of the mole fractions of glucose, mono-*O*-carboxymethyl-, di-*O*-carboxymethyl-, and 2,3,6-tri-*O*-carboxymethylated glucose from the binomial distribution. The CMC were synthesized starting from cellulose acetate **4** via induced phase separation. ▨ water-free (DS_{HPLC} 0.45), ▤ 1%, v/v water (DS_{HPLC} 1.70).

induced concentration gradient. It has to be mentioned that this effect is less drastic in case of more hydrophobic cellulose intermediates like, e.g., TMS-cellulose. The TMS-cellulose dissolved in DMSO yields products of significant deviation of the amount of the mole fractions even in the presence of 1 % of water.

The addition of phase transfer catalysts does not influence the deviation of the amount of the monomeric units from the statistics. Carboxymethylation via phase separation starting from a TMS-cellulose (DS 1.1) in the presence of dibenzo-18-crown-6 yields a polymer of DS_{CMC} 1.56 and with a significant deviation from statistically calculated mole fractions.

It was important to investigate the influence of the size of the NaOH particles. The results obtained with CTFA 2 dissolved in DMSO (5.7 %, w/v) are summarized in Figure 5. The phase separation and carboxymethylation reactions were carried out under equal conditions. By decreasing the size from < 1 mm to < 0.63 mm to < 0.25 mm the total DS_{CMC} values increased from 0.62 to 0.97 to 1.12. The mole fractions of the monomeric units determined by means of HPLC show that a deviation from the statistic values occurs as expected. However, as can be seen from Figure 5, the decreasing particle size has mainly an influence on the total DS_{CMC}. The influence on the amount of the monomeric units within the chains is rather small.

Carboxymethylation reactions via the cellulose intermediates mentioned of appropriate DS in unpolar solvents does not yield CMCs. Thus, the conversion of TMS-cellulose (DS 2.8, DP 460) dissolved in methylene chloride with dispersed solid NaOH (20 mol/mol modified AGU) and monochloroacetic acid (10 mol/mol modified AGU) under reflux for 16 h gives practically no CMC (DS_{CMC} determined by means of HPLC are less than 0.01).

Besides the determination of the amounts of monomeric units that built up the polymer, the distribution of carboxymethyl groups within the AGU was a goal of our studies. For this purpose the polymers were degraded with deuterated sulfuric acid and the corresponding hydrolysates were analyzed by means of ^{1}H-NMR spectroscopy. The partial degree of substitution calculated from the spectra indicates a substitution in the order O-6 > O-3 > O-2 in case of samples synthesized starting from CTFA and CF. The CMCs prepared in DMA/LiCl possess a distribution O-6 > O-2 > O-3. In comparison the carboxymethyl groups of conventionally prepared samples are distributed in the order O-2 ≥ O-6 > O-3. Consequently, the phase separation process does not just effect the regioselectivity on the level of the polymer chain. It has also an remarkable effect on the distribution of functional groups on the level of the monomeric unit.

An important feature due to the new synthesis concept using induced phase separation is the application to other etherification reactions. Preliminary experiments of methylation and ethylation prove the usefulness. For example the conversion of a cellulose acetate (DS 0.8) dissolved in DMSO (5.7 %, w/v, polymer) with methyl iodide (10 mol/mol modified AGU) after induced phase separation with solid NaOH (20 mol/mol modified AGU) yields a methyl cellulose of a DS_{Methyl} of 1.12. The product was analyzed by means of HPLC after degradation with trifluoroacetic acid (15). The absolute deviation of the amount of mole fractions of unsubstituted and methylated glucoses from calculated values according to the binominal distribution is shown in Figure 6. It is obvious that the deviation is comparable with results obtained from CMC synthesized via phase separation. Additionally, analytical results of a commercial methyl cellulose (DS 1.38) and of a methyl cellulose (DS 1.85) prepared

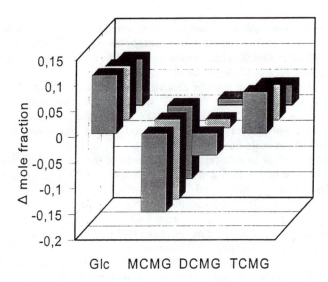

Figure 5. Influence of the NaOH particle size on the absolute deviation of the mole fractions of glucose, mono-*O*-carboxymethyl-, di-*O*-carboxymethyl-, and 2,3,6-tri-*O*-carboxymethylated glucose from the binomial distribution. The CMC were synthesized starting from cellulose trifluoroacetate **2** via induced phase separation. ▤ 1.00 - 0.63 mm (DS_{HPLC} 0.64), ▨ 0.63 - 0.25 mm (DS_{HPLC} 0.97), ▥ < 0.25 mm (DS_{HPLC} 1.12)

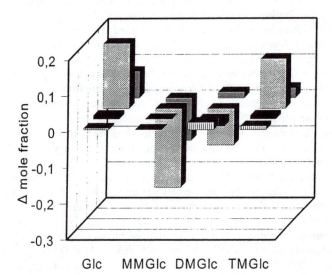

Figure 6. Absolute deviation of the mole fractions of glucose, mono-*O*-methyl-, di-*O*-methyl-, and 2,3,6-tri-*O*-methylated glucose from the binomial distribution of methyl celluloses, ▨ synthesized from TMS cellulose **5** via induced phase separation (DS_{Methyl} 1.12), ▤ from CA **4** via induced phase separation (DS_{Methyl} 0.45), ■ commercial sample (DS_{Methyl} 1.38), and ▥ synthesized via induced phase separation in the presence of 1 % (v/v) of water (DS_{Methyl} 1.85).

via phase separation, however with 1 % water in the reaction mixture are included in Figure 6 as well. Both do not exhibit significant deviations from the calculated values. It is worth mentioning that the water leads to a sample of higher total DS_{Methyl} as already found for the carboxymethylation.

Conclusion

It was shown that induced phase separation processes are useable for the preparation of cellulose products with a non statistic distribution of functional groups within the polymer chains. These samples already show unexpected properties. Thus, CMCs prepared via phase separation dissolve in water starting from DS 1.2 ... 1.6. In comparison conventionally prepared samples are water soluble at about 0.4. On the other hand, solutions of samples described here show a comparable lower decrease in viscosity upon an addition of electrolytes (*12*).

The validity of the synthesis concept was shown recently for the preparation of methyl starches. Products synthesized starting from a solution of the polymer and subsequent phase separation yields methyl starch with a block-like structure as concluded from FAB-MS analysis and statistical calculations (*16*). Moreover, Miyamoto et al. have shown that a methyl cellulose prepared via phase separation possess a very different water solubility as well as thermotropic gel formation tendency compared with a sample prepared by the so-called alkali cellulose process, which is a totally heterogeneous reaction (*17*).

In contrast to common knowledge that the etherification of cellulose in the solvent system DMA/LiCl has no particular advantage over conventional heterogeneous processes for cellulose ether production (*18*) our results show that new cellulosic polymers with both unconventional distribution of functional groups and properties may be designed. A first indication for such alternative molecular structures was already found by us in 1994 (*9*).

Besides the necessary more detailed elucidation of the structural features that are under progress, a very helpful tool for the understanding of the reaction mechanism would be a mathematical model suitable for the simulation of non statistical distributions of substituents along the polymer chains. One approach is the establishment of a series of rate laws and a mathematical handling that yield equations useable for the calculation of the mole fractions of repeating units dependent on the rate constants. By iteration and fitting of the analytical data with the calculated values, essential drawbacks concerning the mechanism of the reaction path are possible. These results will be published elsewhere.

Experimental

Materials. Spruce sulfite pulp, *N,N*-dimethylacetamide (DMA), NaOH, monochloroacetic acid and LiCl were purchased from FLUKA. DMA was dried over CaH and distilled under reduced pressure. All other chemicals were used for the carboxymethylation after drying at 105°C for 5 h in vacuum.

The cellulose intermediates were synthesized and analyzed according to (*13*): cellulose trifluoroacetate, (*11*): cellulose formate, (*14*): trimethylsilyl cellulose.

Carboxymethylation of cellulose in isopropanol/aqueous NaOH. The carboxymethylation was carried out by a standard solvent method. 5 g of air-dry cellulose **1** (spruce sulfite pulp) in 150 ml isopropanol was stirred vigorously, while 13.3 ml of a 5 to 30 % (w/v) aqueous NaOH (see Table I) was added dropwise during 10 min at room temperature. Stirring was continued for 1 h and 6 g of monochloroacetic acid was then added. The mixture was placed for 5 h on a water bath at 55 °C with stirring. The mixture was filtrated, suspended in 300 ml of aqueous methanol and neutralized with dilute acetic acid. The product (CMCs **6a-f**) was washed three times with 80 % aqueous ethanol and with ethanol and dried at 60 °C. DS_{CMC} are given in Table I, IR (KBr): 1630, 1410 cm^{-1} (C=O, carboxylate group).

Carboxymethylation of cellulose in DMA/LiCl. For a typical preparation, 1 g of cellulose **1** (spruce sulfite pulp) and 60 ml DMAc was kept at 130 °C for 2 h under stirring. After the slurry was allowed to cool to 100 °C, 3 g of anhydrous LiCl were added. By cooling down to room temperature under stirring the cellulose dissolved completely. After standing overnight, a suspension of pulverized NaOH (4 - 10 mol/mol AGU) in 20 ml DMA and a suspension of monochloroacetic acid (2-5 mol/mol AGU) in 20 ml DMA were added under vigorous stirring. The temperature was raised to 70°C. After 48 h reaction time the mixture was cooled to room temperature and was precipitated into 300 ml ethanol. The precipitates were filtered off, suspended or dissolved in 75 ml distilled water, neutralized with acetic acid and reprecipitated into 300 ml ethanol. After filtration the products were washed with ethanol and dried in vacuum at 50°C. DS_{CMC} are given in Table II, IR (KBr): 1630, 1410 cm^{-1} (C=O, carboxylate group).

Carboxymethylation of a cellulose intermediates 2-5, typical example. 1g of the intermediate (**2-5**) was dissolved in 17.5 ml DMSO under nitrogen. A suspension of dried pulverized NaOH (10-40 mol/mol modified AGU; dried under vacuum at 45 °C, 5 h; see Table II) in DMSO (2.75 ml per g NaOH) was added to the solution within 10 minutes, followed by sodium monochloroacetate (5-20 mol/mol modified AGU, dried under vacuum at 45 °C, see Table II) under vigorous stirring. The temperature was raised to 70 °C. After various reaction times (Table II) the reaction mixture was cooled to room temperature and precipitated into 75 ml methanol. The precipitate was filtered off, dissolved or suspended (in dependence on the DS_{CMC}) in water, neutralized with acetic acid, and reprecipitated into 100 ml of 80 % (v/v) aqueous ethanol. The products obtained are summarized in Table II. DS_{CMC} are given in Table II, IR (KBr): 1620, 1410 cm-1 (C=O, carboxylate group).

Measurements. The HPLC analysis of the CMC was carried out as described in ref. (*9*). However, the samples were hydrolysed with perchloric acid. 0.1 g of CMC were dispersed in 2 ml HClO$_4$ (70 %) and after 10 min at room temperature dilute with 18 ml distilled water. This mixture was kept at 100 °C for 16 h. The solution obtained was carefully neutralized with 2 M KOH and kept at 4°C for 1 h to guarantee a complete precipitation of the KClO$_4$. The salt was filtered off and washed three times with distilled water. The obtained solution was reduced to approximately 3 ml and dilute with distilled water to give exactly 5 ml sample. A Jasco HPLC equipment with two Bio-Rad Aminex HPX-87 columns (H$^+$ form) was used. The ^{1}H-NMR analyses were carried out according to ref. (*11*).

Table II. Conditions and results of carboxymethylation of cellulose (1) dissolved in *N,N*-dimethylacetamide (DMA)/LiCl as well as cellulose trifluoroacetate (CTFA, 2), cellulose formate (CF, 3), cellulose acetate (CA, 4), and trimethylsilyl cellulose (TMSC, 5) via induced phase separation with NaOH particles (size < 0.25 mm).

Staring cellulosic material	Molar ratio[a]	Reaktion time (h)[b]	Carboxymethyl cellulose		
			No	DS_{HPLC}[c]	Solubility in water
Cellulose (1) in	1:2:4	48	**7a**	1.13	-
DMA/LiCl	1:4:8	48	**7b**	1.88	+
	1:5:10	48	**7c**	2.07	+
CTFA (2)	1:5:10	2	**8a**	0.11	-
	1:10:20	4	**8b**	1.86	+
	1:10:20	16	**8c**	1.54	+
	1:10.20[d]	4	**8d**	1.36	-
	1:20:20[e]	2	**8e**	0.62	-
	1:10:20[f]	2	**8f**	0.97	-
CF (3)	1:10:20	2	**9a**	1.46	+
	1:10:20	4	**9b**	1.91	+
	1:15:30	4	**9c**	1.36	-
	1:20:40	2	**9d**	2.21	+
CA (4)	1:10:20	2	**10a**	0.36	-
	1:10:20	4	**10b**	0.45	-
TMSC (5)	1:10:20	0.5	**11a**	2.04	+
	1:10:20	1	**11b**	1.91	+
	1:10:20	2	**11c**	1.97	+

[a] Molar ratio: Modified anhydroglucose unit (AGU) : $ClCH_2COOH$ (Na) : NaOH.
[b] Reaction temperature 70 °C.
[c] DS_{HPLC}: degree of substitution determined by means of HPLC (see ref. *9*).
[d] First addition of $ClCH_2COONa$ and subsequent phase separation with solid NaOH particles.
[e] NaOH particle size: 0.63 - 1.00 mm.
[f] NaOH particle size: 0.25 - 0.63 mm.

72

Acknowledgment

The general financial support of the German "Deutsche Forschungsgemeinschaft (DFG; Schwerpunktprogramm Cellulose)" is gratefully acknowledged. We would like to thank the "Stifterverband für die Deutsche Wissenschaft" for generous financial support.

Literature Cited

1.   Klemm, D.; Stein, A.; Heinze, Th.; Philipp, B.; Wagenknecht, W. In *Polymeric Materials Encyclopedia: Synthesis, Properties and Applications;* Salamone, J.C. Ed.; CRC Press, Inc., Boca Raton, USA, vol. 2, **1996**, 1043-1053.
2.   Harkness, B.R.; Gray, D.G. *Macromolecules* **1991**, *24*, 1800.
3.   Heinze, Th.; Röttig, K.; Nehls, I. *Macromol. Rapid Commun.* **1994**, *15*, 311.
4.   Kamide, K.; Saito, M. *Macromol. Symp.* **1993**, *83*, 233.
5.   Klemm, D.; Heinze, Th.; Wagenknecht, W. *Ber. Bunsenges. Phys. Chem.* **1996**, *100*, 730.
6.   Kadakowa, J.-I.; Karasu, M.; Tagaya, H.; Chiba, K. *J.M.S.-Pure Appl. Chem.* **1996**, *A33*, 1735.
7.   Baar, A.; Kulicke, W.-M.; Szablikowski, K.; Kiesewetter, R. *Macromol. Chem. Phys.* **1994**, *195*, 1483.
8.   Reuben, J.; Conner, H.T. *Carbohydr. Res.* **1983**, *115*, 1.
9.   Heinze, Th.; Erler, U.; Nehls, I.; Klemm, D. *Angew. Makromol. Chem.* **1994**, *215*, 93.
10. Spurlin, H.M. *J. Am. Chem. Soc.* **1939**, *61*, 2222.
11. Liebert, T.; Klemm, D.; Heinze, Th. *J.M.S.-Pure Appl. Chem.* **1996**, *A33*. 613.
12. Heinze, Th.; Heinze, U.; Klemm, D. *Angew. Makromol. Chem.* **1994**, *220*, 123.
13. Liebert, T.; Schnabelrauch, M.; Klemm, D.; Erler, U. *Cellulose*, **1994**, 249.
14. Klemm, D.; Stein, A. *J.M.S.-Pure Appl. Chem.* **1995**, *A32*. 899.
15. Erler, U.; Mischnick, P.; Stein, A.; Klemm, D. *Polymer Bull. (Berlin)*, **1992**, *29*, 349.
16. Mischnick, P.; Kühn, G. *Carbohydr. Res.* **1996**, *290*, 199.
17. Miyamoto, T.; Donkai, N.; Nishimura, H. *Proceedings of "Japanese-German seminar on future developments of polysaccharides. Fundamentals and applications"*, **1996**, Hokkaido University, Sapporo, Japan, p.35.
18. Dawsey, R.D. In *Polymer Fiber Sci.: Recent Adv.;* Forners, R.E., Gilbert, E.D., Ed.; VCH, New York, USA, **1992**; pp. 157-176.

Chapter 5

Methods for the Selective Oxidation of Cellulose: Preparation of 2,3-Dicarboxycellulose and 6-Carboxycellulose

A. C. Besemer[1], A. E. J. de Nooy[1], and H. van Bekkum[2]

[1]TNO Nutrition and Food Research Institute, P.O. Box 360, 3700 AJ, Zeist, Netherlands
[2]Delft University of Technology, Julianalaan 136, 2628 BL, Netherlands

Three methods for the selective oxidation of cellulose are described. The classical method consists of consecutive oxidation with sodium periodate, leading to 2,3-dialdehyde cellulose and sodium chlorite, giving 2,3-dicarboxy cellulose. This material, which is obtained in high yield and has a high carboxylate content (7.6 mmol COONa/g; 90% of the theoretical value), has a very good calcium sequestering capacity.
The second method is by oxidation of the substrate, dissolved in concentrated phosphoric acid with nitrite/nitrate, leading to the selective oxidation of the substrate at the 6-position of the glucose unit. Generally, the yields are higher than 80%, and the degree of oxidation is 80-90%. However, the reaction is not completely specific, since some oxidation at the secondary hydroxylic groups occurs. Borohydride reduction of the product restores the diol configuration and also ß-elimination is avoided and thereby depolymerization. Oxidation with sodium hypochlorite and bromide as a catalyst and TEMPO as a mediator appears also to be applicable to cellulose. Selectivity of oxidation at the 6-CH_2OH group is somewhat lower than that obtained earlier for glucans like starch and pullulan. Products with a degree of oxidation of 80% are obtained in 90% yield or higher.

Oxidation of (poly)saccharides has been studied in detail by numerous investigators, but, because of the presence of several reactive groups, it is not easy to attain high selectivity, and only a limited number of reagents are available for this purpose (1). Because it is insoluble in water and most common organic solvents, there are especially difficulties with cellulose.
In view of the structure of an anhydroglucose unit in glucans such as cellulose and starch, one has to account for the presence of three reactive groups: one primary and two secondary OH-groups. Usually the oxidation of the secondary hydroxyl

groups in glucans results in ring cleavage. Well-known methods for this conversions are reactions with sodium periodate or with lead(IV) tetraacetate, which lead to the formation of the corresponding 2,3-dialdehyde derivatives (2-4). Upon subsequent oxidation of this material with sodium chlorite the corresponding dicarboxy derivatives are obtained. The conversion of starch has been studied in detail because the oxidation products have excellent calcium binding properties and therefore may be used as a substitute for builders in laundry detergents (3, 4). Good results can also be obtained with cellulose. Floor *et al.* (4) improved the procedure by using hydrogen peroxide in the second step. For the quantitative conversion of dialdehyde derivatives 6 mol instead of 2 mol of sodium chlorite are required:

Dialdehyde polysaccharide + 2 $NaClO_2$ → Dicarboxy polysaccharide + 2 NaOCl

because the reaction product (NaOCl) decomposes sodium chlorite according to

$NaOCl + 2NaClO_2 + H_2O$ → 2 ClO_2 + NaCl + 2 NaOH.

Use of hydrogen peroxide has two advantages: less reagent is needed, since sodium hypochlorite reacts faster with hydrogen peroxide than with sodium chlorite according to

$NaOCl + H_2O_2$ → $NaCl + H_2O + O_2$

and, better products are obtained.

About twenty years ago the most attractive way to selectively oxidize polysaccharides such as starch and cellulose at the 6-position of the anhydroglucose unit was by exposing them to (gaseous) NO_2 or N_2O_4 (5). Satisfactory results can be obtained with cellulose and starch; i.e., the corresponding 6-carboxy polysaccharide can be prepared with a high carboxylate content and with a satisfactory yield.

An alternative, which has been studied by a few authors, is oxidation with nitrous acid (6,7). In this system, the substrate is dissolved in concentrated phosphoric acid and allowed to react with sodium nitrite. In the highly viscous solution a foam develops in which various oxidizing species like N_2O_3 and N_2O_4 are present. A drawback of these methods is that considerable depolymerization occurs and that the degree of oxidation is no higher than approximately 80% (6).

Recently, we improved the latter oxidation and we developed a new method for oxidation of primary alcohol groups in polysaccharides using hypochlorite as the primary oxidant and bromide and 2,2,6,6-tetramethylpiperidine-N-oxyl (TEMPO) as the catalysts. In (8-11) we presented the results of this reaction with respect to selectivity, depolymerization, and scope. Mainly water soluble-polymers, such as starch, inulin, and pullulan, were investigated.

In this study we describe results obtained by applying these oxidation methods to poorly water-soluble polymers such as cellulose. A comparison is made with some model compounds.

Results and discussion

a. Periodate/clorite/hydrogen peroxide oxidation of cellulose

Table I shows the results of the glycol cleavage of starch and cellulose, using 2 mol of sodium chlorite and 2 mol hydrogen peroxide. For comparison we have also presented data on the method in which 6 mol of sodium chlorite are used (and no hydrogen peroxide). Generally, in polysaccharide oxidation, method A gives the best results. However, it is seen that oxidation of cellulose according to method B can give somewhat better results.

Table I. Dicarboxy polysaccharides obtained by sodium periodate/sodium chlorite oxidation

Substrate	Method[a]	Yield(%)	CC(%)	SC (mmol Ca/g)[b]
Cellulose	A	93	82	2.20
Cellulose	B	91	86	2.29
Starch	A	98	79	2.51
Starch	B	93	86	2.39
Amylose	A	87	74	2.39
Amylose	B	95	80	2.29

[a]Method A: 2 mol of sodium chlorite and 2 mol of hydrogen peroxide per anhydro-glucose unit
Method B: 6 mol of sodium chlorite per anhydroglucose unit
[b] SC = sequestering capacity defined as the mmol of Ca bound by 1 gram of material until the concentration is below 10^{-5} M Ca(II). CC = carboxylate content (at 100 % conversion the CC = 8.4 mmol COONa/g)

b. 6-carboxy cellulose (oxidation in phosphoric acid with sodium nitrate/nitrite)

The results of the oxidation of cellulose in the oxidation with sodium nitrate are shown in Table II. Generally, satisfactory results are obtained. Good results were also obtained in the oxidation of amylose (yield 80%; carboxylate content 95%). It can be concluded that this method is very suitable for the oxidation of poorly soluble biopolymers in water. Yield and carboxylate content are high. Although some may occur (see Figure 1), depolymerization appears to be very modest, especially in comparison to former procedures (5) . From the kinetic experiments (see Figure 2) it can be seen that nitrite has a catalytic effect. The reduction of nitric acid by NO, leading to the formation of NO_2, is thought to play an important role. An important advantage of this method over the TEMPO method is that no glycolic oxidation can occur. Table III shows some results of the oxidations of ß-cyclodextrin using the phosphoric acid/nitrate/nitrite method.

From the results presented in Table IV, it appears that only the stoichiometric amount of nitrate is needed, according to the theoretical reaction equation

$$3 -CH_2OH + 4 HNO_3 \rightarrow 3 -COOH + 4NO + 5H_2O,$$

indicating 1.33 $NaNO_3$ per CH_2OH-group.

Table II. Oxidation of cellulose with sodium nitrate/sodium nitrite in phosphoric acid

Substrate	T(°C)	Time (hours)	DO[a] (%)	Yield(%)
cellulose	20	3	23	77
cellulose	20	6	72	85
cellulose	20	9	88	82
cellulose	20	20	99	74
cellulose	4	6	<10	nd
cellulose	4	24	90	85
cellulose	4	40	97	88

[a] DO = Degree of oxidation determined by the Blumenkrantz method (12).

It is important to note that when using only sodium nitrite, as in the literature method (7), a much larger amount of NO will be produced per primary alcohol group:

$$-CH_2OH + 4HNO_2 \rightarrow -COOH + 4 NO + 3H_2O.$$

c. TEMPO-oxidation of starch, pullulan and cellulose

The results of the TEMPO-oxidation of cellulose and amylose are shown in Table IV.

It is observed that the results with cellulose are not quite reproducible; i.e. yield and carboxylate content vary to some extent, and seem to depend on some unknown variables. No clear explanation exists for the fact that some material remains undissolved and that the composition varies. The fact that a yield of higher than 100%

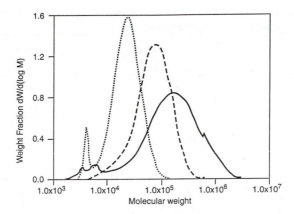

Figure 1. Molecular weight distribution of pullulan (right peak) and completely oxidized samples at 4 °C (middle peak) and at 20 °C (left peak) as obtained by SEC-MALLS (oxidation method sodium nitrate/sodium nitrite in phosphoric acid).

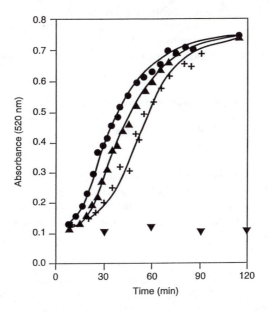

Figure 2. Influence of the initial concentration of sodium nitrite on the reaction rate (measured with the Blumenkrantz assay). Conditions: 0.70 g β-cyclodextrin and 0.70 g sodium nitrate dissolved in 5 ml 85% phosphoric acid with 0 mg (▾), 20 mg (+), 40 mg (▴), 60 mg (●) sodium nitrite, respectively.

Table III. Degree of oxidation of ß-cyclodextrin in the nitrate/nitrite oxidation in relation to the nitrate/substrate molar ratio

Entry	NaNO₃/primary alcohol(molar ratio)	DO(%) [a]
1	0.27	17
2	0.54	41
3	0.81	56
4	1.10	68
5	1.38	77
6	1.93	80

[a] Based on the assumption that under optimum conditions the degree of oxidation per anhydroglucose unit is 80%.

can be obtained with a low uronic acid content of the materials points to the occurrence of a competing oxidation (glycol cleavage). Although the oxidation has not been studied in detail, evidence exists that the reagent sodium hypochlorite should be added gradually to avoid undesired reactions, such as glycol cleavage, due to the presence of excess sodium hypobromite (13).

To overcome this complication, three options should be considered for better results:

- avoiding excess of reagent by adding only small amounts and waiting until the reaction has ceased
- use of a higher TEMPO-concentration
- use of "activated" cellulose.

Use of activated (water-swollen) cellulose will lead to a more accessible substrate, which combined with a higher TEMPO-concentration is expected to favour the desired reaction. Use of a lower hypochlorite concentration will suppress glycol cleavage. It is therefore supposed that all these measures will be effective.

d. Molecular weight

The molecular weights of some representative samples have been measured and, again, the high selectivity is shown; i.e. the molecular weight decreases only from 300.000 to 150.000.

In Figure 3 the molecular weight distribution is given for pullulan converted with the TEMPO/hypochlorite system to varying degrees of oxidation. It can be seen that loss of molecular weight is moderate. Whereas the maximum in pullulan is found at 300.000, the maximum of the TEMPO-oxidized material (100%) is found at 170.000. In the N_2O_4 oxidation (under optimum conditions, 4°C reaction temperature), the maximum shifts from 300,000 to 100,000 (see Figure 3). A

Table IV. Oxidation of cellulose and amylose with the sodium hypo-chlorite/bromide/TEMPO system

Substrate	time(hours)	T(°C)	Yield[a] (%)	Yield[b] (%)	CC[a] (%)	CC[b] (%)
cellulose	20	20	86	13	70	35
cellulose	24	20	82	28	48	20
cellulose	24	20	80	25	30	25
cellulose	24	5(20)[c]	85	10	73	35
amylose	24	5(20)[c]	95	-	52	-

[a] Yield and degree of oxidation (DO) of water-soluble material
[b] Yield and carboxylate content (CC) of non-dissolved material
[c] Initial temperature. The reaction temperature was allowed to rise to 20 °C.

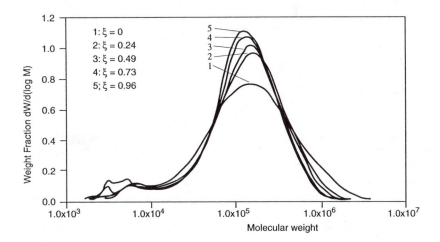

Figure 3. Molecular weight distribution of pullulan with 0 (line 1), 25 (line 2), 50 (line 3), 75 (line 4) and 100% (line 5) degree of oxidation. Oxidation method hypochlorite/NaBr/TEMPO.

higher reaction temperature leads to severe depolymerization (from 300,000 to approximately 30,000). From this result it appears that for retention of a high molecular weight the TEMPO-method is preferred. However, so far, no data are available of the molecular weights of cellulose, oxidized before and after TEMPO-oxidation. The competing glycol cleavage accompanied by depolymerization may especially lead to severe depolymerization. In this respect it is expected that optimization of the conversion of cellulose with the TEMPO-system will lead to better results (see recommendations above).

Conclusion

There is no doubt that the 2,3-oxidation reaction is highly selective, a fact that can be attributed to the first step: it is postulated that a cyclic ester of periodate with the diol group is formed (14).

The behaviour of cellulose in the two oxidation methods directed on oxidation of the primary alcohol group is different, mainly because the substrate is weakly soluble in water. In the TEMPO-oxidation it is seen that a part of the cellulose does not react or does not dissolve.

The yield of the TEMPO-oxidized material is 70-75%, but the conversion is not quantitative (75%). A part of the material remains undissolved and appears to be only partly oxidized. It may be recalled that the TEMPO-oxidation gives better results with water-soluble or swellable polymers. It should also be kept in mind that the NO_2-oxidation is in fact not selective. Some non-specific oxidation can be repaired by sodium borohydride reduction. It is to be concluded that the phosphoric acid/$NaNO_2$/$NaNO_3$ method gives better results, so far.

Up to now no data are available with regard to the molecular weight of TEMPO-oxidized cellulose. During TEMPO-oxidation hypochlorite/hypobromite oxidation may be an important side reaction, which may lead to the glycol cleavage reaction. Some measures to improve the results of the TEMPO-catalysed cellulose oxidation are indicated.

Experimental

Materials. The cellulose used was a highly purified cotton wool; microcrystalline cellulose (Avicel) was purchased from Merck; Amylose-V (amorphous), potato starch, and ß-cyclodextrin were gifts of Avebe, Veendam, The Netherlands. Sodium hypochlorite solution (150 gram of active chlorine/L was a gift of AKZO-Nobel (Hengelo, The Netherlands). Sodium nitrite, sodium nitrate, sodium borohydride, and sodium hydroxide were obtained from Merck. Sodium chlorite (80% purity) was obtained from Aldrich. TEMPO (tetramethylpiperidine-N-oxyl) was a Sigma Chemicals product. Pullulan was purchased from Hayashibara (Japan).

Oxidation methods

2,3-dicarboxy cellulose (sodium periodate/sodium chlorite). Method A. To a suspension of 5.0 g of cellulose in 100 ml of water (pH 5), cooled to 5°C, sodium periodate (6.64 g) was added. The mixture was stirred in the dark for 168 hours. The product (dialdehyde cellulose) was filtered off, washed several times with water, and dried in vacuo. The yield was about 98 %. This material was dispersed in water, and a solution of hydrogen peroxide (6.3 ml 30% w/w) was added. In the course of two hours 6.8 g sodium chlorite was added. The mixture was allowed to react for 48 hours. During the reaction the pH was kept constant at 5 by addition of 0.5 M NaOH-solution. To precipitate the dicarboxy-cellulose, the solution was poured out into ethanol. The product was filtrated, redissolved in water, and again precipitated.

Method B. Procedure as described in method A, however, no hydrogen peroxide was applied and 20 g of sodium chlorite was used (molar ratio starch/NaClO$_2$ = 1:6).

Carboxylate content of the products was measured by ion exchange (strong acid), followed by freeze drying and titration with NaOH. The calcium sequestering capacity was measured using a calcium ion selective electrode. Two other substrates, starch and amylose, were subjected to the same oxidation procedures.

6-carboxy cellulose (phosphoric acid/sodium nitrate/nitrite). Cellulose (1.5 g) was dissolved in 30 ml 85% phosphoric acid at 4 °C, which took about 4-6 hours of stirring. To the viscous solution 1.5 g sodium nitrate was added and stirring was continued for 30 minutes to dissolve the salt. Then sodium nitrite (40-60 mg) was added. After stirring for another 60 minutes, the mixture was left at 4°C for 40 hours. A foam developed from which (the toxic) NO escaped. The product was iso-lated by pouring out the solution while stirring in cold ethanol. A solid precipitated, which was collected by filtration. The material was dissolved in water and the sol-ution was brought to pH 7 with sodium carbonate. To this solution 100 mg sodium borohydride was added to remove any carbonyl function present. After 20 hours the solution was brought to pH 5-6 with acetic acid. Any salts present were removed by nanofiltration (Toray, UTC 260 membrane filter). The final solution was freeze dried. Generally, 6-carboxy-cellulose was obtained in the sodium form as a white product and in a yield of approximately 80%. The uronic acid content was measured with the Blumenkrantz method (12).

Some kinetic measurements were carried out with ß-cyclodextrin in order to prove the catalytic effect of NO (induced by the presence of nitrite). The experiments were conducted in the same way as described above. ß-Cyclodextrin (0.70 g) was dissolved in 5 ml phosphoric acid. To this solution 0.70 g sodium nitrate was added, and after dissolution of this compound, the catalyst sodium nitrite was added. During the reaction the uronic content was measured as a function of time. Another experiment was conducted with the objective to study the influence of the amount of nitrate on the degree of oxidation.

6-carboxy cellulose (oxidation with sodium hypochlorite/bromide/TEMPO). In 100 ml of water 15 mg of TEMPO and 100 mg of sodium bromide were dissolved. In this solution 2 g of the substrate was suspended and 1 ml of an aqueous solution of sodium hypochlorite solution(1 M) was added. Within a few minutes the pH starts to drop. During the reaction the pH was kept constant by the addition of 0.5 M NaOH solution (when the rate of the consumption of NaOH-solution decreased, again a small amount of hypochlorite was added). In this way the competing reac-tion, glycolic oxidation by sodium hypochlorite, is largely prevented. For a complete conversion 12 ml of hypochlorite solution (2M), was used. It was observed that a homogeneous mixture could not be obtained; i.e. a small amount of non-reacted or partially converted material did not dissolve. This material was separated from the solution by centrifugation. The supernatant liquid was desalted through nanofiltration. The solution was freeze dried. A white material was obtained of which weight, yield and carboxylate content were determined. The pellet was washed with ethanol (96%) and dried. From this material the weight and carboxylate content were also determined.

Literature cited

1.   Radley, J.A. *Starch and its derivatives*, Chapter 11, Chapman and Hall, London (1968).
2.   Nieuwenhuizen, M; Kieboom, A.P.G.; Van Bekkum, H.; Starch/Stärke **1985,** 37, 192.
3.   Floor, M; Thesis, Delft University of Technology Delft, The Netherlands (**1989**).
4.   Floor, M; Kieboom, A.P.G.: Van Bekkum, H.; Rec. des Travaux Chim. Pays-Bas **1989**, 108, 384.
5.   Yackel, E.C.; Kenyon, W.O.; J. Am. Chem. Soc. **1942**, 64, 121.
6.   Painter, T.J.; Carbohydr. Res. **1977**, 55, 95.
7.   Painter, T.J.; Cesaro, A.; Delben, F.; Paoletti, S.; Carbohydr. Res. **1985,** 61, 140.
8.   De Nooy, A.E.J.; Besemer, A.C.; Van Bekkum, H.; Rec. des Travaux Chim. Pays-Bas **1994**, 113, 165.
9.   De Nooy, A.E.J.; Besemer, A.C.; Van Bekkum, H; Carbohydr. Res., **1995,** 269, 89.
10   De Nooy, A.E.J.; Besemer, A.C.; Van Bekkum, H; Tetrahedron **1995**, 51, 8023.
11.   De Nooy, A.E.J.; Besemer, A.C.; Van Bekkum, H.; Van Dijk, J.A.P.P.; Smit, J.A.M.; Macromolecules **1996**, 29, 6541.
12.   Blumenkrantz, N.; Asboe-Hansen, G.; Anal. Biochem. **1973,** 54, 484
13.   Besemer, A.C.; Van Bekkum, H; Starch/Stärke **1994,** 46, 95
14.   De Wit, D.; Thesis, Delft University of Technology, Delft, The Netherlands (**1990**).

Chapter 6

Reaction of Bromodeoxycellulose

N. Aoki[1], M. Sakamoto[2], and K. Furuhata[3]

[1]Molecular Engineering Division, Kanagawa Industrial Research Institute, 705-01, Shimo-imaizumi, Ebina-shi, Kanagawa Prefecture 243-04, Japan
[2]College of School Education, Joetsu University of Education, Yamayashiki-machi, Joetsu-shi, Niigata Prefecture 943, Japan
[3]Department of Organic and Polymeric Materials, Faculty of Engineering, Tokyo Institute of Technology, O-okayama, Meguro-ki, Tokyo 152, Japan

The aim of this paper is to show that bromodeoxycellulose, whose C-6 hydroxyl groups are regioselectively and quantitatively substituted with bromine atoms, is useful for the synthesis of cellulose derivatives. The comparison of rate constants of nucleophilic halogen substitution of halogenated methyl glycosides revealed that the rates for bromodeoxysaccharides were about 1000 times higher than those of corresponding chlorodeoxysaccharides. Bromodeoxycellulose was converted effectively to S-substituted deoxymercaptocellulose derivatives by the reaction with thiols under homogeneous conditions. Deoxymercaptocellulose samples having high degrees of substitution were obtained by the reaction of bromodeoxycellulose with thiourea and consecutive alkali treatment.

Chlorodeoxycellulose (Cell-Cl) and cellulose tosylate have been used for the syntheses of many kinds of cellulose derivatives (1). Cell-Cl is potentially more useful than cellulose tosylate because the regioselective and quantitative substitution of hydroxyl groups at C-6 alone or both at C-6 and C-3 is possible (2, 3). However, the use of Cell-Cl is somewhat limited because of its relatively low reactivity. Bromodeoxycellulose (Cell-Br) is considered to be a better cellulose derivative than Cell-Cl for further reactions because bromine is a better leaving group than chlorine. Until recently, however, it was difficult to obtain Cell-Br samples with both excellent regioselectivity and high degree of bromine substitution.

Several organic solvent systems for cellulose have been developed in these 20 years and some of them are used as the reaction media in the chemical modification of cellulose (4). Lithium halide–N,N-dimethylacetamide (DMAc) systems were found to be very suitable for the substitution of hydroxyl groups with halogen atoms because they include high concentrations of halide ion. The degree of substitution (DS) of halodeoxycellulose has been improved remarkably under homogeneous conditions in the lithium halide–DMAc systems. The substitution occurs first at C-6, and next at C-3 but not at C-2. The maximum DS by chlorine achieved for Cell-Cl in LiCl–DMAc was 1.8 (2) or 1.9 (3) while that of Cell-Br in LiBr–DMAc was 0.9 (regioselectively substituted at C-6) with the N-bromosuccinimide–triphenyl-phosphine (TPP) reagent system (5) or 1.6 with tribromoimidazole–TPP (6).

This paper describes three topics relating to the reactions of Cell-Br; (i) evaluation of reactivity of halodeoxycellulose using model saccharides, (ii) reaction of

Cell-Br with thiols in a homogeneous system and (iii) synthesis of deoxymercaptocellulose.

Experimental

Syntheses. Methyl 3,6-dichloro-3,6-dideoxy-β-D-alloside (Me 3,6-Cl$_2$-β-All)(*7*), methyl 6-chloro-6-deoxy-α(β)-D-glucoside (*8*) and methyl 6-bromo-6-deoxy-α(β)-D-glucoside (*9*) were synthesized by the methods described in the literature, as were methyl 3,6-dibromo-3,6-dideoxy-β-D-alloside (Me 3,6-Br$_2$-β-All) and methyl 3,6-dibromo-3,6-dideoxy-β-D-glucoside (Me 3,6-Br$_2$-β-Glc)(*10*).

Bromodeoxycellulose (Cell-Br) used in this study was synthesized with *N*-bromosuccinimide and triphenylphosphine in the LiBr–*N,N*-dimethylacetamide (DMAc) solvent system as described previously (*5*). In the Cell-Br samples obtained by this method, only hydroxyl groups at C-6 were substituted with bromine. The degree of substitution by bromine was 0.85 - 0.97 which was determined based on the elemental analysis (*5*).

For the reaction with a thiol, Cell-Br was dissolved in LiBr–DMAc. After stirring for 1 h at 60°C, the thiol and triethylamine were added to the solution at reaction temperatures. The product was recovered by reprecipitation with excess acetone. The precipitates were treated with a dilute Na$_2$CO$_3$ solution and dialyzed against distilled water and freeze-dried.

Deoxymercaptocellulose (Cell-SH) was synthesized from Cell-Br by a method similar to that from Cell-Cl (*11*). A typical reaction procedure is as follows; Cell-Br (1 g) was treated with thiourea (1.77 g) in 100 mL of dimethyl sulfoxide (DMSO) at 80 °C for 24 h. The mixture became homogeneous as the reaction proceeded. Deoxyisothiouroniumcellulose bromide (Cell-TU) was recovered as a precipitate by pouring the reaction solution into 1 L of acetone. Cell-TU was dissolved into water and the solution was treated with aqueous alkali. Cell-SH was precipitated by neutralizing the solution with hydrobromic acid and dialyzed against distilled water for 3 days. The content of the dialysis tube was freeze-dried.

Analysis. The analysis of saccharide mixtures was carried out by gas chromatography (GC) with a gas chromatograph 4BPMF (Shimadzu Corp.) after trifluoroacetylation (*3*). Cellulose derivatives were hydrolyzed in sulfuric acid. Gas chromatography–mass spectrometry (GC–MS) was applied to support the assignments obtained by the GC analysis with Shimadzu GC-MS LKB 9000S gas chromatograph–mass spectrometer (Shimadzu Corp.). Details of the conditions for GC and GC–MS analyses were described previously (*3*). NMR spectra were recorded with a spectrophotometer JNM-A500 (JEOL, Ltd.).

Halogen Exchange Reactions of Model Saccharides. Halogen exchange reactions were carried out in DMAc. In order to terminate the reaction, the lithium chloride was precipitated by pouring the reaction solution into butyl acetate and the bromide ion was precipitated as silver bromide by adding silver lactate to the solution. After filtration, the filtrate was evaporated, the residue was trifluoroacetylated, and subjected to the GC analysis.

Results and Discussion

Evaluation of Reactivity of Halodeoxycellulose using Model Saccharides. There is a difference in the configuration of repeating units in chlorodeoxycellulose (Cell-Cl) and that in bromodeoxycellulose (Cell-Br) with high DS (Scheme 1). Cell-Cl with DS over 0.8 includes 3,6-dichloro-3,6-dideoxyallose units (*3*) while Cell-Br with high DS contains two different dibromodideoxysaccharide units, that is, 3,6-dibromo-3,6-dideoxyglucose and 3,6-dibromo-3,6-dideoxyallose units (*6*). The mechanism of

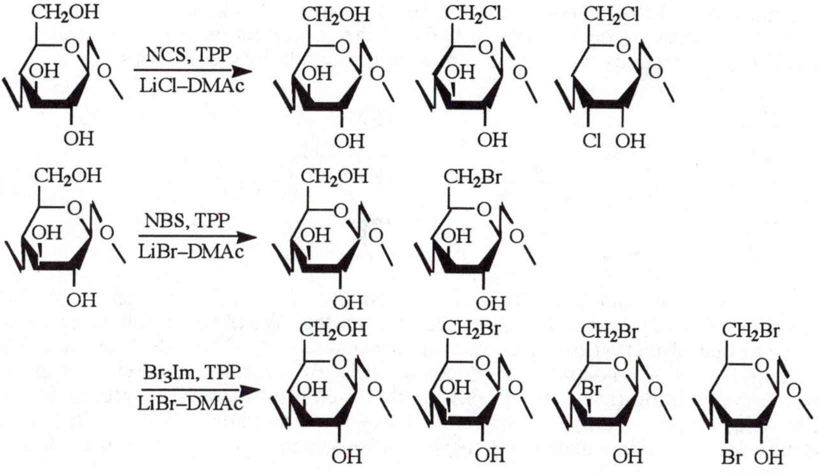

NCS, N-Chlorosuccinimide: NBS, N-Bromosuccinimide: Br₃Im, Tribromoimidazole

Scheme 1

these halogenations of cellulose is considered to be the same; formation of triphenylphosphonium ester followed by S_N2 attack of halide ion (12). We have found that the high reactivity of Cell-Br is the reason for this difference. Three types of model experiments using halogenated methyl glycosides were carried out to examine the reactivities of halodeoxysaccharides.

Type I is the halogen exchange reaction at C-3 of a methyl 3,6-dideoxy-3,6-dihaloglycoside using the ion of the same kind of halogen as included in the starting saccharide (10, 13);

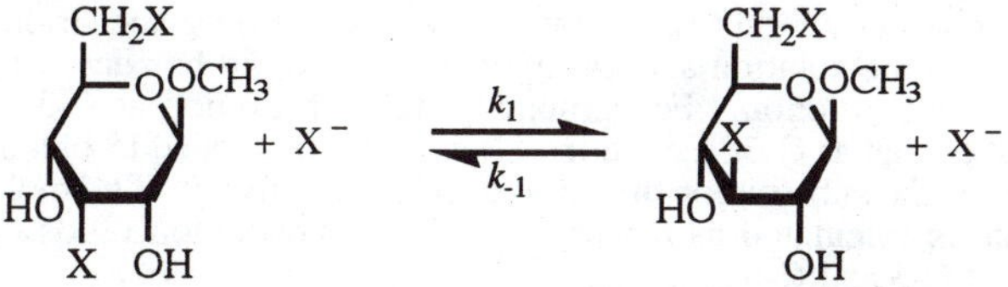

$$y_t = y_e\{1-\exp(-k_1[B]t/y_e)\} \tag{2}$$

where X is Cl or Br. This type of reaction can be monitored by GC analysis of the reaction mixture after trifluoroacetylation. The reversible reaction shown above is considered to be a model for the inversion of configuration at C-3 positions of dideoxydihalo-units which may occur during the substitution of hydroxyl groups of cellulose with halogen atoms in lithium halide–DMA. The rate equation for this reaction is relatively simple because the concentration of halide ion is constant during the reaction.

$$y_t = y_e\{1-\exp(-k_1[B]t/y_e)\} \tag{2}$$

where y_t and y_e are the mole fractions of product saccharide at time t and at equilibrium, respectively, and [B] is the concentration of the halide ion (13).

Type II is the halogen exchange reaction of a methyl 6-deoxy-6-haloglucoside using the ion of halogen not contained in the starting saccharide. In this type of reaction, the exchange occurs only at C-6 and the configuration is not changed. It is possible to follow the exchange at C-6 most conveniently by GC analysis;

$$\text{(structure with } CH_2X,\ OCH_3,\ OH,\ HO,\ OH) + X'^{-} \quad \underset{k_{-1}}{\overset{k_1}{\rightleftharpoons}} \quad \text{(structure with } CH_2X',\ OCH_3,\ OH,\ HO,\ OH) + X^{-} \quad (3)$$

where X is Cl or Br and X' is Br or Cl, respectively. The molar response of the FID detector in GC analysis is different for the saccharides. We obtained following values relative to that of methyl α-D-glucoside (internal standard); 0.76, 0.82, 0.89 and 0.85 for methyl 6-chloro-6-deoxy-α-D-glucoside (Me 6-Cl-α-Glc), methyl 6-bromo-6-deoxy-α-D-glucoside (Me 6-Br-α-Glc), methyl 6-chloro-6-deoxy-β-D-glucoside (Me 6-Cl-β-Glc) and methyl 6-bromo-6-deoxy-β-D-glucoside (Me 6-Br-β-Glc), respectively. The rate equation for these reactions is more complicated than that for Type I reactions;

$$y_t = y_e\{1-\exp(-\alpha t)\}/\{1+\beta\exp(-\alpha t)\} \tag{4}$$

where $\alpha = k_1(2[B]_0/y_e-[A]_0-[B]_0)$, $\beta = 1-y_e([A]_0+[B]_0)/[B]_0$, y_t and y_e are the mole fractions of product saccharide at time t and at equilibrium, respectively, and $[A]_0$ and $[B]_0$ are the initial concentrations of the starting saccharide and halide ion, respectively. In the cases where the equilibrium points are shifted largely to one side, the rate constants are determined from the initial slopes in order to avoid possible errors in iterative calculations.

Type III is the halogen exchange reaction of a methyl 3,6-dideoxy-3,6-dihaloglycoside using the ion of halogen not contained in the starting saccharide. This will give a complex saccharide mixture as shown in Scheme 2 for the reaction of methyl 3,6-dichloro-3,6-dideoxy-β-D-alloside (Me 3,6-Cl$_2$-All) with LiBr as an example. It is very difficult to analyze this mathematically exactly. Eight saccharides can theoretically be present in the solution and 24 rate constants would be necessary to effectively describe the total reaction. For simplicity, the substitution at C-3 is assumed to be independent of that at C-6. Equations (2) and (4) are applied to obtain the average rate constants for the substitution at C-3 and C-6, respectively. The mole fraction of each saccharide is calculated as the product of the substitution ratios of constituent halogen atoms at C-3 and C-6.

A typical time course of Type I reaction between Me 3,6-Cl$_2$-All and LiCl (*13*) is shown in Figure 1. The symbols show the experimental data and the theoretical curves were calculated with the obtained parameters. The curves coincide very well with the data. The equilibrium is on the glucoside side at all temperatures studied. The kinetic parameters for Type I and Type II reactions are summarized in Table I. The rates of interconversion between dibromodideoxyglycosides are about 1000 times higher than those between dichlorodideoxyglycosides. The difference in the configurations at C-3 of repeating units between Cell-Cl and Cell-Br can be ascribed to that in the rates of nucleophilic substitution of halogen atoms in dideoxydihalo-units with halide ions in lithium halide–DMAc (*10, 13*). Most of the 3,6-dibromo-3,6-dideoxyallose units originally formed in Cell-Br are converted quickly to thermodynamically more stable 3,6-dibromo-3,6-dideoxyglucose units in LiBr–DMAc.

Table I Kinetic parameters for Type I and Type II reactions

Starting compound	Temp. (°C)	[Saccharide] (g/L)	[LiX] (mol/L)	$k_1 \times 10^3$ (Lmol⁻¹min⁻¹)	$k_{-1} \times 10^3$ (Lmol⁻¹min⁻¹)	Activation energy (kJ/mol)	
						k_1	k_{-1}
Me 3,6-Cl₂-β-All	80	1.65	2.14	0.19	0.03	175	177
	90	1.58	2.04	0.77	0.13		
	100	1.68	0.99	4.67	0.80		
Me 3,6-Br₂-β-All	60	1.00	0.0826	17.5	2.02	143	145
	70	1.03	0.042	101	17.2		
	80	1.00	0.040	264	36.1		
	90	0.93	0.00415	1470	192		
Me 6-Cl-β-Glc	70	3.07	2.32	0.34	(401)	85	
	80	2.68	2.03	0.93	(191)		
	90	2.33	2.03	1.74	(286)		
Me 6-Br-β-Glc	63	2.78	0.00993	257	(0.49)	100	
	70	2.69	0.00993	373	(0.34)		
	80	2.93	0.00993	1370	(0.27)		
Me 6-Cl-α-Glc	70	2.67	2.31	0.50	(67)	59	
	80	2.88	2.09	0.95	(144)		
	90	2.73	2.21	1.53	(132)		
Me 6-Br-α-Glc	60	2.37	0.00993	152	(15.8)	102	
	70	2.34	0.00993	455	(89.4)		
	80	2.33	0.00993	1230	(57.3)		

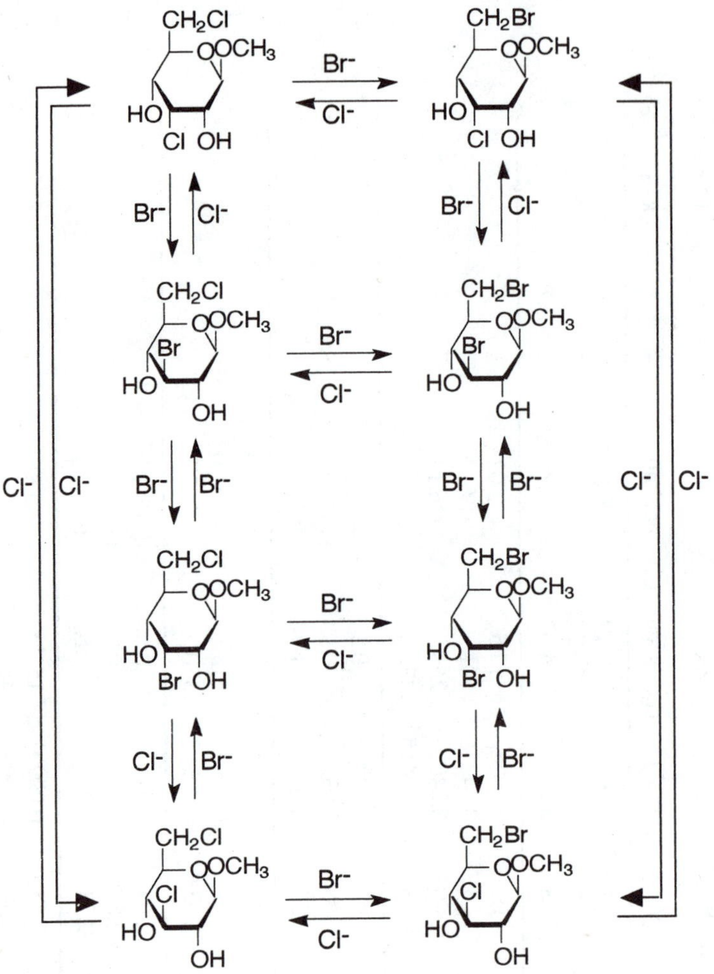

Scheme 2

The kinetic parameters are considered to be affected by the leaving group, nucleophile, anomeric structure, position of halogen substitution, and if halogen is on C-3, configuration at C-3. For example, it is possible to estimate the effect of anomeric structure on the rate of substitution at C-6 from data shown in Table I; the difference in rates is very small between the anomers. The halogen exchange between Me 3,6-Cl$_2$-All and LiBr (Type III) was carried out to estimate the effects of other factors. Only five dideoxydihalosaccharides were detected on the chromatograms. An additional assumption is introduced to simplify the analysis, that is, molar response values of these saccharides are equal. Figure 2 shows the time course of the Type III reaction. The calculated curves fit with the experimental data very well except two at 48 h. Table II compares the rate constants obtained with those for Type I and Type II reactions. Although the values for Type III may contain some errors, it

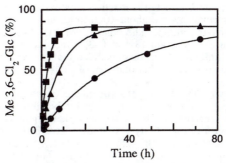

Figure 1. Time course of reaction between Me 3,6-Cl₂-β-All and LiCl. Reaction temperature: ●, 80°C; ▲, 90°C; ■, 100°C.

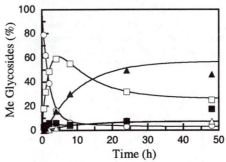

Figure 2. Time course of reaction between Me 3,6-Cl₂-β-All and LiBr. ○, Me 3,6-Cl₂-β-All; □, methyl 6-bromo-3-chloro-3,6-dideoxy-β-alloside; ■, methyl 3-bromo-6-chloro-3,6-dideoxy-β-glucoside; △, Me 3,6-Br₂-β-Glc; ▲, Me 3,6-Br₂-β-Glc.

Table II   Kinetic parameters of halogen exchange reactions at 90°C

	Type of exchange[a]		$k \times 10^3$ (Lmol^{-1}min^{-1})	
Position	Configuration	Halogen	Type III[b]	Type I and II
C–6	—	Cl → Br	2.97	1.74
	—	Br → Cl	116	1370 (80°C)
C–3	*allo* → *gluco*	Cl → Br	0.476	—
		Cl → Cl	—	0.77
		Br → Br	—	1470
	gluco → *allo*	Br → Cl	85.4	—
		Br → Br		192
		Cl → Cl	—	0.13

[a] Changing from the left-side one to the right-side one.
[b] The differences in relative molar responses were not taken into account.

is clearly shown which factor affects the reaction most strongly; when bromine is the leaving group, the rate constants are about 1000 times larger than those for the case where chlorine is the leaving group. The other factors, kind of nucleophile, position of halogen substitution and configuration at C-3, affect much less than the kind of halogen. From these findings, it can be concluded that the reactivity of Cell-Br for nucleophilic substitution is much higher than that of Cell-Cl.

Reaction of Cell-Br with Thiols in a Homogeneous System. Above results show that Cell-Br, whose hydroxyl groups are regioselectively substituted with bromine, is potentially very useful for the regioselective modification of cellulose derivatives through nucleophilic substitution. Only few studies have been reported in this area. We reported the reactions of Cell-Br with amines (*14, 15*), and with thiols under heterogeneous (*16*) and homogeneous (*17*) conditions. The reactivities of amines in nucleophilic substitution are relatively low while the reactions with thiols under homogeneous conditions are convenient for the introduction of a variety of functional groups into cellulose, especially under alkaline conditions. The LiBr–DMAc solvent system was used where no pretreatment was necessary for the dissolution of Cell-Br (*17*). Structures of reaction products of Cell-Br (bromine substitution at C-6) with thiols were studied with GC–MS and NMR analyses which confirmed that bromine atoms were substituted with thiol moieties. The amino groups of cysteine and 2-aminoethanethiol did not react We tried the reaction of Cell-Cl but the substitution did not occur under the reaction conditions studied.

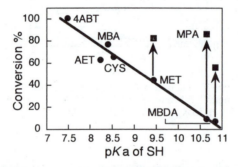

Figure 3. Relation between pKa of mercapto group and conversions (filled squares show the data obtained at higher concentrations of triethylamine and/or thiols). Thiol: 4ABT, 4-aminobenzenetiol; MBA, 2-mercaptobenzoic acid; AET, 2-aminoethanethiol; CYS cysteine; MET, 2-mercaptoethanol; MPA, 3-mercaptopropionic acid; MBDA, 2-mercaptobutanedioic acid.

Figure 3 shows the relation between pKa values of mercapto groups of used thiols and achieved conversions of Cell-Br to S-substituted deoxymercaptocellulose derivatives (*17*). The data plotted with filled circles were obtained under the same reaction conditions. It is clear that the reactivity of thiol depends strongly on the pKa value of mercapto group. This relation could not be observed in the case of the heterogeneous reactions in aqueous alkali (*16*). The pKa values of mercapto groups for thiols having carboxyl group(s) are low and the conversions are low as shown in the figure. Even these thiols give higher conversions of Cell-Br when the concentrations of triethylamine and/or thiols are increased (shown with filled squares). Table III compares the solubilities of the products (*17*). Some of the products obtained under homogeneous conditions are soluble in aqueous solutions while all the

Table III   Solubilities of cellulose derivatives obtained from Cell-Br and thiols

Thiol used	DS	HCl			NaOH	
		1 N	pH 5.0	H$_2$O	pH 9.0	1 N
Aminoethanthiol	0.72	–	–	–	–	–
Mercaptopropionic acid	0.92	–	–	–	–	±
Meracaptobutandioic acid	0.47	–	+	+	+	+
Cysteine	0.55	+	–	–	–	+
4-Aminobenzenethiol	0.84	–	–	–	–	–
2-Mercaptobenzoic acid	0.64	–	+	+	+	+

+, Soluble.   ±, Slightly soluble or swallen.   –, Insoluble

products obtained under heterogeneous conditions were insoluble.   The carboxyl group is effective in improving the solubilities of samples.   We used four carboxyl-bearing thiols and three of them gave samples soluble in alkaline solutions.   The most part of sample having cysteine moiety (Cell-CYS) is soluble both in acidic and alkaline solutions, and the dissolved Cell-CYS is precipitated by neutralization of the filtered solution.   This cycle of dissolution and precipitation can be repeated many times.

Synthesis of Deoxymercaptocellulose (Cell-SH).   As another example of the applications of Cell-Br, we studied the introduction of mercapto groups at C-6. Mercapto-bearing cellulose derivatives are interesting materials.   They are expected to be useful, for example, as a specific sorbent for Hg$^+$ and Ag$^+$ and also as starting materials for the introduction of functional groups through reactions with various organic halides.   The syntheses of Cell-SH from Cell-Cl (11) or cellulose tosylate (18) were reported previously.

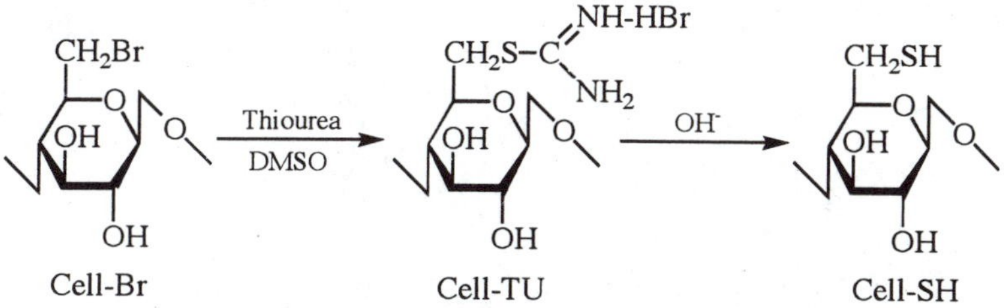

Scheme 3

In this paper, Cell-Br was converted to Cell-SH through the reaction with thiourea (Scheme 3) where DMSO was used as a solvent   Other organic solvents such as DMAc with or without lithium halide were examined but DMSO gave the best result.   The DS by bromine (DS$_{Br}$) of starting Cell-Br ranged from 0.85 to 0.97.   The time course of the formation of deoxyisothiouroniumcellulose bromide (Cell-TU) from Cell-Br and thiourea is shown in Figure 4.   The degree of substitution by isothiouronium moiety (DS$_{TU}$) and DS$_{Br}$ for Cell-TU were calculated with nitrogen and bromine contents.   The figure shows that DS$_{TU}$ gradually increases with increasing time (DS$_{Br}$ decreases at a corresponding rate).   The maximum conversion of bromine

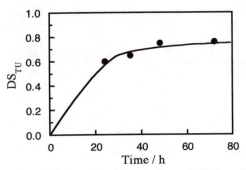

Figure 4. Time course of reaction between Cell-Br and thiourea.

into isothiouronium moiety achieved is higher than 70%. Approximately 20% of bromine in the original Cell-Br was lost within 24 h, probably by dehydrobromination. It is impossible to compare these results with those reported for Cell-Cl (*11*) because the chlorine contents of Cell-SH samples from Cell-Cl were not given. Cell-TU samples obtained in this study are soluble in water while those from Cell-Cl and thiourea were only swollen in water (*11*). The increased solubility can be explained in terms of the homogeneous distribution of isothiouronium residues which is due to the fact that both the synthesis of Cell-Br and reaction with thiourea proceeded under homogeneous conditions.

Cell-TU was treated next with alkali to convert isothiouronium groups into mercapto groups. Cell-TU was dissolved in water and Cell-SH was precipitated by neutralizing the solution after alkali treatment. Two conditions were examined for the

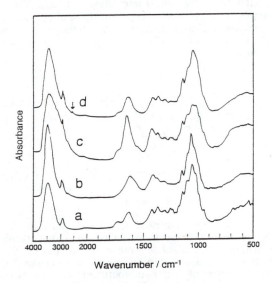

Figure 5. IR spectra of Cell-TU and Cell-SH. a, Cell-Br (DS_{Br} 0.85); b, cellulose; c, Cell-TU (DS_{TU} 0.60); d, Cell-SH (DS_{SH} 0.57) obtained under condition B.

alkali treatment in water; treating with a Na_2CO_3 solution (pH 11.0) in a dialysis tube for 15 h (condition A) and stirring in 2% NaOH solution for 1.5 h (condition B). With regard to the conversion and reaction time, condition B is better than condition A. The conversion of isothiouronium moieties to mercapto groups reached 95% under condition B. Much harsher conditions, for example, boiling in an alkaline solution for several hours, were applied to the synthesis of mercapto-bearing polymers (19). Both of our alkali treatment conditions are much milder but effective enough to obtain Cell-SH. This may be ascribed to the higher hydrophilicity of cellulose chains.

Figure 5 shows IR spectra of Cell-TU and Cell SH together with those of Cell-Br and cellulose. A weak peak due to S-H stretching vibration is clearly observed at 2560 cm^{-1} in the spectrum of the Cell-SH sample. This peak was not reported before for the cellulose derivatives containing mercapto groups.

Cell-SH and other deoxymercaptocellulose derivatives were stored under reduced pressure in the dark. Their IR spectra showed no detectable change after storage for several months indicating that no oxidation by air had occurred under the storage conditions.

We showed here the reactions between 6-bromo-6-deoxycelllulose and some nucleophiles under homogeneous conditions. We believe that this method is generally applicable to other nucleophiles and is useful for the regioselective introduction of functional groups to cellulose.

Literature Cited.

1. Ishizu, A., In *Wood and Cellulosic Chemistry*; Hon, D.N.-S.; Shiraishi, N., Eds.; Marcel Dekker: New York, 1991, chap. 16.
2. Furubeppu, S.; Kondo, T.; Ishizu, A. *Sen'i Gakkaishi*, **1991**, *47*, 592.
3. Furuhata, K.; Chang, H.-S.; Aoki, N.; Sakamoto, M. *Carbohydr. Res.*, **1992**, *230*, 151.
4. Dawsey, T. R.; McCormick, C. L. *JMS-Rev. Macromol. Chem. Phys.*, **1990**, *C30*, 405.
5. Furuhata, K.; Koganei, K.; Chang, H.-S.; Aoki, N.; Sakamoto, M. *Carbohydr. Res.*, **1992**, *230*, 165.
6. Furuhata, K.; Aoki, N.; Suzuki, S.; Sakamoto, M.; Saegusa, Y.; Nakamura, S. *Carbohydr. Polym.*, **1995**, *26*, 25.
7. Dean, D.M.; Azarek, W.A.; Jones, J.K.N. *Carbohydr. Res.*, **1974**, *33*, 383.
8. Evans, M.E.; Long, Jr., L.; Parrish, F.W. *J. Org. Chem.* **1968**, *33*, 1074.
9. Hannesian, S.; Plessas, N.R. *J. Org. Chem.*, **1969**, *34*, 1035.
10. Furuhata, K.; Aoki, N.; Suzuki, S.; Arai, N.; Ishida, H.; Saegusa, Y.; Nakamura, N.; Sakamoto, M. *Carbohydr. Res.*, **1995**, *275*, 17.
11. Tashiro, T.; Shimura, Y. *J. Appl. Polym. Sci.*, **1982**, *27*, 747.
12. Hodosi, G.; Podányi, B.; Kuszmann, J. *Carbohydr. Res.*, **1992**, *230*, 327.
13. Furuhata, K.; Aoki, N.; Suzuki, S.; Arai, N.; Sakamoto, M.; Saegusa, Y.; Nakamura, N. *Carbohydr. Res.*, **1994**, *258*, 169.
14. Aoki, N.; Taniguchi, T.; Arai, N.; Furuhata, K.; Sakamoto, M. *Sen'i Gakkaishi*, **1993**, *49*, 563.
15. Saad, G.R.; Sakamoto, M.; Furuhata, K. *Polym. Intern.*, **1996**, *41*, 293.
16. Aoki, N.; Koganei, K.; Chang, H.-S.; Furuhata, K.; Sakamoto, M. *Carbohydr. Polym.*, **1995**, *27*, 13.
17. Aoki, N.; Furuhata, K.; Saegusa, Y.; Nakamura, S.; Sakamoto, M. *J. Appl. Polym. Sci.*, **1996**, *61*, 1173.
18. Sakamoto, M.; Yamada, Y.; Ojima, N.; Tonami, H. *J. Appl. Polym. Sci.*, **1972**, *16*, 1495.
19. Chanda, M.; O'Druscill, K. F.; Rempel, G. L. *React. Polym.*, **1986**, *4*, 11.

Chapter 7

Cationization of Cellulose Fibers in View of Applications in the Paper Industry

E. Gruber, C. Granzow, and Th. Ott

Institute of Macromolecular Chemistry, University of Technology Darmstadt, Alexanderstrasse 10, D-64283 Darmstadt, Germany

Three methods were presented to render pulps cationic:

- direct reaction of epichlorohydrin and a tertiary amine
- coupling of oligo-ionomers
- grafting of cationic monomers

Advantages of such cationic pulps are:

- more effective than soluble cationic celluloses
- not sensitive towards pH-changes
- very versatile
- more economic

The direct reaction of epichlorohydrin and a tertiary amine is catalyzed by hindered tertiary amines (possible auto catalysis). A wide range of different products carrying various chemical groups (different polarity, accessibility, charge density) can be achieved by this method. Disadvantages of this reaction are, that the reaction proceeds also within the fiber and that it causes cross linking.

The method of coupling ionomers to the fiber yields higher charge densities, but surface selectivity is still poor. Surprisingly a higher surface charge has an adverse effect on retention of anionic filler particles.

For radical grafting besides charged monomers neutral comonomers have to be used (e.g. acrylamide). This method exhibits the best surface selectivity.

As paper aids cationic pulps excel at

- high total retention effectiveness
- good strength properties
- good drainability

Cellulose fibers normally are negatively charged. This has a strong bearing on properties of a paper stock, which predominately consists of anionic species like pulp fibers, fillers, and additives. Repelling interactions between such negatively charged particles hamper the process of flock formation, which may cause problems for water drainage and particle retention.

It is common practice in paper making to add cationic polymers to improve flock formation and particle retention. However such soluble polyelectrolytes may also exhibit some disadvantages. One draw back is the viscosity contribution of soluble polymers, which in most cases also depends on pH and salt concentration in the aqueous medium. Polymeric aids are also very sensitive to dosage. They should be applied in stoichiometric concentrations, over dosage leads to a change from a negative to a positive net charge for all species present, resulting in another repelling interaction.

This is demonstrated on a soluble cationic cellulose. Figure 1 shows, that the effectiveness of soluble ammonium alkyl cellulose as a retention aid passes a maximum as a function of dosage[1]. The maximum shifts to higher dosages, when the DS increases.

A similar effect is found with most other cationic polymers used as paper additives[2;3].

The reason is, that a flexible polyelectrolyte may easily approach anionic charges on the surface of either a fiber or filler particle, thus effectively neutralizing their electrical charge. A stiff polyelectrolyte on the other hand will only form a minor fraction of ion pairs, such aggregates will still carry naked anions as well as „unused" cations. *Figure 2* demonstrates schematically this different situation. If there are only stiff charge carriers flock formation is more efficient, as there are always free ions available.

Bridging capacity of cationic compounds will also depend on the accessibility of the cationic groups. Evidently they should sit on the surface and have some mobility to be able to form contact ion pairs. The chemical nature of the cationic groups will also be of importance as it will control the degree of hydration.

Based on such considerations there should be a potential for cationic fibers for paper making. Such fibers could offer some benefits over soluble polyelectrolytes:

- improving filler retention
- support drainage
- adsorb anionic trash
- do not disperse into white and waste water
- are more easily biologically degradable

Stone and Rutherford[4] have already described cationization of cellulose fibers by using glycidyl ammonium salts in 1969. Krause and Käufer suggested to use such cationic pulps as paper making additives and investigated their basic properties[5].

This paper describes some investigations based on such previous work extending it by using different chemical routes to obtain cationic pulp fibers.

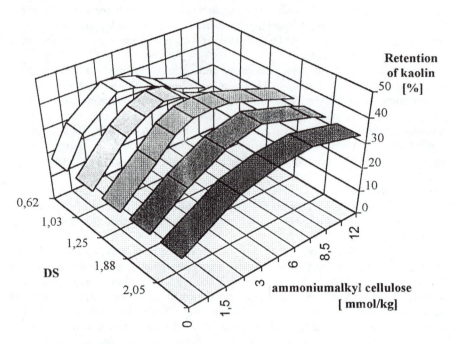

Figure 1: Retention effect of a soluble ammonium - alkyl – cellulose (trimethyl glycidyl ammonium cellulose)

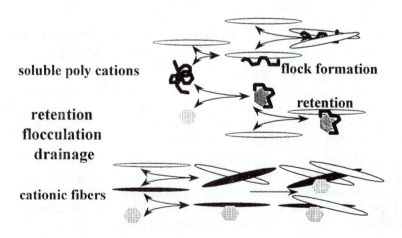

Figure 2: Interactions among electrically charged fibers, pigments, and polymers

Different types of cationic pulp fibers

In order to study morphological effects we prepared three different kinds of grafted fibers illustrated schematically in Figure 3. Direct addition reaction yields glycidyl ammonium derivatives of cellulose, where charged groups are tightly bond to the cellulose material. By coupling oligomers short side chains carrying cations can be introduced. Finally by grafting cationic polymer chains somewhat longer and more accessible side chains of ion carriers could be attached to the cellulose fibers.

Chemical routes to cationic pulp fibers

Simultaneous reaction of epichlorohydrin and a tertiary amine

Normal cationization uses glycidyl trimethyl ammoniumchloride as a reactant. To obtain a wider range of different cationic functionalities a one step method can be applied, which starts from epichlorohydrin and any tertiary amine (Equation 1).

Equation 1: One step cationization of a polysaccharide

$$\text{Posac}-\text{OH} + \text{(Epichlorohydrin)} + \text{(Tertiary Amine)} \longrightarrow \text{Cationic Polysaccharide}$$

| Polysaccharide | Epichlorohydrin | Tertiary Amine | Cationic Polysaccharide |

As there are also some carboxylic functions present in natural polysaccharides, besides ethers, esters and salts may be formed (Equation 2).

Equation 2: Possible side reactions with carboxylic groups

To get some insight in the mechanism of these reactions, methyl-α-D-glucopyranoside and glucuronic acid were used as model substances. These were reacted with epichlorohydrin and triethyl amine and the resulting products were analyzed by NMR and MALDI mass spectroscopy. The conclusions, drawn from these investigations are listed in *Table 1*.

Table 1: Features of one step cationization

Feature	Method	Result
yield	MALDI	if stoichiometrically applied, epichlorohydrin is consumed quantitatively
formation of ethers	NMR	C2, C3, C6 rather similar in reactivity
formation of uronic esters	NMR	insignificant
formation of uronic ammonium salts	NMR	positive
formation of advancement products	MALDI	few oligomers
crosslinking	reaction in solution of dimethyl acetamide / LiCl	some crosslinking occurs

We should expect, that the reaction proceeds predominately on the surface of the fiber. However the kinetics of the reaction (see *Figure 4*) show that both, a surface reaction and penetration into the depth of the fiber occurs. A slow diffusion controlled reaction of accessible areas within the fibers follows a swift reaction on the surface.

The ratio between surface and bulk reaction depends on catalysts applied and the size of the ligands of the amine. The degree of surface cationization can be determined by polyelectrolyte titration (see *Figure 5*).

Cationization by oligo - ionomers

By coupling a oligo-ionomer to the fiber higher charge densities can be achieved. Such oligomers are synthesized according to *Equation 3*:

Equation 3: Synthetic route for preparing oligo - ionomers

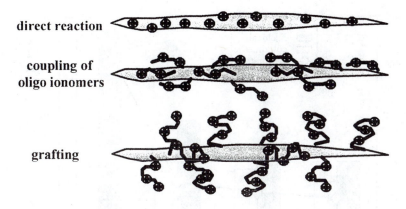

Figure 3: Types of cationic pulp fibers, prepared by different methods

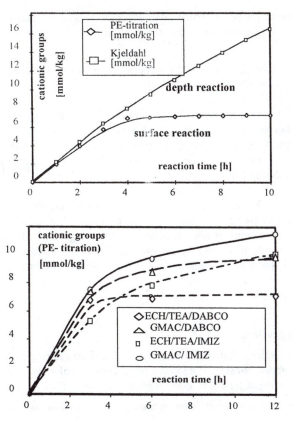

Figure 4: Kinetics of amine/epichlorohydrin cationization
(ECH = epichlorhydrin; TEA = triethyl amine; DABCO = 1,4-
diazabicyclo[2.2.2]octane; GMAC = glycidyl trimethyl ammonnium chloride;
IMIZ = 4-methylimidazole)

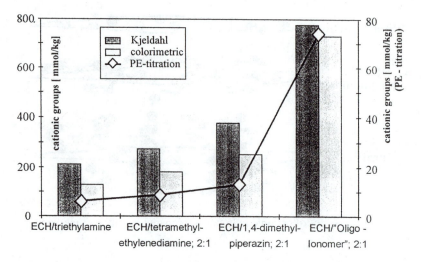

Figure 5: Comparison of some different degrees of cationization obtained (colorimetric titration according to reference [6])

Figure 5 shows, that cationization by oligo-ionomers is most effective, the surface selectivity however is still poor (app. 10%).

As an example of technical features of such pulps the retention of filler ($CaCO_3$) was studied. The fiber stock used as model of paper stock contained unmodified pulp, modified pulp, and filler. *Figure 6* shows the results obtained with pulp slurries, which contained the same amount of cationic charge equivalent but different charge densities of the cationic pulps added.

It can be seen, that filler retention capacity decreases with the charge density applied. Oligomer grafted pulps had the highest surface charges but were specifically least effective. To achieve optimal filler retention a rather high feed of charged fibers is needed, each of which should carry only a moderate surface charge. Among differently substituted ammonium ions trimethyl compounds are most effective.

Grafted cationic pulp

Heterogeneous grafting

Another way to render cellulose cationic was pursued by grafting cationic monomers to fibers. The reaction consists of a radical polymerization of an unsaturated monomer (*Equation 4*) starting at a radical, generated on the fiber surface.

Equation 4: Grafting on cellulose (Cell· = cellulose radical)

Table 2: Substances used for grafting

Name	Short	Formula
Diallyl dimethyl ammoniumchloride	DADMAC	
[3-(Methacryloylamino)-propyl]-trimethyl-ammonium chloride	MAPTAC	
Acrylamide	AAM	

The cellulose radical may be generated by chemical activated initiation of grafting:

- redox systems (e.g. Ce IV or Fe II)
- chemical oxidation (e.g. ozone, peroxide)

In these investigations Ce (IV) was used as a matter of convenience.

The substances used for grafting are listed in *Table 2*.

As fiber material bleached beech sulfite pulp was applied.

Trials to graft charged monomers alone to cellulose however were not successful. With DADMAC practically no grafting reaction was observed. But even with the neutral monomer acrylamide only very low grafting yields could be achieved. Surprisingly both monomers together reacted vigorously to a graft copolymer with copolymeric side chains. The reaction is summarized in *Equation 5*.

Equation 5: Cationic grafting onto cellulose

Pure (meth)acryl ammonium compounds like MAPTAC also do not graft to cellulose but contrary to allyl ammonium salts, they do not copolymerize easily with acrylamide (*Figure 7*).

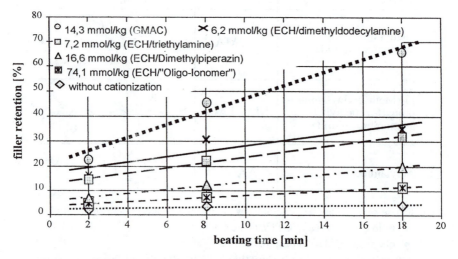

Figure 6: Filler retention as a function of beating time at various degrees of cationization
(ECH = epichlorhydrin; GMAC = glycidyl trimethyl ammonnium chloride)

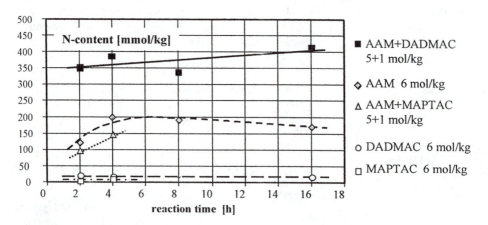

Figure 7: Influence of monomer type and reaction time on grafting yield
(AAM = acrylamide; DADMAC = diallyldimetyl ammoniumchloride; MAPTAC = [3-(methacryloylamino) propyl] trimethylammoniumchloride)

Technical features of grafted cationic pulp fibers

Retention effects

As expected cationic pulps operate as retention aids to negatively charged filler particles. The grafted cationic pulp was tested in a fiber stock containing calcium carbonate. The effect on retention of such filler particles was compared with a soluble cationic polymer (PDAMAC), which contained the same charged groups like the grafted pulp. As seen in Figure 8 at lower dosage of the polyelectrolyte the soluble polymer (PDADMAC) is more effective than the grafted pulp (BuSiKat). At higher dosages however grafted fibers are more active and do not show any sign of saturation and over dosage.

From this experiment we may conclude, that cations fixed to fibers are specifically less effective than chain bond ones, but there is no self inhibition and a higher total amount of filler can be fixed.

Mechanical properties

Each chemical modification may damage the pulp fibers leading to losses of mechanical strength. In addition to that, the modified surface charge will have an influence on the fiber to fiber interactions and on the formation of the sheet.

To test the influence of modified pulps, sheets were made from a mixture of modified and unmodified pulp fibers. As shown in Figure 9 the strength of the paper is augmented by the content of grafted pulp. This may be caused by stronger fiber flocks formed and may be also an indication, that the strength of the fibers as such is not hampered.

In practice however this positive effect may be camouflaged by the retention effect, which will lead to a higher filler content in papers containing cationic pulp. To evaluate this issue, filled papers were made using either grafted pulp or a synthetic polyelectrolyte (PDADMC) as a retention aid. Figure 10 shows, that the mechanical strength decreases as normal, when the filler content is increased. However this decline is comparable with fibers and soluble polyelectrolytes.

Drainage effects

As drainage of a paper slurry depends among others on flock formation, it has to be expected, that grafted pulp fibers will influence drainability. As a general information on drainage properties Schopper-Riegler freeness was measured as a function of feed of grafted pulp and beating time. The results are summarized in Figure 11.

As can be seen, cationic pulps drain more readily and exhibit slightly higher beating resistance. This may suggest, that flocks of higher densities are formed by fibers of opposite charge. This leads to slightly more substantial inhomogeneities in the paper. In fact this can be confirmed by image analysis of microphotographs. Opacity and, as shown, mechanical strength however does not suffer by these stronger variations of fiber density.

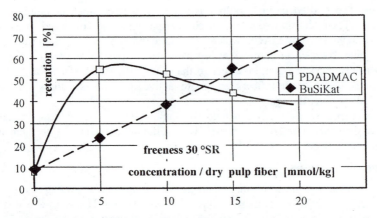

Figure 8: Retention of CaCO$_3$ by cationic graft pulps compared to a soluble
cationic polymer
(PDADMAC = poly diallyldimetyl ammoniumchloride; BuSiKat = beech sulphite
pulp grafted by PDADMAC)

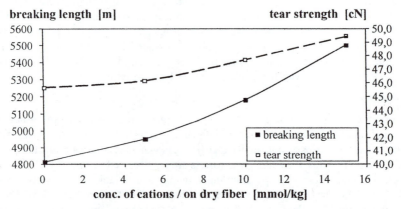

Figure 9: Mechanical strength of lab sheets without filler
(made from grafted + ungrafted beech sulphite pulp)

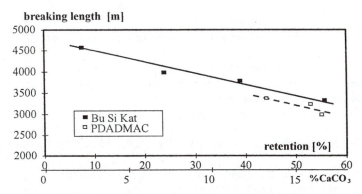

Figure 10: Breaking length as a function of filler concentration
(PDADMAC = poly diallyldimetyl ammoniumchloride; BuSiKat = beech sulphite
pulp grafted by PDADMAC)

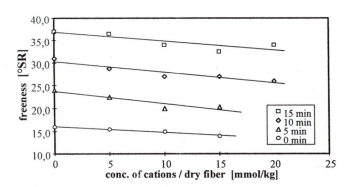

Figure 11: Influence of cationic pulps on the drainability and freeness of fiber stock
suspensions

106

Acknowledgment

This work was sponsored by a grant AIF 10210 N by the German „Arbeitsgemeinschaft Industrieller Forschungseinrichtungen" and by the German Papermakers Association („Verband der Deutschen Papierfabriken").

Literature Cited

[1]   Ott, G., Thesis Darmstadt 1992

[2]   Swerin, A.; Sjödin, U.; Ödberg, L.; Nordic Pulp Paper Res. J. **8** (1993) No. 4, 389 - 396

[3]   Klix, J., Thesis Darmstadt 1991

[4]   Stone, F.W., Rutherford, J.M.; US-Patent 3 472 840 (1969)

[5]   Käufer, M.; Thesis, Darmstadt 1982;

Käufer, M., Krause, Th.; Schempp, W.; Das Papier **34** (1980), Nr. 12, 575-579

Käufer, M., Krause, Th.; Schempp, W.; Das Papier **35** (1981) Nr. 10A, V33-V38

Käufer, M., Krause, Th.; Das Papier **37** (1983), Nr. 5, 181-185

Käufer, M., Das Papier **35** (1981) Nr.12, 555-562

[6]   Gruber, E.; Ott, Th.: Das Papier **49** (1995), Nr. 6, 289 - 296

Chapter 8

A Preliminary Study of Carbamoylethylated Ramie

C. C. L. Poon[1], Y. S. Szeto[1,3], W. K. Lee[1], W. L. Chan[2], and C. W. Yip[2]

[1]Institute of Textiles and Clothing and [2]Department of Applied Biology and
Chemical Technology, Hong Kong Polytechnic University, Hong Kong

Ramie fibres were treated with acrylamide in the presence of alkali
acting as a catalyst; the add-on ranged from 0.8% to 11.0%. Both the
amount of char produced during thermal degradation and the flame
resistance in terms of Limiting Oxygen Index were found to have
increased from 18 to 24. The properties of the resulting textile were
compared with those of the intact ramie fibre, and the relationship
between the yield of reaction and the application parameters –
temperature, catalyst concentration, and duration of the reaction – is
described.

The chemical modification of cellulose and methods for modifying cotton have
been studied extensively in recent decades. One method for modifying cotton, the
Michael addition reaction, uses a vinyl compound to react with the cellulose chain.
Frick et al. (*1-2*) reported that acrylamide readily reacted with cotton under base-
catalysed conditions. The carbamoylethyl ether derivative of cellulose fibres prepared
by this reaction has good fabric properties and modified dyeing characteristics. This
carbamoylethylated cotton can be further modified; the resulting fabric can be dyed
with different classes of dyestuffs that have very little or no affinity to cellulose (*3*).
    Ramie, which is also cellulosic in nature, is a bast fibre obtained from the
stems of the plants *Boehmeria nivea* or *Boehmeria tenacisseama*. The fibre of ramie
possesses many superior properties, which make it popular in North American and
European countries – it is strong, white, lustrous, and durable (*4*). Nevertheless, there
are few academic reports on the chemical modification of ramie as a means of further
improving its properties.
    Based on previous work on cotton, this study aimed to determine some of the
factors involved in the application of the Michael addition to the preparation of
carbamoylethylated ramie. The physical and thermal properties of the treated and

[3]Corresponding author.

the untreated ramie were compared. The thermal behavior of the treated samples was improved, and their textile properties, including tensile strength, were retained.

Experimental

Materials. The ramie fabric used was plain weave that had been desized and scoured. The ramie fibre had 96-98% α-cellulose on dry basis with a small amount of lignin. Commercial acrylamide (97%) was used as received. Reagent grade sodium hydroxide was used as catalyst.

Carbamoylethylation. Ramie fabrics weighing 5.0 grams were carbamoylethylated by immersing them in a 200 ml aqueous solution containing acrylamide and sodium hydroxide at specified temperatures and durations (Table 1). Thorough water washing was followed by Soxhlet extraction, using water as a solvent to remove the unreacted chemicals.

Measurements

Weight Gain. The weight gain of the fabric after carbamoylethylation was calculated using the following equation:

$$\text{Weight Gain (\%)} = \frac{W_f - W_i}{W_i} \times 100$$

where W_f and W_i are the weights of the treated and the untreated fabrics, respectively. All the fabrics were held at 21°C, 65% relative humidity, overnight before measurement.

Moisture Regain. The moisture regain of sample fibres was evaluated according to the standard procedure, ASTM D2654-89a.

Thermogravimetric Analysis. The thermal properties of the reacted fabrics were studied by a Mettler TA2000 thermal analysis system; scanning ranged from 100°C to 600°C at 30K/min in nitrogen atmosphere with a flow rate of 200 cm^3/min. The onset temperature and the residual amount were evaluated.

Limiting Oxygen Index. The flammability of the fabrics was determined as Limiting Oxygen Index according to the procedures stated in ASTM D2863-87.

Tensile Properties. Tensile strength of the fabrics was measured by the standard strip test method (ASTM D 5035) using the Instron tensile tester.

Results And Discussion

Influence of Reaction Conditions. Figure 1 shows that increasing the acrylamide concentration under 5% sodium hydroxide was accompanied by an increase in weight

Table 1. Experimental Conditions of Carbamoylethylation of Ramie

Experiment Number	Composition of Solution Acrylamide %	NaOH %	Reaction Parameters Temp. °C	Duration Hrs	Weight Gain (%)
CAA001	5	5	40	4	2.60%
CAA002	10	5	40	4	4.80%
CAA003	15	5	40	4	7.40%
CAA004	20	5	40	4	6.60%
CAA005	25	5	0	4	0.80%
CAA006	25	5	10	4	1.00%
CAA007	25	5	20	4	4.00%
CAA008	25	5	30	4	7.00%
CAA009	25	5	40	4	8.40%
CAA010	25	5	50	4	5.80%
CAA011	25	5	60	4	4.60%
CAA012	25	5	40	0.25	2.00%
CAA013	25	5	40	0.5	3.60%
CAA014	25	5	40	0.75	4.60%
CAA015	25	5	40	1	5.00%
CAA016	25	5	40	1.5	5.60%
CAA017	25	5	40	2	7.00%
CAA018	25	5	40	3	8.40%
CAA019	25	5	40	4	8.80%
CAA020	25	5	40	6	8.00%
CAA021	25	5	40	8.33	7.00%
CAA022	25	5	40	10	7.20%
CAA023	25	0.5	40	4	0.60%
CAA024	25	2.5	40	4	3.40%
CAA025	25	5	40	4	8.60%
CAA026	25	7.5	40	4	11.00%

gain. At lower concentration, 0-10% of acrylamide, the weight gain increased almost linearly with the increase in concentration. At higher concentrations, as the molar ratio of catalyst to acrylamide became smaller, the increase in weight gain became non-linear and less significant.

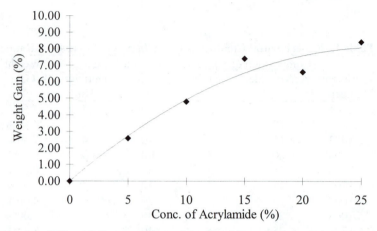

Figure 1. Effect of Concentration of Acrylamide on Weight Gain of Fabric.

Khalil et al. (*5*) reported the following reactions when starch was carbamoylethylated in a mixture of cellulose, acrylamide, sodium hydroxide, and water:

$$\text{CELL}-\text{OH} + \text{CH}_2=\underset{\underset{\text{CONH}_2}{|}}{\text{CH}} \underset{}{\overset{\text{NaOH}}{\rightleftharpoons}} \text{CELL}-\text{O}-\text{CH}_2-\underset{\underset{\text{CONH}_2}{|}}{\text{CH}_2} \tag{1}$$

$$\text{CELL}-\text{O}-\text{CH}_2-\underset{\underset{\text{CONH}_2}{|}}{\text{CH}_2} + \text{H}_2\text{O} \xrightarrow{\text{NaOH}} \text{CELL}-\text{O}-\text{CH}_2-\underset{\underset{\text{COONa}}{|}}{\text{CH}_2} + \text{NH}_3 \tag{2}$$

$$\text{CH}_2=\underset{\underset{\text{CONH}_2}{|}}{\text{CH}} + \text{H}_2\text{O} \xrightarrow{\text{NaOH}} \text{CH}_2=\underset{\underset{\text{COONa}}{|}}{\text{CH}} + \text{NH}_3 \tag{3}$$

$$\text{CELL}-\text{OH} + \text{CH}_2=\underset{\underset{\text{COONa}}{|}}{\text{CH}} \xrightarrow{\text{NaOH}} \text{CELL}-\text{O}-\text{CH}_2-\underset{\underset{\text{COONa}}{|}}{\text{CH}_2} \tag{4}$$

Khalil et al. (5) also noted that the extent of the reactions depended upon the temperature of the reaction, the concentration of the catalyst, and the duration of the reaction. Our experiments were designed to study the effect of these same three parameters on the carbamoylethylation of ramie in terms of weight gain of fabric after reaction.

Figure 2 shows the effect of reaction temperature on the weight gain of ramie fabrics treated with 25% acrylamide and 5% sodium hydroxide for 4 hours. It is apparent that an optimum reaction temperature of 40°C gave the maximum fabric weight gain. In the range of 0 to 40°C, the higher the temperature employed, the faster the reaction rate, and the greater the weight gain.

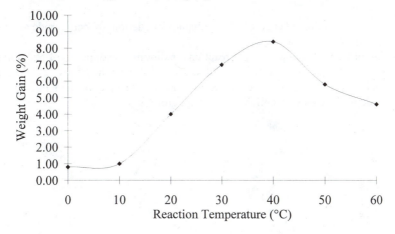

Figure 2. Effect of Reaction Temperature on Weight Gain of Fabric.

When the temperature was higher than 40°C, the weight gain was lower because of hydrolysis of the ether linkages of carbamoylethylated ramie (6) and the loss of beta and gamma cellulose of ramie in the presence of alkali. The activation energy of the reaction was found by plotting the logarithm value of percentage weight gain versus 1/T (Figure 3).

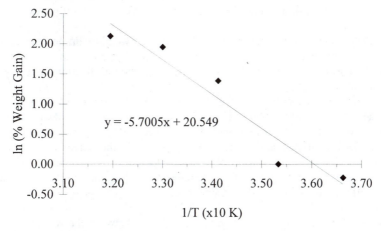

Figure 3. A Plot of ln (% Weight Gain) vs 1/T.

The activation energy, E_A, was calculated by the equation:

$$E_A = -R \left[\frac{\delta \ln (\% \text{ Weight Gain})}{\delta \frac{1}{T}} \right]$$

The activation energy of the carbamoylethylation of acrylamide on ramie cellulose was 47.40 kJmol-1.

In order to develop efficient reaction conditions, the duration of reaction time was studied (Figure 4). As reaction time increased to 4 hours, weight increased significantly, obviously because of the increase in the reaction and interaction of acrylamide molecules with cellulose hydroxyl groups.

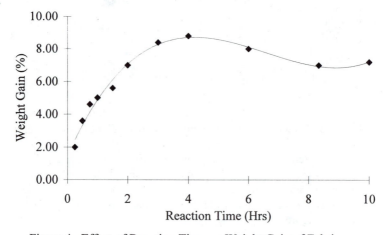

Figure 4. Effect of Reaction Time on Weight Gain of Fabric.

As reaction time increased beyond 4 hours, weight gain decreased, probably because of the hydrolysis of acrylamide in the solution and the partial hydrolysis of ether linkages of carbamoylethylated ramie, i.e. de-etherification, as suggested by Ibrahim et al. (7)

Figure 5 shows the relationship between the concentration of the catalyst used and the percentage weight gain. The higher the concentration of NaOH, the greater the weight gain, until the NaOH concentration reached 7.5%, after which gelling occurred on the surface of the fabric. The gelling is due to the extensive crosslinking of acrylamide with cellulose, as well as to self-polymerization under highly alkaline conditions (8-9). Thus, in this study the highest possible concentration of catalyst was 5.0%, offering the least significant side reactions.

Moisture Regain. Figure 6 shows the relationship between the weight gain and the moisture regain of carbamoylethylated ramie. Although the increase of moisture regain was not very significant when compared to the untreated ramie, the carbamoylethylated ramie generally showed an enhancement of moisture regain.

Thus, the hydrophilic properties of cellulose increased with the addition of amide groups, whereas the addition of other vinyl monomers caused a hydrophobic reaction (*10*).

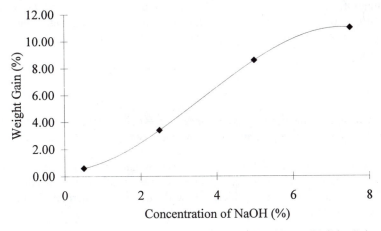

Figure 5. Effect of Concentration of Sodium Hydroxide on Weight Gain of Fabric.

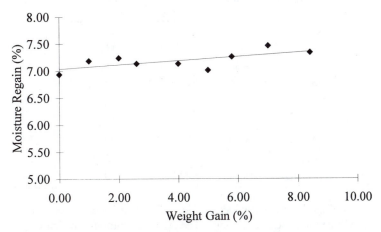

Figure 6. Relation of Moisture Regain of Carbamoylethylated Ramie with Weight Gain.

Thermal Behavior of Carbamoylethylated Ramie.   The results of the thermogravimetric analysis of thermal behavior are given in Table 2.

As the weight gain of the fabric increased, the onset temperature decreased. This resulted in early decomposition of the fibre and delay of the decomposition rate, so the residual amount of the treated ramie was higher than that of the untreated ramie. The addition of amide groups to the cellulose chain thus improved the thermal behavior of ramie, as was reported by Shimada and Nakamura (*11*) for the graft copolymerization of acrylamide for cotton and for the carbamoylethylation of cotton.

Evaluation of Flammability by Limiting Oxygen Index (LOI). The LOI test determines the minimum concentration of oxygen in a flowing mixture of oxygen and nitrogen that will just support a flaming combustion of the material. The limiting oxygen index, expressed as volume percentage, is calculated as follows:

$$n\% = \frac{100 \times O_2}{O_2 + N_2}$$

where n% = the limiting oxygen index, O_2 = volumetric flow rate of oxygen in mm^3/s, N_2 = corresponding volumetric flow rate of nitrogen in mm^3/s.

As the weight of the fabric increased, its flammability became correspondingly higher, as shown in Table 2. These results confirm the results of the thermogravimetric analysis, which showed improved thermal stability for carbamoylethylated ramie.

Table II. Thermal Behavior of Carbamoylethylated Ramie

Experiment No.	Weight Gain (%)	Onset Temperature (°C)	Residue Amount (%)	LOI (%)
Control	0	360.10	14.51	18
CAA005	0.80	357.90	16.43	19
CAA013	3.60	357.20	15.07	20
CAA011	4.60	341.10	17.08	21
CAA002	4.80	336.30	18.22	21
CAA022	7.20	337.70	19.41	22
CAA009	8.40	336.80	22.86	22
CAA019	8.80	330.30	21.19	22
CAA026	11.00	325.40	30.04	24

During the pyrolytic degradation of carbamoylethylated ramie, the amide group will decompose to ammonia. The released ammonia, which is an energy-poor fuel, exhibits flame retardant effects by several modes (*12-13*). The flame inhibitory action is a result of the concentration dilution of the released volatile combustible material from the decomposed cellulose, i.e. the levoglucosan, during the combustion.

Tensile Properties of Carbamoylethylated Ramie Fabrics. Tensile strength of the fabrics was measured by standard method ASTM D 5035. The breaking strength of carbamoylethylated fabrics gradually decreased with increasing weight gain. The strength loss ranged from 0 to a maximum of 23% (Figure 7). This result was similar to that of the carbamoylethylated cotton (*14*).

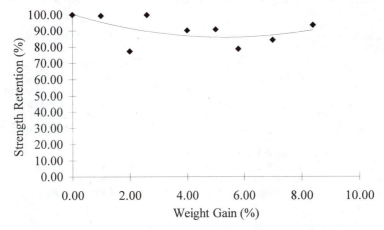

Figure 7. Strength Retention of Carbamoylethylated Ramie.

The loss of tensile strength is due to the addition of amide groups, which are bulkier than the hydroxyl groups, thus reducing the extent of hydrogen bonding between cellulose chains. Nevertheless, the loss of strength is not very significant, as amide groups can also form hydrogen bonds between cellulose chains.

Conclusions

The application of Michael addition was useful in modifying ramie fibre by using the all-in exhaustion method. The best yield of the reaction between ramie cellulose and acrylamide was obtained at 40°C, a 4-hour reaction, in the presence of 5% NaOH. The carbamoylethylated ramie had better hydrophilicity and thermal properties without significant loss of fabric strength after the reaction.

Acknowledgment

The authors thank The Hong Kong Polytechnic University for the financial support of this project.

References

1.   Frick, J.W.; Reeves, W.A.; Guthrie, J.D. *Text. Res. J.* **1957**, *27*, No.2, pp.92-99.
2.   Frick, J.W.; Reeves, W.A.; Guthrie, J.D. *Text. Res. J.* **1957**, *27*, No.4, pp.294-299.
3.   Abou-zeid, N.Y.; Anwar, W.; Hebeish, A. *Cell. Chem. Technol.* **1981**, *15*, pp.321-330.
4.   How, Y.L.; Cheng, K.P.; Lau, M.P. *Textile Asia* **1991**, *22*, No.5, pp.74-78.

5.   Khalil, M.I.; Bayazeed, A.; Farag, S.; Hebeish, A. *Starch/Stärke* **1987**, *39*, No.9, pp.311-318.
6.   Feit, B.A.; Zilkha, H. *J. Org. Chem.* **1963**, *28*, p.406.
7.   Ibrahim, N.A.; Haggag, K.; Abo-Shosha, M.H. *Am. Dyestuff Reptr.* **1988**, *77*, No.7, pp.34-42.
8.   Thomas, W.M. and Wang, D.W. In *Encyclopedia of Polymer Science and Engineering*, Herman F. Mark. Ed., 2nd Edition, John Wiley & Sons: New York, **1985**, p.185.
9.   Kurenkov, V.F. and Myagchenkov, V.A. In *Polymeric Materials Encyclopedia*, Joseph C. Salamone. Ed., CRC Press, Inc., **1996**, Vol.1, pp.47-54.
10.   Hebeish, A.; Guthrie, J.T. *The Chemistry and Technology of Cellulosic Copolymers*, Springer-Verlag, Berlin-Heidelberg-New York, **1981**, p.295.
11.   Machiko Shimada and Yoshio Nakamura In *Inititation of polymerization*, ACS symposium series 212, Frederick E. Bailey, Jr. Ed., Bailey, March/April, **1982**, pp.237-248.
12.   Miller, D.R.; Evans, R.L.; Skinner, G.B. *Comb. Flame* **1963**, *7*, pp.137-142.
13.   Haynes, B.S.; Jander, H.; Mätzing, H.; Wagner, H.G. *19th Symp. (Intl.) on Combustion*, The Combustion Institute, **1982**, pp.1379-1385.
14.   Grant, J.N.; Greathouse, L.H.; Reid, J.D.; Weaver, J.W. *Text. Res. J* **1955**, *25*, No.1, pp.76-83.

CHEMICAL AND MOLECULAR STRUCTURE

Chapter 9

Characterization of Polysaccharide Derivatives with Respect to Substituent Distribution in the Monomer Unit and the Polymer Chain

Petra Mischnick

Department of Organic Chemistry, University of Hamburg, Martin-Luther-King-Platz 6, D-20146 Hamburg, Germany

Besides other structural parameters the distribution of substituents greatly influences the properties of polysaccharide derivatives. To characterise the substitution pattern in the monomer unit and in the polymer chain, different approaches can be used. Strategies of monomer analysis by gas chromatography and mass spectrometry after chemical degradation will be presented. Examples of alternative separation methods will be given. The distribution along the polysaccharide chain was studied on methyl amyloses as model compounds. Oligomeric mixtures obtained after partial degradation were analysed by mass spectrometry. The results were interpreted by comparison with a homogeneous distribrion calculated from the monomer composition. First applications on cellulose sulfates and acetates are reported.

Polysaccharides are very well suited for their functions in nature, e.g. to build up mechanically stable cell walls in plants. However, to get new materials with new properties cellulose and starch as the main sources of renewable polymers have been modified in different ways since more than 150 years now. The primary structure of those derivatives is responsible for the formation of higher molecular architectures as helices, aggregates in solution, liquid crystalline phases or monomolecular layers. Properties as thickening, gelation, film building or flocculation are mainly influenced by the distribution of substituents in the polymer chain, while effects involving molecular recognititon also show a dependence on the regioselectivity of substitution in the monomer unit. Therefore, analysis in this field has to aim at a determination of all structural features for a better understanding of the relationship between the reaction conditions, the primary structure and the resulting properties.

Monomer Analysis of Cellulose Derivatives. Cellulose is a linear homopolymer of 1,4-linked β-D-glucose residues. By a partial derivatisation of the OH functionalities it becomes a "copolymer" of up to eight different monomers, if one type of substituent is introduced. Due to the high DP the ends of the chains can be neglected. Then, 3 positions (C2, C3, and C6, m = 3) per anhydro glucose unit (AGU), which can be either OH or OR (n = 2), result in $n^m = 2^3 = 8$ different monomer patterns. If two types of substituents are present (n = 3) as in mixed esters as acetates/butyrates $3^3 = 27$ different monomer units are possible.

The distribution in the AGU, especially the distribution on the positions 2, 3 and 6, have been determined for a number of different cellulose derivatives by ^{13}C NMR spectroscopy (see chapter x). The alternative approach includes degradation of the polymer to monomers without any discrimination, to yield a mixture which is representative for the original copolymer. The constituents of this mixture in an appropriate form are then separated by an efficient chromatographic method. After peak assignment by combined mass spectrometry or comparision with authentic standard compounds, the relative molar ratios of all components are calculated from their peak areas after correction for their individual detector response (Figure 1).

How can this be achieved in practice? The polysaccharide derivative is usually permethylated in the first step to protect and to sign all free hydroxy groups. Furthermore, by this step all hydrogen bonds are broken, the derivative can be really dissolved, and all AGU should be accessible. Then, this mixed derivative is submitted to subsequent hydrolysis, reduction and acetylation to get partially methylated, partially substituted glucitol acetates (Figure 2, standard methylation analysis). An alternative type of methylation analysis is shown in Figure 3. In the reductive-cleavage-method, introduced by Rolf and Gray (1), the glucosidic linkages are cleaved under promotion of a Lewis-acid. The cyclic carboxonium ions formed are *in situ* reduced to yield the corresponding 1,5-anhydroglucitols which are usually acetylated in this one-pot-reaction sequence. A mixture of volatile compounds is obtained, appropriate for capillary gas chromatography with its high separation efficiency. A further advantage of GC is the combination with a flame ionization detector (FID). The response of this detector in principle corresponds to the number of C-atoms present in the compound. Electronegative atoms reduce the response of a carbon for a certain amount. Therefore, the response values of the individual components can be calculated by an experimentally established increment system [effective-carbon-concept, (2-4)]. In addition, the combination with a mass spectrometer (GC/MS) allows the elucidation of the structure and the differentiation of regioisomers. Therefore, the tedious synthesis of standard compounds is not essential. These procedures have been applied to alkyl ethers (5,6) hydroxyethyl- and hydroxyproypl derivatives (7-10).

If two or more types of substituents are present in the cellulose derivative, the situation becomes much more complex, e.g. for ethylhydroxyethylcellulose [EHEC, (11)] or hydroxypropylmethylcellulose (HPMC). Without tandem substitution (oligo-ether formation of the hydroxyalkyl groups) 4 different types of OR are possible in the AGU: OH, OMe, OHP, and OMP (methoxypropyl). That means that $4^3 = 64$ monomer patterns can theoretically occur. Due to the asymmetric C in the HP residue, HP derivatives occur in two diastereomeric forms, di-O-HP (or MP-) substituted AGU give four diastereomers and so on. At higher MS values, tandem-substitution will cause additional products of theoretical unlimited number. We have investigated a series of HPMC with a DS_{Me} of 2 and a MS_{HP} of about 0,2.

To avoid that small amounts of O-HP and O-MP ethers, which are spread over glucose derivatives with all different methylation patterns, do not escape detection, we used a twofold approach (Mischnick, P.; Dönnecke, J., unpublished results). To analyse the methyl pattern, which was responsible for about 90% of the total DS, the cellulose ethers were directly hydrolysed, reduced and acetylated. The partially methylated glucitol acetates are eluted first in the gas chromatogram (Figure 4a) and could easily be assigned and calculated. The main peaks of glucitols bearing hydroxy-propyl groups *and* methyl groups could be identified by GC-MS. For the analysis of the HP pattern the sample was permethylated prior to degradation to focus all monomers with a HP/MP group in a certain position in one peak: for example 1,4,5-tri-O-acetyl-2-O-(2-methoxy)propyl-3,6-di-O-methyl-D-glucitol includes the 2-O-HP

120

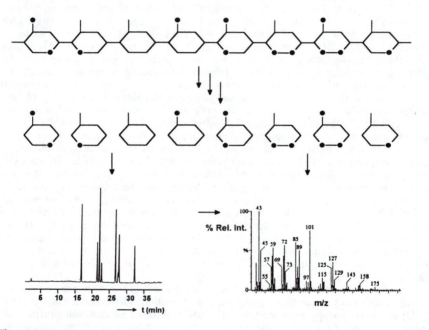

Figure 1
General Approach for the determination of the substituent distribution in the monomer unit of polysaccharide derivatives

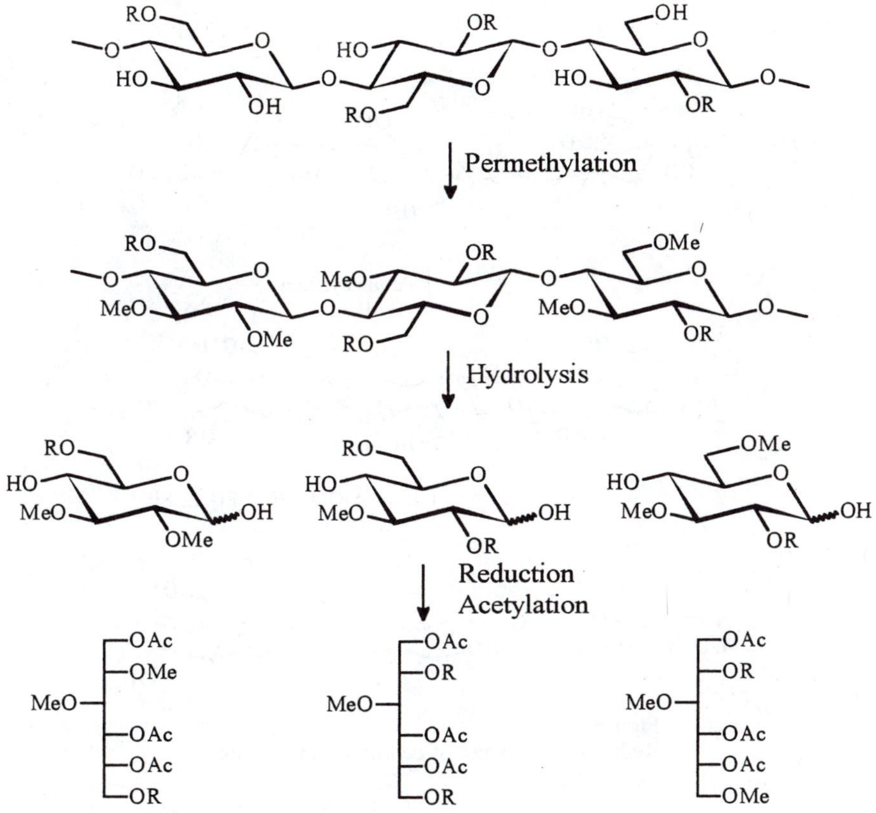

Figure 2
Standard methylation analysis of cellulose derivatives

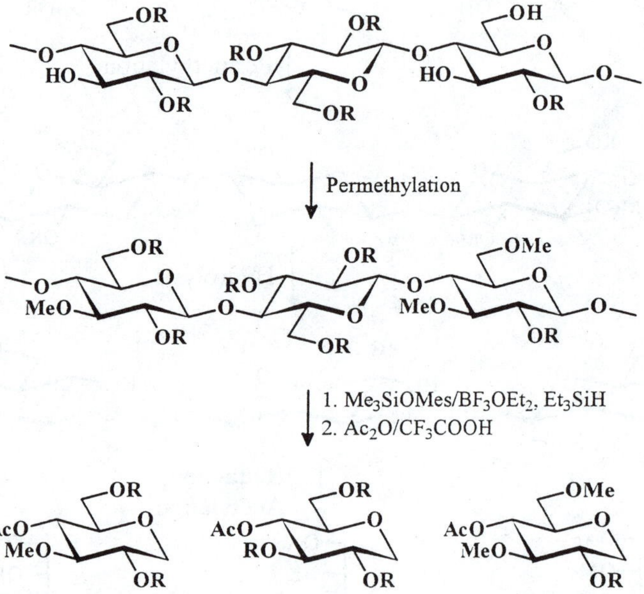

Figure 3
Reductive-cleavage of cellulose derivatives

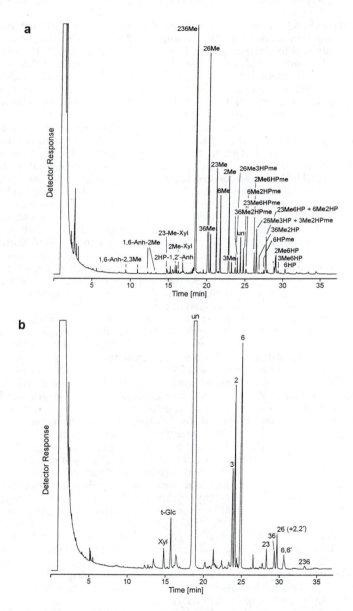

Figure 4
(a) GC of HPMC after hydrolysis, reduction and acetylation and (b) after permethylation, hydrolysis, reduction and acetalytion according to Figure 2. Peaks are assigned according to the substituted position in the AGU

ethers with eight different methyl patterns (Figure 4b). So the sensitivity is significantly enhanced.

Figure 5 shows the results obtained for the methyl pattern and the distribution of the HP groups summerized for the positions 2, 3 and 6. The complete monomer compositon only slightly deviates from the model of Reuben (*12-13*) (Figure 6). This kinetical model already includes the enhanced reactivity of 3-OH in 2-O-methylated AGU. Prefered 2-O-methylation is observed, which is favoured by low alkali concentration during the reaction. Then the 2-OH as the most acidic one is the most reactive. In the competition with the methyl iodide the HP groups are directed to the steric less hindered primary 6-OH.

The procedures just outlined require, that the substituents are stable under the reaction conditions applied. If this is not the case, as for trialkylsilyl ethers, acetates or sulfates, an indirect determination of the substitution pattern by the analysis of the complementary methyl pattern is performed. Trialkylsilyl ethers of cellulose as TBDMS- or THxDMS-derivatives are important intermediates in the synthesis of regioselectively substituted cellulose derivatives (see chapter 1). During the hydrolysis or the reductive-cleavage of the permethylated samples the silyl groups are cleaved and partially methylated (anhydro)glucitol acetates are obtained. However, during the analysis of a series of cellulose-, cyclodextrin-, and amylose derivatives it turned out, that the silyl groups nearly quantitatively migrate from O-2 to O-3 in α-glucans under the alkaline methylation conditions, while only 5% or less rearrangement was observed for the β-linked cellulose (*14-15*) This different behaviour may be caused by stereoelectronic effects (α *versus* β) and conformational differences. Methyl 2-O-TBDMS-β-D-glucopyranoside shows about 60% rearrangement of the silyl group. With methyl triflate as the methylating agent no migration occured as expected.

Sulfation of several polysaccharides as laminarin, schizzophyllan or cellulose plays an important role in the research for heparinoidic materials or drugs with antiviral properties. The ionic sulfate groups shall also help to get soluble products. Nehls *et al.* reported, that a DS of 0,3 was efficient to achieve water solubility for cellulose sulfates from a homogeneous reaction, while a DS of 1,8 was necessery for samples produced under heterogeneous conditions (*16*). Methylation analyis (Figure 2) can be applied (Gohdes, M.; Mischnick, P. *Carbohydr. Polym.* in press). However, problems arise from the residues sulfated in position 2, which can be displaced by an intramolecular nucleophilic attack of the vicinal 3-O⁻ anion under the alkaline methylation conditions. In addition, these residues are preferably hydrolysed and partially lost during long hydrolysis times. For such samples the method of Stevenson and Furneaux reported for polysaccharides from algae (*17*) is the method of choice.

Acyl groups are labile under both alkaline methylation conditions and acid hydrolysis. Methylation can be achieved with methyl triflate or trimethyloxonium tetrafluoroborate and 2,6-di-*tert.*-butyl-pyridine (*18*). In a similar etherification reaction with benzyl triflate neither migration nor cleavage of acyl groups was observed (*19-20*). In a subsequent alkaline alkylation step the acyl residues are exchanged against ethyl or higher alkyl groups. The resulting mixed ether can be analysed as described above. Lee and Gray have reported a direct analysis of the methylated cellulose acetates by reductive-cleavage (*21*). With a high excess of the reagents, the acyl groups are reduced to the corresponding alkyl ethers (OAc ⟶ OEt) (*22*). This method also allows the analysis of mixed esters (Gray, G. R., University of Minneapolis, unpublished data).

Besides gas chromatography HPLC has been used for the determination of the molar fractions of un-, mono-, di- and trisubstituted glucose residues (*23*), which complements the data for 2 : 3 : 6 substitution obtained bei NMR spectroscopy. Anionic deriviates as carboxymethyl cellulose (CMC) and sulfoethyl cellulose (SEC) have been analysed by High-pH anion exchange chromatography (HPAEC) after hydrolysis (*24*). We recently succeeded in the separation of partially methylated glucoses, which are still soluble in water (*25*). In contrast to gas chromatography all

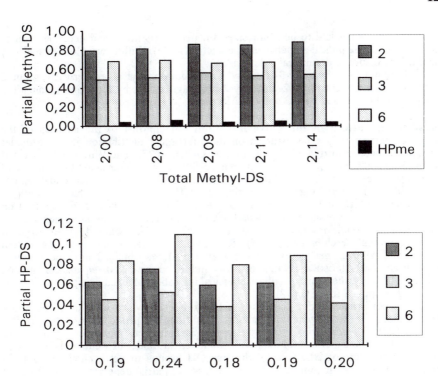

Figure 5
Distribution of methyl groups and hydroxypropyl groups in five HPMC

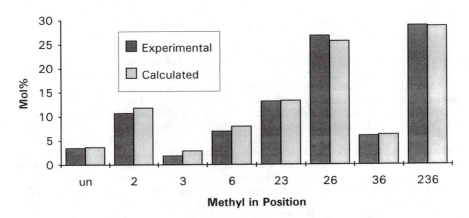

Figure 6
Comparison of the monomer composition (methyl pattern) of one HPMC with the
calculated ratios according to the model of Reuben (*13*)

standard compounds had to be synthesized for peak assignment and determination of the response in the pulsed amperometric detector (PAD). With every blocked hydroxy function the response dramatically decreases.

Cationic O-(2-hydroxy-3-trimethyl ammonium)propyl ether derivatives could be transformed to neutral O-(2-methoxy-2-propenyl) ethers by Hofmann-elimination, which could be analysed by GC and GC/MS (26).

Substitution Pattern in the Polymer Chain. In contrast to the analysis of the monomer pattern only a few studies on the substituent distribution in the polymer chain have been reported in the literature. Gelman (27) and Wirick (28) treated hydroxyethylcellulose (HEC) and CMC with cellulases and determined the number of chain cleavages/1000 AGU. These data were compared with theoretical values calculated from the amount of unsubstituted AGU ($\longrightarrow$ average block length of unsubstituted AGU) and the assumed enzyme specifity. Steeneken *et al.* reported on different enzyme accessibility of methyl starches produced under homogeneous or heterogeneous conditions (29) and on topochemical effects caused by the layered structure of the starch granules (30). Recently, a new approach involving partial hydrolysis, FAB-MS analysis of the oligomeric mixtures obtained after perdeutero-methylation, and statistical evaluation was published by Arisz *et al.* for methyl celluloses (31). Furthermore, pyrolysis chemical ionisation MS and pricipal component analysis was applied to HEC (32).

We performed model studies on methyl amyloses, which were prepared under homogeneous or heterogeneous reaction conditions in protic or aprotic solvent systems (33). The analytical strategy applied included (a) monomer analysis and comparison of the experimental data with the kinetical model of Reuben as described above, (b) perdeuteromethylation of the free OH groups to get a chemical uniform product and prove a random degradation process, (c) partial depolymerisation by mild methanolysis or mild reductive cleavage, (d) mass spectrometric analyis by FAB-MS or MALDI-TOF-MS, and (e) comparison of the substitution patterns in the oligomeric mixtures with a calculated statistical arrangement of the monomers. The average DS of each oligomeric mixture has to agree with the average DS of the sample, to confirm, that its substituent pattern is representative for the whole sample. For a methyl amylose, which was prepared in a homogeneous solution in water, the calculated statistical composition and the experimental data fitted very well. Samples from a heterogeneous reaction showed a broader distribution. When amylose was dissolved in DMSO and treated with powdered NaOH and methyl iodide a bimodal distribution pattern was observed as the result of two competing methylation processes. The amylose molecules are adsorbed from the solution on the solid surface of the NaOH particles according to the model of Fleer and Scheutjens (34). There, the deprotonation and subsequent methylation occured very fast while the not adsorbed molecules are nearly randomly methylated with a lower rate. Filtration and titration of the NaOH suspensions in DMSO showed that the NaOH is not really dissolved in DMSO. The smallest particles were found to have a diameter of about 280 nm by light scattering experiments (Mischnick, P., Burchard, W., unpublished data). Larger aggregates of these particles were also present. While amylose was completely recoverd from a solution in DMSO after filtration (16 μm), it was strongly retained in the presence of powdered NaOH. The extent depended on the ratio of amylose and NaOH and the particle size of the latter. Liebert and Heinze (see chapter y) also reported on an enhanced ratio of the trisubstituted fraction in CMC from such reaction systems, which indicates the fast etherification process of the adsorbed polymer.

For a quantitative MS analysis the oligomers of a certain DP should be as similar as possible with respect to the molecular mass and the polarity, since these parameters mainly influence the desorption and ionisation properties. This requirement is best met by mixed CH_3/CD_3 ethers. As already described above acetates, sulfates or cellulose silyl ethers can in principle be transformed to their complementary methyl ethers and subsequent to the oligomeric mixed ethers. Figure 7 shows the results for a

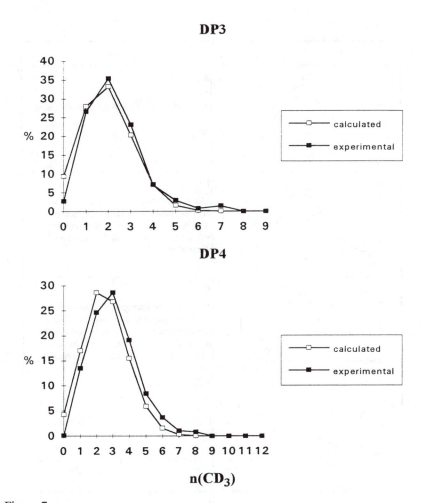

Figure 7
Distribution of CD$_3$ groups (corresponding to the original sulfate groups) in the
tri- and tetrameric fraction of a cellulose sulfate from a homogeneous reaction

cellulose sulfate with preferred 6-sulfation from a homogeneous reaction in the
N$_2$O$_4$/DMF system (35). The average DS of the oligomers deviated about 10%,
indicating that the partial hydrolysis with simultaneous cleavage of the sulfate group
was not really random. Therefore, desulfation and perdeuteromethylation should be
performed prior to partial degradation. However, a relative good agreement with the
calculated homogeneous statistical distribution pattern along the chain can already be
realised. These investigations are in progress. First results for a commercial cellulose
acetate (DS 2) show a heterogeneous distribution of the type **B** in Figure 8.

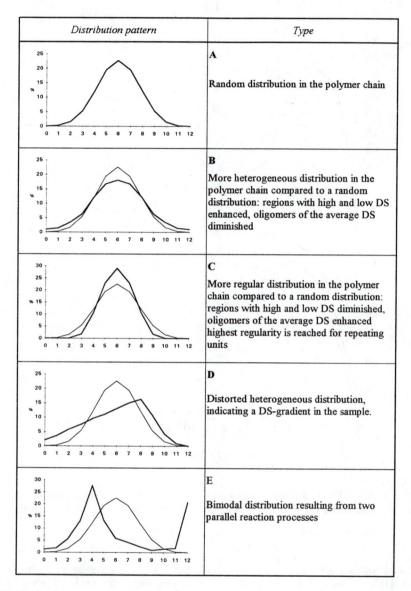

Figure 8
Classification of substitution patterns demonstrated for the tetrameric fraction of a theoretical polysaccharide derivative with a DS 1,5. The plane graph shows the random distribution as a reference. 0, 1, 2, 3,....12 is the number of substituents in 2-, 3-, and 6-position. (Reproduced with permission from ref. 33, Copyright 1996, Elsevier Science Ltd)

The random substituent distribution in the polysaccharide chain, which is calculated for a certain monomer composition, always serves as a defined reference structure. The typical deviations have been classified in a qualitative sense (Figure 8). Mathematical models have to be developed, which include the positive or negative intermonomeric effects, e.g. the enhanced rate of the reaction due to a local enhanced solubility and consequently accessibility after the introduction of the first substituents, or a real cooperative effect on the vicinal AGUs, which propagates along the chain, when a higher structure of the polymer in the solution has been disturbed.

Conclusion

While the monomer analysis of polysaccharide derivatives is well established, the determination of the substitution pattern on higher structural levels and therefore a molecular understanding of the macroscopic properties is still in the beginning. Analytical strategies for the determination of the monomer compositon of cellulose derivatives have been outlined for alkyl and hydroxyalkyl ethers, chemically less stable substituents as silyl ethers, sulfates or organic esters, and also for anionic and cationic species. With respect to the substituent distribution in the polymer chain model studies with methyl amyloses have given promising results and a first insight in the influence of the reaction system on the primary structure of the derivatives obtained. The analytical approach applied to the methyl ethers is also appropriate for acetates sulfates or silyl ethers.

Literature Cited

(*1*)   Rolf, D.; Gray, G. R. *J. Am. Chem. Soc.* **1982,** *104*, 3539-3541.
(*2*)   Sweet, D. P.; Shapiro, R. H.; Albersheim, P. *Carbohydr. Res.* **1975,** *40*, 217-225.
(*3*)   Scanlon, J. T.; Willis, d. E. *J. Chromatogr. Sci.* **1985,** *23*, 333-339.
(*4*)   Jorgensen, A. D.; Picel, K. C.; Stamoudis, V. C. *Anal. Chem.* 1990, *62*, 683-689.
(*5*)   D'Ambra, A. J.; Rice, M. J.; Zeller, S. G.; Gruber, P. R.; Gray, G. R. *Carbohydr. Res.* **1988,** *177*, 111-116.
(*6*)   Mischnick-Lübbecke, P.; Krebber, R. *Carbohydr. Res.* **1989,** *187*, 197-202.
(*7*)   Lindberg, B.; Lindquist, U.; Stenberg, O. *Carbohydr. Res.* **1987,** *170*, 207-214.
(*8*)   Mischnick, P. *Carbohydr. Res.* **1989,** *192*, 233-241.
(*9*)   Steeneken, P. A. M.; Woortman, A. J. J.; Tas, A. C.; Venekamp, J. C. *Carbohydr. Netherlands* **1993,** *9*, 31-34.
(*10*) Arisz, P.W.; Lomax, J.A.; Boon, J.J. *Carbohydr. Res.* **1993,** *243*, 99-114.
(*11*) Lindberg, B.; Lindquist, U.; Stenberg, O. *Carbohydr. Res.* **1988,** *176*, 137-144.
(*12*) Reuben, J. *Macromolecules* **1984,** *17*, 156-161.
(*13*) Reuben, J.; Casti, T.E. *Carbohydr. Res.* **1987;** *163*, 91-98.
(*14*) Mischnick, P.; Lange, M.; Gohdes, M.; Stein, A.; Petzold, K. *Carbohydr. Res.* **1995,** *277*, 179-187.
(*15*) Icheln, D.; Gehrcke, B.; Piprek, Y.; Mischnick, P.; König, W.A.; Dessoy, M.A.; Morel, A. F. *Carbohydr. Res.* **1996,** *280*, 237-250.
(*16*) Nehls, I.; Philipp, B.; Wagenknecht, W.; Klemm, D.; Schnabelrauch, M.; Stein, A.; Heinze, T. *Das Papier* **1990,** *44*, 633-640.
(*17*) Stevenson, T.T.; Furneaux, R.H. *Carbohydr. Res.* **1991,** *210*, 277-298.
(*18*) Mischnick, P. *J. Carbohydr. Chem.* **1991,** *10*, 711-722.
(*19*) Lemieux, R. U.; Kondo, T. *Carbohydr. Res.* **1974,** *35*, C4-C6.
(*20*) Berry, J. M.; Hall, L. D. *Carbohydr. Res.* **1976,** *47*, 307-310.
(*21*) Lee, C. K.; Gray, G. R. *Carbohydr. Res.* **1995,** *269*, 167-174.

130

(22) Mischnick, P. *Min. 5th Intern. Symp. Cyclodex.* Ed. de Santé, Paris **1990**, 90-94.
(23) Erler, U.; Mischnick, P.; Stein, A.; Klemm, D. *Polym. Bull.* **1992**, *29*, 349-356.
(24) Kragten, E. A.; Kamerling, J. P.; Vliegenthart, J. F. G. *J. Chromatogr.*, **1992**, *623*, 49-53.
(25) Heinrich, J.; Mischnick, P. *J. Chromatogr. A* **1996**, *749*, 41-45.
(26) Wilke, O.; Mischnick, P. *Carbohydr. Res.* **1995**, *275*, 309-318.
(27) Gelman, R. A. *J. Appl. Polym. Sci.* **1982**, *27*, 2597-2964.
(28) Wirick, M. G. *J. Polym. Sci.*, Part A **1968**, 6, 1705-1718 and 1965-1974.
(29) Steeneken, P. A. M.; Woortman, A. J. J. *Carbohydr. Res.* **1994**, *258*, 207-221.
(30) Steeneken, P. A. M.; Smith, E. *Carbohydr. Res.* **1991**, *209*, 239-249.
(31) Arisz, P.W.; Kauw, H. J. J.; Boon, J. J. *Carbohydr. Res.* **1995**, *271*,1-14.
(32) Arisz, P. W. *Thesis: Mass spectrometric analysis of cellulose ethers*, Amsterdam, **1995.**
(33) Mischnick, P.; Kühn, G. *Carbohydr. Res.* **1996**, *290*, 199-207.
(34) Scheutjens, J. M. H. M.; Fleer, G. J. *J. Phys. Chem.* **1980**, *84*, 178-190.
(35) Wagenknecht, W.; Nehls, I.; Philipp, B. *Carbohydr. Res.* **1993**, *240*, 245-252.

Chapter 10

Characterization of Cellulose Esters by Solution-State and Solid-State NMR Spectroscopy

Douglas W. Lowman

Research Laboratories, Eastman Chemical Company, Kingsport, TN 37662–5150

The characterization of organic cellulose esters, including acetyl, propionyl, and butyryl esters, mixed cellulose esters, and acylated cellulose ethers by modern solution-state and solid-state NMR techniques over the past 10 - 15 years is reviewed. Modern 1D and 2D NMR techniques enable detailed structural elucidation of these heteropolymers. The importance of molecular characteristics determined by NMR, such as total and site-specific degree of substitution, solution conformation, and molecular dynamics as well as crystalline and amorphous content, are discussed in terms of structure-property relationships for these cellulosic polymers.

The major organic cellulose ester derivatives, including acetate, propionate, and butyrate, have been important commercial products for many years. They find applicability in plastics as well as in biodegradable polymers. Study of these acylated polymers in relation to various properties has not been possible until recently due to the lack of detailed structural information on these very complex, heterogeneously substituted homopolymers and heteropolymers. With the development of new structure elucidation techniques in nuclear magnetic resonance spectroscopy (NMR), these detailed analyses are becoming possible.

Cellulose esters are polymers resulting from acylation of cellulose. Cellulose ($\underline{1}$) is a linear 1,4-β-D-glucan with three hydroxyl groups per anhydroglucose unit (AGU). Each AGU contains hydroxyl functions at the 2-, 3- and 6-positions. Acylation can occur at none of the hydroxyl positions, at any one of the three hydroxyl positions, at any two of the three hydroxyl positions, and at all three hydroxyl positions, resulting

$\underline{1}$ R = H, Cellulose
$\underline{2}$ R = Acetyl, CTA
$\underline{3}$ R = Propionyl, CTP
$\underline{4}$ R = Butyryl, CTB

in the possible formation of 8 different AGU's in a cellulose ester polymer (Figure 1). If all three hydroxyl positions are acylated, the triacylated cellulose ester homopolymer has a degree of substitution (DS) of 3. Depending on how the acylation is accomplished, either by direct acylation or back-hydrolysis after acylation from a DS = 3 homopolymer, the possible range of DS in the heteropolymer containing any combination of the 8 possible AGU's is between 0 and 3. The complexity of the chemical structure of the cellulose ester heteropolymer, and thus its properties, conformation and dynamics, is related to the polymer DS.

NMR techniques have been successfully developed and applied to the analysis of structure-property relationships, sequence determinations, molecular dynamics, and conformations in biological heteropolymers, such as peptides, proteins and enzymes. These same techniques are now being used for similar analyses of cellulose esters. Structural details available from these modern NMR techniques have proven very useful in aiding our understanding of structure-property relationships, conformational properties, and the molecular dynamics of these polymers.

It is the intent of this chapter to discuss the characterization of organic cellulose esters, including acetyl, propionyl, and butyryl esters, by modern solution-state and solid-state NMR techniques in terms of ^{1}H and ^{13}C chemical shifts, total and site-specific DS, solution conformation, molecular dynamics, and crystalline allomorphs. In this chapter, inorganic cellulose esters, such as nitrate, sulfate, and phosphate, will not be considered. The application of NMR to related studies on acylated cellulose

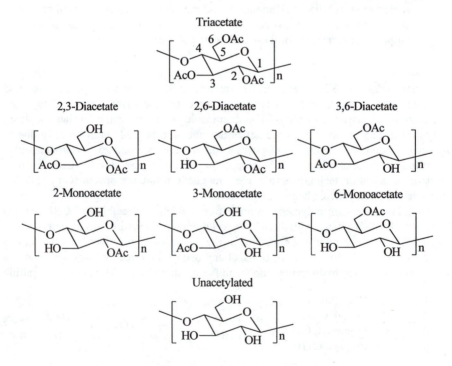

Figure 1. Eight AGU's present in Cellulose Acetate

ethers will be discussed. The chapter will begin with an analysis of applications of 1D and 2D solution-state NMR techniques to CTA (**2**), CTP (**3**), CTB (**4**), cellulose mixed esters, and cellulose ethers. The chapter will conclude with an analysis of solid-state NMR techniques applied to cellulose esters. A discussion of the NMR analysis of all known cellulose esters is outside the scope of this chapter. However, a review appeared recently discussing the characterization of cellulose and cellulose derivatives, including esters, by ^{13}C NMR (*1*). Kamide and Saito (*2*) recently reviewed the research from the authors' laboratories between 1985 and 1993 relative to the molecular and supramolecular characterization of cellulose and cellulose derivatives, including cellulose acetates.

Solution-State NMR

Cellulose acetate (CA) is the most common commercial cellulose ester. The largest amount of detailed information is available about this polymer's structure, DS, conformation, and dynamics based on NMR analysis relative to the other cellulose derivatives discussed in this chapter. The classic work of Goodlett and coworkers (*3*) in 1971 provided the first method for direct determination of the acetyl distribution in CA. They reacted the unacetylated hydroxyl groups in CA with DS less than 3 with acetyl-d3 chloride. The proton NMR spectrum in CD_2Cl_2 of this fully acetylated CA presented 3 acetyl methyl proton resonances with chemical shifts of 2.09, 1.99, and 1.94 ppm assigned to substitution at the 6-, 2- and 3-positions, respectively. These assignments were confirmed by Shiraishi and coworkers (*4*). Using a digital computer method, Goodlett and coworkers (*3*) determined the relative DS at each site to a standard deviation of about 0.03. This simple measurement allowed the correlation of the site-specific DS with methods of preparation for CA.

In 1984, Shibata and coworkers (*5*) used ^{13}C NMR to measure site-specific acetylation based on the ring carbons of CA in DMSO-d6. Acetylation at C-6 results in deshielding of C-6. Resonances for C-2 and C-3 are not separated enough from the C-5 resonance to allow direct observation of the impact of acetylation on these carbon resonances. Instead, substitution at C-2 and C-3 results in shielding of the C-1 and C-4 resonances, respectively. Using the assignments of Goodlett and coworkers (*3*) and comparing signal intensities, Shibata and coworkers (*5*) assigned the carbonyl carbon resonances of the acetyl groups also. The three carbonyl carbon resonances at 169.9, 169.5 and 168.8 ppm were assigned to acetyl groups substituted at the 6-, 3- and 2-positions, respectively. Later, Kowsaka, Okajima and Kamide (*6*) made similar assignments, correcting previous assignments by Kamide and Okajima (*7*).

Chemical Shift Assignments. In order to correlate structure and properties, chemical shift assignments must be known in detail. Proton and ^{13}C NMR assignments depend on a 2D homonuclear correlated spectrum (COSY) to assign the ring protons, then a 2D [13]C-^{1}H direct-detected (HETCOR) or inverse-detected (HMQC) heteronuclear correlated spectrum to assign the protonated ring carbons, followed by a long-range heteronuclear correlated spectrum (COLOC) or multiple-bond HETCOR or a multiple-bond inverse-detected heteronuclear correlated spectrum (HMBC) to assign acyl carbon and proton resonances including carbonyl carbons.

Table I. Proton NMR Chemical Shifts[a] and Coupling Constants[b]
for CTA, CTP, CTB and Cellulose

	CTA			CTP	CTB	Cellulose
	DMSO-d_6 25°C (8)	DMSO-d_6 80°C (8)	CDCl$_3$ 25°C (8)	CDCl$_3$ 25°C (8)	CDCl$_3$ 25°C (8)	DMSO-d_6 80°C (26)[c]
H-1	4.65 (7.9)	4.65 (7.9)	4.42 (7.9)	4.35 (7.9)	4.34 (7.9)	4.35
H-2	4.52 (7.3)	4.55 (8.6)	4.79 (8.6)	4.77 (8.6)	4.76 (8.6)	3.10
H-3	5.06 (9.2)	5.04 (9.2)	5.07 (9.0)	5.07 (9.1)	5.06 (9.2)	3.38
H-4	3.65 (9.2)	3.68 (9.2)	3.71 (9.2)	3.66 (9.1)	3.61 (9.2)	3.38
H-5	3.81	3.77	3.53	3.47	3.48	3.38
H-6S	4.22 (10[d])	4.26 (11[d])	e	e	e	3.78
H-6R	3.98	4.04	4.06	4.03	4.03	3.60

[a] In ppm relative to CDCl$_3$ at 7.24 ppm or DMSO-d_6 at 2.49 ppm

[b] Shown in parentheses, in Hz; digital resolution 0.2-0.26 Hz

[c] Unsubstituted AGU from a low DS CA

[d] $^2J_{6S,6R}$

[e] H-6S and H-1 overlap

Table II. Carbon-13 NMR Chemical Shifts[a] and Coupling Constants[b]
for CTA, CTP, and CTB

	CTA		CTP		CTB
	DMSO-d_6 90°C [c]	CDCl$_3$ 25°C [c]	CDCl$_3$ 25°C		CDCl$_3$ 25°C [c]
C-1	99.8 (167)	100.4 (165)	100.3 (163) [c]	100.37 [d]	100.1 (163)
C-2	72.2 (152)	71.7 (153)	71.7 (150)	71.81	71.4 (150)
C-3	72.9 (151)	72.5 (148)	72.2 (148)	72.25	71.8 (147)
C-4	76.4 (151)	76.0 [e]	75.8 (153)	75.90	75.8 [e]
C-5	72.5 (146)	72.7 (139)	73.0 (138)	73.08	73.1 (143)
C-6	62.8 (151)	61.9 (151)	61.9 (147)	62.04	61.9 (145)

[a] In ppm relative to CDCl$_3$ at 77.0 ppm or DMSO-d_6 at 39.5 ppm.

[b] Shown in parentheses, in Hz; digital resolution 0.52 Hz.

[c] Reference 8.

[d] Reference 19.

[e] The coupled resonance overlaps with the solvent resonance.

The first detailed, unambiguous chemical shift assignments for the ring protons (Table I) and carbons (Table II) of CTA , CTP and CTB dissolved in DMSO-d_6 or CDCl$_3$ using 2D NMR techniques were provided by Buchanan and coworkers (8). Proton assignments were accomplished with COSY (9,10) (CTA, Figure 2) while ^{13}C assignments were accomplished with HETCOR (11,12) (CTA, Figure 3) for all three cellulose esters. Two- and 3-bond correlations (CTA, Figure 4) from HETCOR (13), COLOC (13,14), and INAPT (15) experiments extended these assignments from the ring protons across the ester oxygen to the acetyl carbonyl carbons (in CDCl$_3$: C6: 170.1 ppm, C3: 169.6 ppm, C2: 169.1 ppm) (15) and methyl protons (in 1,1,2,2-tetrachloroethane-d_2: H6: 2.07 ppm, H2: 1.97 ppm, H3: 1.94 ppm) (14). Acetyl methyl carbon assignments are taken from the HETCOR spectrum. Ring proton 3-bond coupling constants are presented in Tables I and III from the homonuclear 2D J-Resolved experiment (16) and 1D proton NMR (17) using resolution enhancement post-processing (8). Kowsaka and coworkers (18) employed similar 2D correlation techniques to accomplish these assignments in DMSO-d_6 and CDCl$_3$ for CTA. Shuto and coworkers (19) accomplished these assignments by a similar protocol for CTP in CDCl$_3$ at 25°C. Tezuka and Tsuchiya (20) determined carbonyl carbon assignments (C6: 173.6 ppm, C3: 173.1 ppm, C2: 172.7 ppm) for CTP in CDCl$_3$ by INAPT.

For a series of peracetylated oligomers from dimer (degree of polymerization, DP = 2) to nonamer (DP = 9), the ring protons and carbons as well as carbonyl carbons were assigned (21). The ring carbons of oligomers DP = 2 through DP = 5 were also assigned by Capon and coworkers (22).   Chemical shift assignments were accomplished by COSY, 2D homonuclear Hartman-Hahn correlated (HOHAHA) spectroscopy (23-25), HETCOR, and INAPT. Subspectra of the reducing terminus,

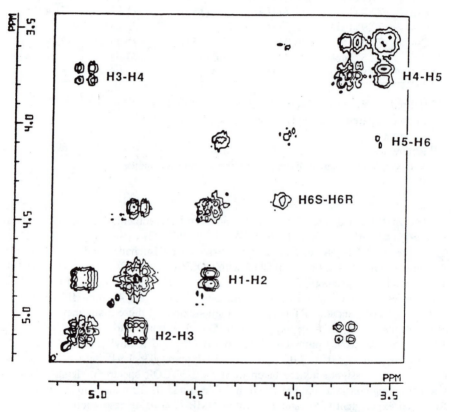

Figure 2.   COSY spectrum of CTA in CDCl₃ at 25°C. (Adapted from ref. 8.)

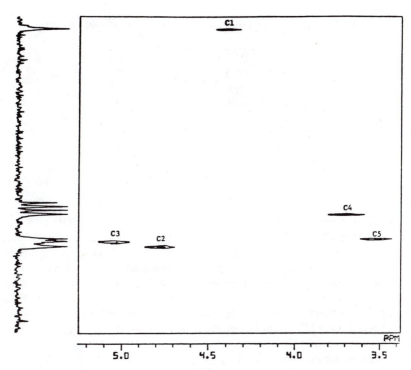

Figure 3. C-H Heteronuclear Correlated 2D NMR spectrum of CTA in CDCl3 at 25°C. (Reproduced with permission from ref. 8. Copyright 1987 American Chemical Society.)

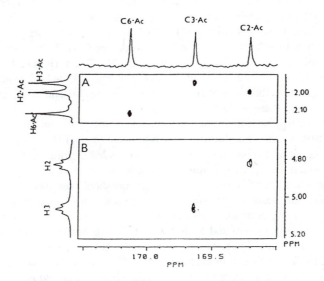

Figure 4. 2-Bond (A) and 3-Bond (B) Heteronuclear Correlated 2D NMR Spectrum of CTA. (Reproduced with permission from ref. 13. Copyright 1990 John Wiley and Sons, Inc.)

Table III. 3-Bond Coupling Constants (in Hz) for CTA in CDCl3

	50°C (17)	50°C (16)	25°C (8)
H1,H2	7,6	7.6	7.9
H2,H3	9.0	9.7	8.6
H3,H4	9.0		9.0
H4,H5	9.5	9.5	9.2
H6S,H5		2.0	
H6R,H5		5.5	

nonreducing terminus and internal AGU's were extracted from the HOHAHA spectra of the series (Figure 5). The AGU adjacent to the nonreducing terminus was also demonstrated to be unique for DP = 3 through 6 oligomers. Therefore, the termini of these lower oligomers must be viewed as being composed of three monomer units -- the reducing terminus, the nonreducing terminus, and the monomer unit adjacent to the nonreducing terminus.

A high level of chain rigidity is demonstrated by comparing T_1 values for these oligomers and CTA (21). At a DP of about 7, the critical DP is reached since T_1 values from the internal monomer ring (Figure 6) and carbonyl (Figure 7) carbons in the oligomers are similar to that of CTA.

For cellulose esters with DS < 3, interpretation of individual resonances for each of the 8 AGU's (Figure 1) displaying 12 magnetically different acetyl groups becomes much more complicated. Buchanan and coworkers (26) synthesized a series of carbonyl carbon ^{13}C-enriched cellulose acetates with DS = 2.54, 2.03, 1.99, 1.61, 1.06 and 0.42. A total of 16 carbonyl carbon resonances (Table IV) were observed in the series of cellulose acetates including CTA. All of these resonances were correlated with acetyl groups in either the 2-, 3- or 6-position using INAPT. Specific monomers in the heteropolymer backbone were identified by 2D COSY (Figure 8). Cellulose acetates with DS < 2.85 are not soluble in chloroform. DMSO-d6 exhibits a wide solubility range for cellulose acetates with DS between 3.0 and 0.4, making it the solvent of choice for these studies.

To assign the ring protons of the individual monomers, several observations from the proton NMR spectra of cellulose esters were used (8): 1) H3 protons attached to carbons bearing an acetyl group are the most deshielded, 2) ring protons attached to carbons that do not bear an acetyl group resonate between 3.8 and 2.9 ppm, and 3) H2, H3 and H6 attached to carbons bearing acetyl groups should resonate between 5.1 and 3.9 ppm. On the basis of proton NMR chemical shift assignments and these observations, coupling network A (Figure 8) was assigned to CTA, the triacetylated monomer. Assuming preferential hydrolysis of the C6 acetyl and that loss of this acetyl causes only a minor perturbation in the rest of the monomer structure, network B is assigned to the 2,3-diacetyl monomer. Since H3 in network C connects to an H2 above 3.9 ppm, network C was assigned to overlapping 3,6-diacetyl and 3-monoacetyl monomers. Buchanan and coworkers (26) suggest that network D results from either

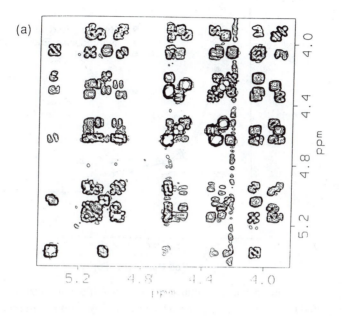

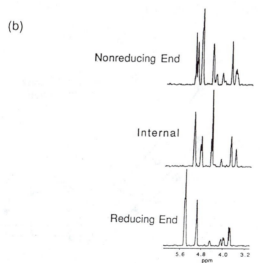

Figure 5. (a) 2D HOHAHA spectrum of cellotriose hendecaacetate. (b) Subspectrum for each monomer is obtained by taking slices through the F_1 dimension. (Reproduced with permission from ref. 21. Copyright 1990 American Chemical Society.)

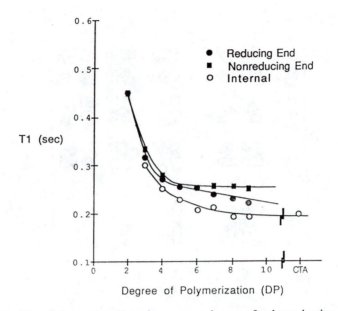

Figure 6. Plot of ring carbon T_1 values versus degree of polymerization for CTA oligomers with DP = 2 through 9. (Reproduced with permission from ref. 21. Copyright 1990 American Chemical Society.)

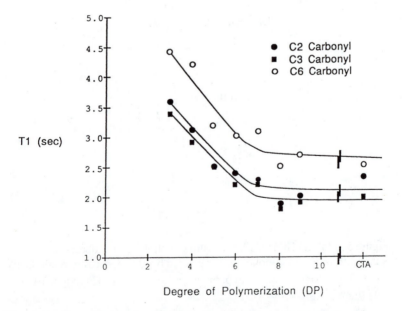

Figure 7. Plot of internal carbonyl carbon T_1 values versus degree of polymerization for CTA oligomers with DP = 2 through 9. (Reproduced with permission from ref. 21. Copyright 1990 American Chemical Society.)

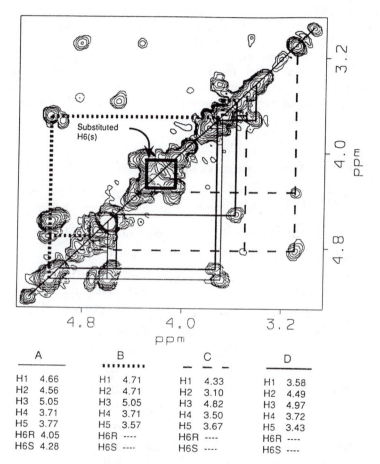

Figure 8.   COSY spectrum of cellulose acetate with DS = 2.54. Four coupling networks, found in this spectrum, are identified by the solid and dashed lines and by their chemical shifts. (Reproduced with permission from ref. 26. Copyright 1990 American Chemical Society.)

intra-chain or inter-chain hydrogen bonding between triacetyl monomer carbonyl carbons and a hydroxyl group either on an adjacent AGU or on a neighboring chain. By similar analyses, other coupling networks were assigned. From these coupling network assignments and the INAPT experiment, 15 of the 16 observed carbonyls were assigned to monomers (Table IV). The remaining carbonyl carbon resonance at 169.86 ppm could only be assigned in general terms to a C6 carbonyl carbon. This assignment resulted from the observation that C3 carbonyl carbons resonate between 169.56 and 169.05 ppm. All C6 carbonyl carbon resonances are deshielded relative to C3 carbonyl carbon resonances. All C2 carbonyl carbon resonances are shielded relative to C3 carbonyl carbon resonances. The limitation of this approach to a detailed analysis of monomer content in heteropolymeric cellulose esters is spectral resolution due to peak overlap and the absence of correlation for H5 and H6.

Table IV. Carbon-13 NMR Chemical Shifts for the Carbonyl Resonances from CTA and the Series of Cellulose Acetates with DS < 3 (26)

Chemical Shift, ppm	Monomer
170.01 (0.01, 0.02)[a]	6-mono or diacetyl
169.92 (0.01, 0.02, 0.03)	6-triacetyl
169.86	C6 carbonyl carbon[b]
169.83 (0.01, 0.02, 0.03)	6-mono or diacetyl
169.78 (0.01)	6-mono or diacetyl
169.56	3-monoacetyl[c]
169.48 (0.01, 0.02, 0.04)	3-monoacetyl
169.44	2,3-diacetyl[c]
169.36 (0.01, 0.02, 0.03)	2,3-diacetyl
169.25	3,6-diacetyl[c]
169.14	3-triacetyl[c]
169.13 (0.06)	3,6-diacetyl
169.05 (0.02, 0.03, 0.05)	3-triacetyl
168.83 (0.01, 0.02)	2,3-diacetyl
168.72 (0.01)	2-triacetyl
168.63 (0.01)	2-monoacetyl, 2,6-diacetyl

[a] The numbers in parentheses represent the downfield shift for that resonance with increasing hydroxyl level (decreasing DS).

[b] See text.

[c] This resonance has been shifted by hydrogen bonding.

Source: Adapted from ref. 26.

From Figure 1, it is clear that only 4 carbonyl carbon resonances should be observed for C3 acetylated cellulose esters. In Table IV however, there are 8 carbonyl carbons assigned to C3 acetylated monomers. Buchanan and coworkers (26) suggested that the extra 4 monomers result from hydrogen bonding similar to that described for network D. Also, the chemical shift of the observed carbonyl carbon resonance is impacted by DS. As the hydroxyl content increases, i.e., DS goes down, carbonyl carbon resonances assigned in spectra from higher DS samples shift downfield by the amounts indicated in parentheses (Table IV). This downfield shift is attributed to hydrogen bonding between AGU's or chains.

Kowsaka and coworkers (27) also attempted to assign the individual carbonyl carbon resonances in a series of CA's with DS ranging from 2.92 to 0.43. The three resonances for CTA (DS = 2.92) were easily assigned. For the three resonances observed in the DS = 0.43 sample, it was assumed that the major composition would contain a distribution of monoacetylated AGU's with substitution at the 6-, 3- and 2-positions. Assignments for the monosubstituted AGU's in DS = 0.43 CA were based on the same order of chemical shifts as observed for CTA. Assignments for the remaining closely spaced carbonyl carbon resonances in the diacetylated AGU's were postulated based on substituent additivity relationships only.

Hikichi and coworkers (28) analyzed proton COSY and relayed COSY (29,30) 2D NMR spectra of cellulose diacetate with DS = 2.46 to assign 9 different spin systems. These spin systems were assigned to individual acetylated AGU's. The acetylated AGU's identified were four different 2,3,6-triacetylated AGU's, two 2,3-diacetylated AGU's, a 2,6-diacetylated AGU, a 3,6-diacetylated AGU, and a 6-monoacetylated AGU. The authors suggest that the four 2,3,6-triacetylated AGU's result from chemical shift variation due to the nature of the acetylated AGU's on either side of the 2,3,6-triacetylated AGU. Since the 2,3,6-triacetylated AGU can be flanked by any of 7 different acetylated AGU's or 1 unacetylated AGU, there are 64 different species possible. The four spectra identified by the authors result from the four AGU triads centered on a 2,3,6-triacetylated AGU with the greatest population. Confirmation of these assignments is provided by comparing the spectrum of the ring proton spectral region to a spectrum resulting from the weighted combination of simulated spectra from each of the 9 assigned acetylated AGU's.

Conformation and Molecular Dynamics. Buchanan and coworkers (31) used ^{13}C T_1 measurements to probe molecular dynamics for CTA, CTP and CTB. The authors also used absorption-mode phase-sensitive 2D nuclear Overhauser enhancement (nOe) exchange spectroscopy (NOESY) (32,33) to probe the solution conformation of these polymers. A plot of T_1 values as a function of temperature (Figure 9) for the carbonyl carbons of CTA in DMSO-d_6 and CDCl$_3$ indicates a transition near 53°C. A similar transition has been observed (34) for CTA in acetic acid on the basis of intrinsic viscosity, specific volume, optical rotation, absolute viscosity, and sorption of solvent measurements. The occurrence of this transition has been attributed to a conformational change in the macromolecular structure of CTA by several groups (17,31,34). Ogura and coworkers (35) observed a similar transition near 50-55°C in variable temperature lineshape studies on CTA in CDCl$_3$ which they attributed to a chair-boat conformational change in the AGU. A transition within the temperature range of 40-60°C was not observed in similar NMR studies on CTP and CTB by Buchanan and coworkers (31).

The cause of this transition at 53°C is difficult to assign since it is difficult to differentiate between microscopic conformational changes and supramolecular structure changes based on these results. Microscopic conformational changes can result from changes in either monomer conformation or the virtual angle about the glycosidic bond between repeat units. Changes in interactions between polymer chains result in supramolecular structural changes. Chair-boat conformational changes and rotational changes about the glycosidic bond would result in changes in the internuclear distances between ring protons and between protons on neighboring AGU's. The NOESY experiment provides an excellent means of probing internuclear distances between protons and solution conformational changes.

Buchanan and coworkers (31) used the NOESY experiment to examine the solution conformation of CTA, CTP and CTB in CDCl3. Several NOESY spectra were collected at different mixing times between 40 and 200 msec at 25 and 60°C for CTA concentrations of 0.1M and 0.35M in CDCl3. Similar data were collected for CTP (0.09 M) and CTB (0.08 M) in CDCl3 at 25°C. A typical NOESY spectrum for 0.35 M CTA in CDCl3 is shown in Figure 10 with all resonances labeled. From peak volume measurements at each mixing time, the initial portion of the nOe build-up curve was used to measure individual nOe build-up rates. These individual rates were then used to calculate internuclear distances by comparison of these rates to the build-up rate for the nOe between the geminal protons on H6.

The average distance calculated for each NMR observable interaction is presented in Table V. Internuclear distances from x-ray analysis (36,37) of CTA crystalline forms I and II in fiber are shown in the second column of Table V. There are no significant differences in the NMR data for the five systems. Keeping in mind the small energy differences between chair conformations, the similarity of internuclear distances, and the consistency of the proton coupling constants (Tables I and III), the data are consistent with retention of the $^4C_1$ conformation for the AGU. Substituent, concentration, and solvent do not perturb this chair conformation. Vicinal coupling constants around 9 Hz (Tables I and III) are consistent with the $^4C_1$ conformation for a β-D-glucopyranosyl residue. However, there may be some flattening of the AGU ring based on the smaller size of the H1-H2 coupling constant (7.9 Hz) (8).

The virtual angle between H1 and H4' (Figure 11) can be estimated from the NOESY spectra. CTA in crystalline forms I and II can be represented as a 2/1 helix (a syn planar conformation) where H1 and H4' are aligned with a virtual angle of 0°. Using models, the measured distance between H1 and H4' in a 2/1 helix is 1.8 Å. By rotating one AGU until the distance between H1 and H4' equals the distance measured in the NOESY experiments, the virtual angle between H1 and H4' in solution for these cellulose esters is estimated to be 30 - 34°. This virtual angle is consistent with a 5/4 helix suggesting that 5 glucose monomers are required to rotate through 360°. This virtual angle is consistent with studies on helical conformation in cellulosic chains from circular dichroism (38,39).

The conformation about the C5-C6 exocyclic bond has also been examined (8). Based on the observed coupling constant between H6R and H5 as well as the smaller coupling constant between H6S and H5 (Table III), the preferred rotamer population

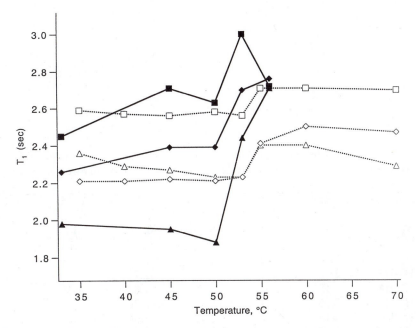

Figure 9. CTA (0.35 M) carbonyl carbon T_1 values versus temperature in CDCl$_3$ (solid lines) and DMSO-d$_6$ (dashed lines) for acetyl groups substituted at C6 (squares), C3 (triangles) and C2 (diamonds). (Adapted from ref. 31.)

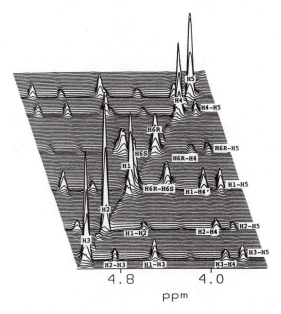

Figure 10. Stack plot of a NOESY spectrum for CTA (0.35 M, 25°C) at a mixing time of 100 ms. The F$_1$ and F$_2$ axes have equal dimensions. (Reproduced with permission from ref. 31. Copyright 1989 American Chemical Society.)

Table V. Average Internuclear Distances (in Angstroms) Obtained by the NOESY Experiment for CTA, CTP and CTB Compared to X-Ray Crystal Structure Distances

Interactions	X-Ray	CTB 25°C 0.08 M	CTP 25°C 0.09 M	CTA 25°C 0.1 M	CTA 60°C 0.1 M	CTA 25°C 0.35 M
H1-H4'	a	2.15	2.11	2.17	2.16	2.08
H1-H3	2.76	2.22	2.25	2.39	2.28	2.24
H3-H5	2.53	2.58	2.73	2.76	2.85	2.57
H2-H4	3.13	2.63	2.79	2.88	3.05	2.63
H1-H5	2.15	2.11	2.12	2.26	2.17	2.09
H3-H4	3.16	3.13	3.35	3.46	b	3.01
H4-H5	2.95	2.22	2.80	2.68	b	2.66
H1-H2	3.00	3.34	3.70	3.83	b	3.48
H3-H2	2.99	3.07	3.51	3.47	b	3.27
H2-H5	3.89	b	3.89	b	b	3.62
H6R-H5	2.45 c	2.69	3.07	3.14	b	3.04
H6S-H5	2.29 c	2.69	3.08	3.15	b	3.17
H6R-H4	2.79 c	3.08	3.50	3.84	b	3.45
H6S-H4	3.58 c	3.08	3.52	3.87	b	3.40
H6R-H6S	1.70					

[a] See the text.

[b] Cross-peaks for these interactions were not observed.

[c] Assignments of R and S for the X-ray data may be reversed relative to the NMR data.

Source: Adapted from ref. 31.

about the C5-C6 bond is the RCS rotamer. The ROS rotamer contributes minimally (Figure 12) in agreement with Perlin and coworkers (*17*).

Tezuka (*40*) examined dipolar interactions between the acetyl methyl protons of CTA in DMSO-d_6 and CDCl3. Using a mixing time of 500 msec in the NOESY experiment, Tezuka observed different interactions in the two solvents. In CDCl3, methyl protons on the acetyl group attached to C3 (Ac3, labeled H3 in Figure 13) exhibited a through-space interaction with Ac6 (Figure 13). In DMSO-d_6, both Ac2 and Ac3 exhibited a through-space interaction with Ac6 (Figure 14). Tezuka concluded that the solution conformations for CTA in these two solvents are different. A 2/1 helical conformation suggests interactions between Ac6 and both Ac2 and Ac3 could be observed which would be consistent with the results in DMSO-d_6. The 5/4 helix suggested earlier (*31*) in CDCl3 would result in increased through-space separation between Ac2 and Ac6 while the distance between Ac3 and Ac6 would become closer. This analysis is consistent with the conformation observed in CDCl3 by the NOESY experiment (*31*).

Morat and Taravel (*41,42*) provide another view of the solution conformation of CTA (DP = 30) in CDCl3. The selective heteronuclear 2D J-Resolved experiment (*43*) was used to determine dihedral angles across the glycosidic bond. The dihedral angles across the fragments C1'-O-C4-H4 and across the fragment H1'-C1'-O-C4 follow a Karplus relationship. By measuring the long-range heteronuclear coupling constants across these dihedral angles, the torsional angles are calculated to be 0°, leading to a 2/1 helix in CDCl3 solution. This result is inconsistent with results presented above unless there is a DP-dependence to solution conformation about the glycosidic bond.

Structure-Property Relationships. A careful monomer compositional analysis enables an analysis of composition versus the properties of water solubility and water absorbency (*44*) for a series of cellulose monoacetates (CMA's). This structure-property relationship study is the first case for which a detailed analysis has been accomplished. The relative degree of substitution (RDS) at C6, C3 and C2 (*45,46*) and 3-monoacetyl content (*45*) have been suggested as important parameters for water solubility. It has also been suggested (*46*) that there is no evidence for a correlation between RDS and water solubility or absorbency.

Several clear differences in water solubility and water absorbency properties can be correlated to monomer composition and molecular weight for CMA's in the DS range of 0.5-0.9 (*44*). In general, for water absorbent CMA's, the monomer ratio of 3-monoacetyl-to-2,3-diacetyl is less than 0.5, the monomer ratio of 3-monoacetyl-to-3-triacetyl is less than 1.1, the monomer ratio of 3-monoacetyl-to-total 3-acetyl content is less than 0.25, and the ratio of the DS for 3-monoacetyl monomer to the total DS is generally less than 0.11. For water-soluble CMA's, each of the monomer ratios above is greater than the number shown. In general, a CMA with a low molecular weight (IV < 1.4; M_W < 2.0 X 10^5) that contains more than 10-11 3-monoacetyl monomers per 100 repeat units will be water soluble. The importance of 3-monoacetyl monomer content and molecular weight in determining water solubility and water absorbency for CMA's is clearly illustrated.

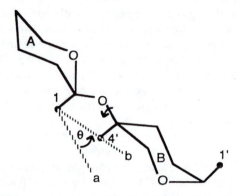

Figure 11. Cellobiose subunit where the virtual angle, θ, between H1 and H4' represents the deviation of H4' from a syn planar relationship with H1. (Reproduced with permission from ref. 31. Copyright 1989 American Chemical Society.)

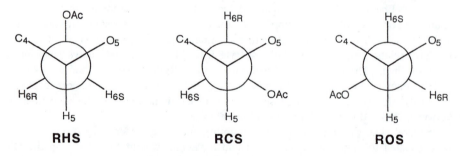

Figure 12. Schematic representation of possible rotamers about the exocyclic C5-C6 bond. (Adapted from ref. 8.)

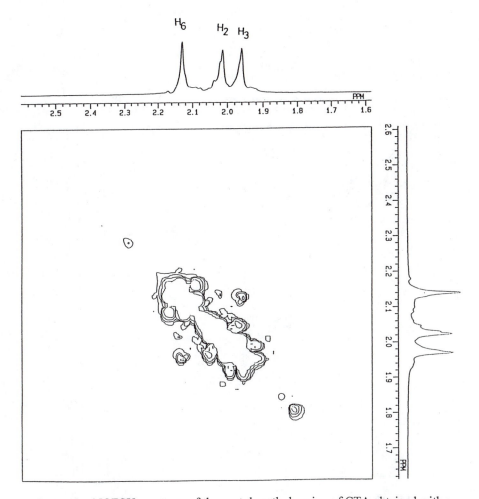

Figure 13. NOESY spectrum of the acetyl methyl region of CTA obtained with a mixing time of 500 ms in CDCl3 at 40°C. (Reproduced with permission from ref. 40. Copyright 1994 John Wiley and Sons, Inc.)

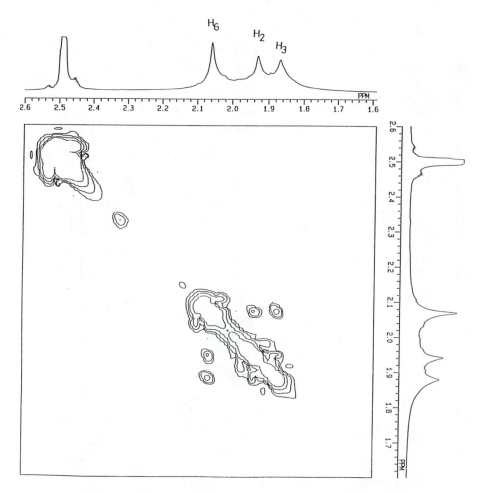

Figure 14.   NOESY spectrum of the acetyl methyl region of CTA obtained with a mixing time of 500 ms in DMSO-d$_6$ at 40°C.   (Reproduced with permission from ref. 40.   Copyright 1994 John Wiley and Sons, Inc.)

Hydroxyl content is also important for these properties. For the water absorbent CMA's, the 3-acetyl of the 2,3-diacetyl monomer was assigned to a non-hydrogen bonding monomer. For three of five water-soluble CMA's, the same resonance was assigned to a hydrogen bonding monomer. These data suggest that a minimum level of 3-acetyl substitution must be present to disrupt intra- and/or intermolecular hydrogen bonding. The delicate balance of C2 hydroxyl content must be maintained to ensure hydrophilicity of the polymer backbone. The C6 hydroxyl is too remote from the polymer backbone to have a significant impact on hydrophilicity.

Cellulose Ethers. As with cellulose esters, knowledge of the specific location of etherification on the AGU is important for understanding properties of the ether derivatives. Unreacted hydroxyl groups in cellulose ethers have been acetylated so that the acetyl distribution can be examined as a means of determining the ether distribution indirectly. Tezuka and coworkers have used this approach extensively for studying ether substituent distribution in O-(methyl)cellulose (47), O-(2-hydroxy-ethyl)cellulose (48), O-(2-hydroxypropyl)cellulose (49), O-methyl-O-(2-hydroxy-ethyl)cellulose (50), O-methyl-O-(2-hydroxypropyl)cellulose (50,51), O-(carboxy-methyl)cellulose (52), and O-(2-hydroxypropyl)-O-methylcellulose acetate succinate (53). Guo and Gray (54) also examined acetyl-derivatized O-ethylcellulose and O-methylcellulose. Generally esterification provides a sensitive indirect probe of ether substitution patterns by examination of the chemical shifts of the ester carbonyl carbon resonances. Assignments are made for carbonyl carbon resonances at 6-, 3- and 2-positions. No attempt is made in these studies to determine detailed esterifi-cation/etherification patterns for individual monomers from the carbonyl carbon resonance patterns.

For methylcellulose (47), high ether DS (above 2.0) polymers are difficult to analyze in DMSO by the acetyl carbonyl carbon resonance approach due to poor solubility. However acetylated methylcellulose is soluble in CDCl$_3$ over the entire range of DS values. The methyl ether carbon resonance splits into multiple resonances with sensitivity to the nature of the substituent at each location as well as neighboring substituents. Assignment of these methyl ether carbon resonances was accomplished by comparison of the NMR results with results from GLC analysis of the polymer hydrolyzed back to the individual monomer units. For example, in Figure 15, C2(3) labels the resonance of the methoxy carbon at the C2 position with a methoxy group at C3 and an acetoxy group at C6.

For acetylated hydroxyalkyl-celluloses, four carbonyl carbon resonances were observed corresponding to acetyl carbonyl carbons in the 6-, 3- and 2-positions as well as the carbonyl carbon from the acetyl group attached to the hydroxyalkyl group. For hydroxyethylcellulose, these four resonances overlap in DMSO-d$_6$ while they are well separated in CDCl$_3$ (48). The four resonances also overlap for O-methyl-O-(2-hydroxyethyl)cellulose in DMSO-d$_6$ while they are a broad, unresolved resonance in CDCl$_3$ (50).

In hydroxypropylcellulose (49), the site-specific degree of substitution at C2 can be determined. The C1 ring carbon chemical shift is sensitive to the nature of the substituent at the 2-position. For 2-O-hydroxypropyl substitution, the C1 chemical shift is 104.8 ppm while, for 2-O-acetyl substitution, this chemical shift is 102.8 ppm.

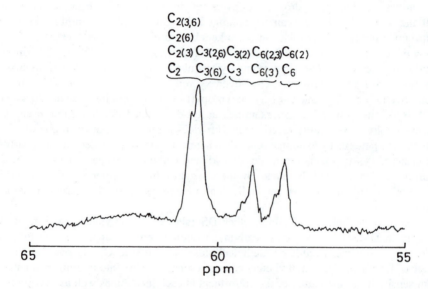

Figure 15.   Proposed methoxy methyl resonance assignments of acetylated methylcellulose in CDCl3 at 30°C. (Reproduced with permission from ref. 47. Copyright 1987 American Chemical Society.)

Mixed Esters of Cellulose. The two cellulose mixed esters that have received detailed NMR analysis by modern NMR techniques are cellulose acetate propionate (CAP) and cellulose acetate butyrate (CAB). Knowledge of the relative and total acyl concentration for each acyl group and the site-specific distribution of the various acyl groups might allow optimization of the final properties.

Iwata and coworkers (55) assigned all proton and ^{13}C resonances for the stereoregular CAP's 6-O-acetyl-2,3-di-O-propanoylcellulose (CADP) and 2,3-di-O-acetyl-6-O-propanoylcellulose (CPDA). By using homonuclear COSY and HETCOR for short-range correlations, all of the ring protons and carbons as well as acyl carbons were identified in the two CAP's. The J-coupling networks could not provide any information about specific sites of acylation. Correlations over 3-bonds between ring protons and acyl carbonyl carbons were accomplished by HMBC (23-25,56,57). This work was the first published application of the HMBC experiment for a cellulose derivative. Figure 16 shows the result of the HMBC experiment for CADP. Even though correlations to H5 and H6 were not clearly observed, assignments to the other ring protons could easily be made by the HMBC experiment to provide site-specific assignments. The HMBC experiment provides a rapid and reliable method for unambiguous site-specific assignments for cellulose esters.

In order to determine acetyl distribution in CA's with DS < 3, Tezuka and Tsuchiya (20) derivatized unreacted hydroxyl groups in the CA's making mixed acetate propionate esters. Using the derivatized mixed ester, they observed the acetyl and propionyl ester carbonyl carbon resonances to analyze the acyl distribution. Proof that acetyl ester exchange during the derivatization step did not occur was provided by monitoring the total acetyl content before and after the derivatization and by failure to exchange any acetyl groups in CTA with propionyl groups.

Tezuka (58) assigned the backbone and sidechain protons of cellulose acetate butyrate (CAB) by COSY. The acetyl carbonyl carbon assignments agreed with previous assignments (14,15). Butyryl carbonyl carbon assignments (C6: 172.6 ppm, C3: 172.1 ppm, C2: 171.6 ppm) were accomplished by the INAPT experiment.

Solid-State NMR

Solid-state ^{13}C NMR spectra of several cellulose esters, including CA's at various DS levels, in different crystalline forms and with different crystalline-to-amorphous ratios, CTP in different crystalline forms and CTB, are reported. NMR chemical shifts for these different cellulose esters are collected in Table VI. In these studies, solid-state NMR was used to address questions relative to hydrogen-bonding, spectral differences due to DS, determination of the composition of crystalline and amorphous components, ester identification, and conformation.

As discussed earlier (5), acylation of the C2 and C3 positions on cellulose results in an upfield shift for adjacent C1 and C4 (γ effect) ring carbons, respectively, while acylation at the C6 position causes a downfield shift for C6 (β effect) in solution-state ^{13}C NMR spectra. Interestingly, all three carbons (C1, C4, and C6) are shifted upfield on acylation in solid-state ^{13}C NMR spectra (compare Tables II and VI).

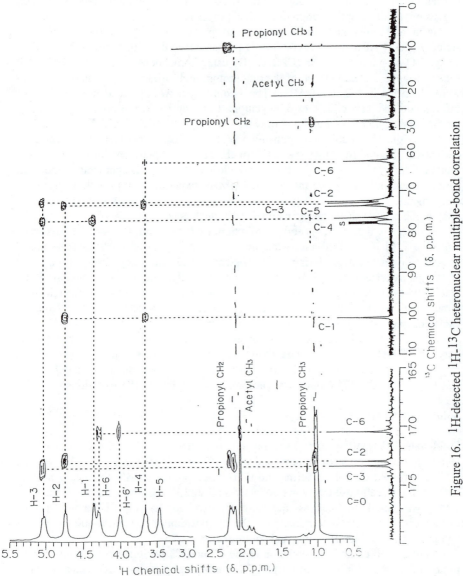

Figure 16. ^{1}H-detected ^{1}H,^{13}C heteronuclear multiple-bond correlation (HMBC) spectrum of cellulose acetate dipropionate (CADP). (Reproduced with permission from ref. 55. Copyright 1992 Elsevier Science Publishers)

Cellulose acetates with 0.48 to 2.92 (CTA) DS have received the greatest amount of attention compared to cellulose propionates and butyrates. VanderHart and coworkers (59) observed that ^{13}C NMR spectra of CTA oligomers with DP between 5 and 9 have the spectral characteristics of high DP CTA in crystalline form I, CTA (I), by noting the consistency of the dominant backbone resonances from these oligomers (Figure 17). ^{13}C NMR spectra of CTA crystalline forms I and II are different (Figure 18) (59-61). The NMR spectrum of CTA (I) is simpler than the corresponding spectrum of CTA (II) in agreement with crystal structure effects observed in the ^{13}C NMR spectra of the two cellulose crystalline allomorphs I and II. In the CTA (I) unit cell, all CTA residues are magnetically equivalent and symmetry, by implication, is high. In CTA (II), there are two magnetically inequivalent sites. In the methyl carbon spectral region, the presence of a unique chemical shift for one of the six methyl carbons is indicative of this magnetic inequivalence as well as the much larger number of resonances observed in the spectrum of CTA (II) relative to the spectrum of CTA (I).

Takai and coworkers (61) prepared CTA (I) from cellulose (I), designated CTA (I$_I$), and from cellulose (II), designated CTA (I$_{II}$), for comparison of these polymers to CTA (II). From these samples, the authors were able to differentiate crystal structures for the three CTA samples, CTA (I$_I$), CTA (I$_{II}$) and CTA (II) .

Subtraction of the spectrum of a related amorphous polymer provided the subspectrum for the crystalline component of the nonamer. By this approach, the nonamer was estimated to contain 47% crystallinity (Figure 18). Subtraction of the ^{13}C NMR spectrum of amorphous CTA from spectra of CTA prepared by different synthetic routes was used to estimate the CTA (I) and CTA (II) crystalline content resulting from these preparations (59).

While ^{13}C NMR spectra of CTA (I) and CTA (II) are quite different, ^{13}C NMR spectra of CTP prepared from either cellulose (I) or cellulose (II), both under low swelling heterogeneous acylation conditions, are similar (60).

Doyle and coworkers (62) also observed amorphous and crystalline CTA. Proton spin-lattice relaxation time, T_1, measurements in the laboratory frame ranged from 0.3 to 1.5 s with a single exponential decay. Proton spin-lattice relaxation time, $T_{1\rho}$, measurements in the rotating frame exhibited a biexponential decay. The short $T_{1\rho}$ values ranged from 3-5 ms and were assigned to amorphous CTA while the long $T_{1\rho}$ values ranged from 15-25 ms and were assigned to crystalline CTA.

At DS = 2.5, the C1 resonance of CTA splits into three resonances. These three resonances have been assigned to resonances from the two crystalline forms Iα and Iβ (62), as observed in cellulose (63).

Buchanan and coworkers (26) discussed the importance of hydrogen bonding between AGU's based on solution-state ^{13}C NMR results. Differences in C2 and C3 carbonyl carbon resonance intensities in the solution- and solid-state ^{13}C NMR spectra of CTA have been suggested to result from hydrogen bonding (64). Similar effects due to hydrogen bonding were not observed for CTP.

Pines and coworkers (65) suggested that ^{13}C shielding tensors should provide a sensitive tool for studying hydrogen-bonding schemes in the solid-state. The ^{13}C shielding anisotropy in the carbonyl group is strongly dependent on hydrogen bonding to the oxygen atom. For CTA and CTP, similar chemical shift anisotropies are

Table VI. Solid-State Carbon-13 NMR Chemical Shifts for Cellulose Esters

Sample	C=O	C1	C4	C2-C5	C6	C-α,β,γ	Ref
Cellulose		104.8	88.5	74.4-71.6	64.9		66
Cellulose		105.1 104.2 103.6	88.8 (+ low frequency shoulder at about 84)	74.3 73.7 71.8 70.8	64.6 (+ low frequency shoulder at about 62)		62
CTA (I)	169.4 171.9	102.2		71.5 75.0 79.2	61.6	α: 20.9 α: 22.0	59
CTA (I) Crystalline	169.00 170.00 171.45	101.37		71.10 74.36 78.51	61.01	α: 20.44 α: 21.42	61
CTA (I) Crystalline	168.87 169.97 171.41	101.56		71.01 74.39 78.52	60.90	α: 20.41 α: 21.56	61
CTA Amorphous DS = 2.90	170.35	100.69		72.85	63.48 63.36 61.40 60.44	α: 20.51	62
CTA (I) DS = 2.92	169.83 171.42	101.34		71.11 74.47 78.60	61.12	α: 20.47	60
CTA (II)	170.0 172.2 173.2	100.0		71.1 72.1 73.3 74.5 74.9 77.9 78.7	63.5 66.4	α: 20.8 α: 23.4	59
CTA (II)	169.87 170.62 171.08 181.54 172.89	99.31		71.70 72.99 73.84 76.73 78.08	60.64 63.05 65.23 66.90	α: 20.50 α: 22.86	61
CTA (II) DS = 2.90	169.79 171.43 172.81	99.06		70.58 71.41 72.91 73.66 76.46 78.02	63.17 65.15	α: 20.45 α: 22.70	60

Table VI. *Continued*

Sample	C=O	C1	C4	C2-C5	C6	C-α,β,γ	Ref
CTA DS = 2.93	169.8 170.3	99.8		72.4	62.8 63.8	α: 19.9 α: 20.4	62
CA DS = 1.5	169.7	100.0 100.6 103.7	87.8 (+ low frequency shoulder)	71.1 72.2 73.6	64.5 (+ low frequency shoulder)	α: 19.9 α: 20.3	62
CA DS = 0.97	169.8 170.5	100.3 103.5 104.4 105.2	87.8 88.4 (+ low frequency shoulder)	71.1 72.1 73.0 73.9 74.6	64.6 (+ low frequency shoulder)	α: 20.1	62
CA DS = 0.48	170.2	103.4 104.1 105.0	88.7 (+ low frequency shoulder)	70.7 71.8 73.5 74.2	65.2 (+ low frequency shoulder)	α: 20.3	62
CTP (Cell I) DS = 2.92	172.68	98.61 101.74		72.03 73.83	63.57	α: 27.53 β: 9.13	60
CTP (Cell II) DS = 2.80	171.98	98.47 101.22 104.32 106.53		72.06 73.83	62.27	α: 26.93 β: 8.83	60
CTB DS = 2.80	170.92 172.54	101.11		72.26 76.51	61.42 62.95	α: 36.21 β: 18.42 γ: 14.07	60

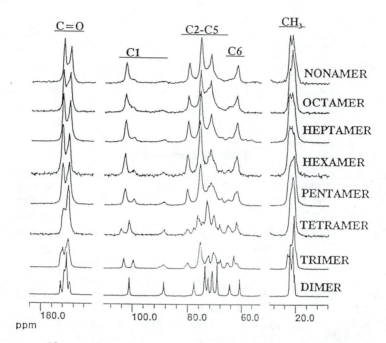

Figure 17. ^{13}C CP/MAS spectra (25.2 MHz) of the indicated oligomers of CTA. (Reproduced with permission from ref. 59. Copyright 1996 American Chemical Society.)

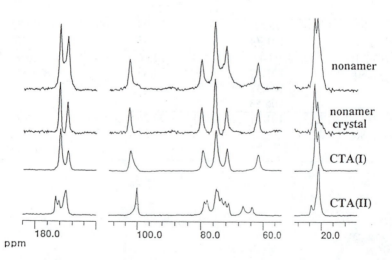

Figure 18. Solid-state ^{13}C CP/MAS NMR spectra of the nonamer of CTA, the crystalline component of the nonamer, and the two CTA allomorphs. (Reproduced with permission from ref. 59. Copyright 1996 American Chemical Society.)

observed for the aliphatic and ring carbons while a much larger chemical shift anisotropy is observed for the carbonyl carbons based on examination of spinning sidebands for ordered, layered polymer film. Based on the measurement of chemical shift tensors σ_{11}, σ_{22} and σ_{33} of the carbonyl resonances in several CTA and CTP films with different physical properties, a correlation between shielding anisotropy and film transparency was observed (64). The largest shielding anisotropy and asymmetry factor for the shielding tensors was found for the film with the least transparency. These observations of tensor anisotropy and reduced transparency are consistent with a more ordered polymer. This study demonstrated the usefulness of probing polymeric ordered domains with carbonyl shielding tensor measurements.

Future Directions

Over the past 10-15 years, considerable progress has been made in understanding the importance of detailed knowledge relating properties, structures, molecular conformations and dynamics for ester- and ether-derivatized cellulosic polymers. The application of new NMR techniques for elucidating detailed polymeric structures has made the analysis of structure-property relationships possible. In this chapter we have brought together relevant solution- and solid-state NMR literature from the past 10-15 years on the structure of acylated cellulosic polymers, including acetate, propionate, and butyrate, as well as acylated cellulose ethers. These studies provide a rich knowledge-base on site-specific substitution as well as specific monomer composition for individual esterified or etherified anhydroglucose units. Also, they begin to make possible an understanding of the impact of structural modifications as they relate to properties, possibly enabling fine-tuning of these structural modifications to approach desired properties.

As other powerful NMR techniques are applied to questions related to structure in cellulosic polymers, more structural information will emerge to aid in our understanding of important property relationships. Determination of the site-specific degree of substitution will undoubtedly benefit from the application of higher field NMR spectrometers as well as application of 3D NMR techniques.

The development of methods for sequencing esterified and etherified anhydroglucose units is an important problem that needs to be solved. Understanding the heterogeneous nature of substitution along the AGU backbone of cellulose derivatives can give useful insight into the solution conformation and molecular dynamics of these derivatives. Some minimal information is available now based on hydrogen bonding, but much more detail is needed to enable further analysis of the structure of these complex heteropolymers.

In studies presented here, specific acetylated monomer assignments from the 8 possible monomers have provided a detailed look at the impact on structure of acetylation at C2 and C3. Only general assignments could be made for C6 acetylated monomers. A monomer-specific analysis of acetylation at C6 is needed and may become possible by application of 3D NMR techniques, zero-quantum coherence 2D NMR techniques, or higher field NMR spectrometers.

Analyses of the solution conformation of cellulose acetates have provided information about the helical nature of the polymer. Some of these studies suggest differing views on the solution conformation about the glycosidic bond. The solution conformation about this bond possibly changes with increasing degree of polymerization. A study of solution-state conformation as a function of degree of polymerization in $CDCl_3$ as well as DMSO-d_6 with emphasis on the virtual angle across the glycosidic bond might resolve this question. In addition, the transition between 50 - 55°C measured for several properties needs to be examined in sufficient detail to determine its source.

Monomer-specific assignments for the cellulose acetate carbonyl carbon resonances have provided detailed knowledge about structure-property relationships pertaining to water solubility and absorbency for this polymer. Similar knowledge based on ^{13}C-enrichment in ester-derivatized cellulose ethers should improve our understanding of important structure-liquid-crystalline property relationships in hydroxyalkylcellulose and methylcellulose derivatives that form thermotropic cholesteric liquid-crystalline phases.

The future looks very bright for applying NMR to solve these tough structure-property relationship problems in cellulose ester and ether chemistry. The results of such studies should aid in the design of cellulose esters and ethers with predictable properties.

Acknowledgments

The author thanks John A. Hyatt, Eastman Chemical Company, for helpful discussions and Yasuyuki Tezuka, Tokyo Institute of Technology, for his assistance.

Literature Cited

1. Nehls, I.; Wagenknecht, W.; Philipp, B.; Stscherbina, D. *Prog. Polym. Sci.*, **1994**, *19*, 29.
2. Kamide, K.; Saito, M. *Macromol. Symp.*, **1994**, *83*, 233.
3. Goodlett, V. W.; Dougherty, J. T.; Patton, H. W. *J. Polym. Sci., Pt. A-1*, **1971**, *9*, 155.
4. Shiraishi, N.; Katayama, T.; Yokota, T. *Cellulose Chem. Technol.*, **1978**, *12*, 429.
5. Miyamoto, T.; Sato, Y.; Shibata, T.; Inagaki, H.; Tanahashi, M. *J. Polym. Sci., Polym. Chem. Ed.*, **1984**, *22*, 2363.
6. Kowsaka, K.; Okajima, K.; Kamide, K. *Polym. J.*, **1986**, *18*, 843.
7. Kamide, K.; Okajima, K. *Polym. J.*, **1981**, *13*, 127.
8. Buchanan, C. M.; Hyatt, J. A.; Lowman, D. W. *Macromolecules*, **1987**, *20*, 2750.
9. Nagayama, K.; Kumar, A.; Wuthrich, K.; Ernst, R. R. *J. Magn. Reson.*, **1980**, *40*, 321.
10. Bax, A.; Freeman, R. *J. Magn. Reson.*, **1981**, *44*, 542.
11. Maudsley, A. A.; Muller, L.; Ernst, R. R. *J. Magn. Reson.*, **1977**, *28*, 463.
12. Bax, A.; Morris, G. A. *J. Magn. Reson.*, **1981**, *42*, 501.
13. Uhrinova, S.; Petrakova, E.; Ruppeldt, J.; Uhrin, D. *Magn. Reson. Chem.*, **1990**, *28*, 979.
14. Dais, P.; Perlin, A. S. *Carbohyd. Res.*, **1988**, *181*, 233.

15. Buchanan, C. M.; Hyatt, J. A.; Lowman, D. W. *Carbohyd. Res.*, **1988**, *177*, 228.
16. Gagnaire, D. Y.; Taravel, F. R.; Vignon, M. R. *Macromolecules*, **1982**, *15*, 126.
17. Rao, V. S.; Saurio, F.; Perlin, A. S.; Viet, M. T. P. *Can. J. Chem.*, **1985**, *63*, 2507.
18. Kowsaka, K.; Okajima, K.; Kamide, K. *Polym. J.*, **1988**, *20*, 1091.
19. Shuto, Y.; Murayama, M.; Azuma, J.; Okamura, K. *Bull. Inst. Chem. Res., Kyoto Univ.*, **1988**, *66*, 128.
20. Tezuka, Y.; Tsuchiya, Y. *Carbohyd. Res.*, **1995**, *273*, 83.
21. Buchanan, C. M.; Hyatt, J. A.; Kelley, S. S.; Little, J. L. *Macromolecules*, **1990**, *23*, 3747.
22. Capon, B.; Rycroft, D. S.; Thomson, J. W. *Carbohyd. Res.*, **1979**, *70*, 145.
23. Bax, A.; Davis, D. G. *J. Amer. Chem. Soc.*, **1985**, *107*, 2820.
24. Bax, A.; Davis, D. G. *J. Magn. Reson.*, **1985**, *65*, 355.
25. Summers, M. F.; Marzilli, L. J.; Bax, A. *J. Amer. Chem. Soc.*, **1986**, *108*, 4285.
26. Buchanan, C. M.; Edgar, K. J.; Hyatt, J. A.; Wilson, A. K. *Macromolecules*, **1991**, *24*, 3050.
27. Kowsaka, K.; Okajima, K.; Kamide, K. *Polym. J.*, **1988**, *20*, 827.
28. Hikichi, K.; Kakuta, Y.; Katoh, T. *Polym. J.*, **1995**, *27*, 659.
29. Wagner, G. *J. Magn. Reson.*, **1983**, *55*, 151.
30. Bax, A.; Drobny, G. *J. Magn. Reson.*, **1985**, *61*, 306.
31. Buchanan, C. M.; Hyatt, J. A.; Lowman, D. W. *J. Amer. Chem. Soc.*, **1989**, *111*, 7312.
32. States, D. J.; Haberkorn, R. A.; Ruben, D. J. *J. Magn. Reson.*, **1982**, *48*, 286.
33. Olejniczak, E. T.; Hoch, J. C.; Dobson, C. M.; Poulsen, F. M. *J. Magn. Reson.*, **1985**, *64*, 199.
34. Ryskina, I. I.; Vakilenko, N. A. *Polym. Sci. USSR*, **1987**, *29*, 340.
35. Ogura, K.; Sobue, H.; Kasuga, M. *Polym. Letters Edit.*, **1973**, *11*, 421.
36. Stipanovic, A. J.; Sarko, A. *Polymer*, **1978**, *19*, 3.
37. Steinmeier, H.; Zugenmaier, P. *Carbohyd. Res.*, **1987**, *164*, 97.
38. Ritcey, A. M.; Gray, D. G. *Biopolymers*, **1988**, *27*, 479.
39. Stipanovic, A. J.; Stevens, E. S. *J. Appl. Polym. Sci., Appl. Polym. Symp.*, **1983**, *37*, 277.
40. Tezuka, Y. *Biopolymers*, **1994**, *34*, 1477.
41. Morat, C.; Taravel, F. R. *Bull. Magn. Reson.*, **1989**, *11*, 321.
42. Morat, C.; Taravel, F. R. *Tetrahedron Letters*, **1990**, *31*, 1413.
43. Bax, A.; Freeman, R. *J. Amer. Chem. Soc.*, **1982**, *104*, 1099.
44. Buchanan, C. M.; Edgar, K. J.; Wilson, A. K. *Macromolecules*, **1991**, *24*, 3060.
45. Miyamoto, T.; Sato, Y.; Shibata, T.; Tanahashi, M.; Inagaki, H. *J. Polym. Sci., Polym. Chem. Ed.*, **1985**, *23*, 1373.
46. Kamide, K.; Okajima, K.; Kowsaka, K.; Matsui, T. *Polym. J.*, **1987**, *19*, 1405.
47. Tezuka, Y.; Imai, K.; Oshima, M.; Chiba, T. *Macromolecules*, **1987**, *20*, 2413.
48. Tezuka, Y.; Imai, K.; Oshima, M.; Chiba, T. *Polymer*, **1989**, *30*, 2288.
49. Tezuka, Y.; Imai, K.; Oshima, M.; Chiba, T. *Carbohyd. Res.*, **1990**, *196*, 1.
50. Tezuka, Y.; Imai, K.; Oshima, M.; Chiba, T. *Makromol. Chem.*, **1990**, *191*, 681.
51. Tezuka, Y.; Imai, K.; Oshima, M.; Chiba, T. *Polym. J.*, **1991**, *23*, 189.
52. Tezuka, Y.; Tsuchiya, Y.; Shiomi, T. *Carbohyd. Res.*, **1996**, *291*, 99.

53. Tezuka, Y.; Imai, K.; Oshima, M.; Ito, K. *Carbohyd. Res.*, **1991**, *222*, 255.
54. Guo, J.-X.; Gray, D. G. *J. Polym. Sci.: Pt. A: Polym. Chem.*, **1994**, *32*, 889.
55. Iwata, T.; Azuma, J.-I.; Okamura, K.; Muramoto, M.; Chun, B. *Carbohyd. Res.*, **1992**, *224*, 277.
56. Frey, M. H.; Leupin, W.; Sorensen, O. W.; Denny, W. A.; Ernst, R. R.; Wuthrich, K. *Biopolymers*, **1985**, *24*, 2371.
57. Bax, A.; Summers, M. F. *J. Amer. Chem. Soc.*, **1986**, *108*, 2093.
58. Tezuka, Y. *Carbohyd. Res.*, **1993**, *241*, 285.
59. VanderHart, D. L.; Hyatt, J. A.; Atalla, R. H.; Tirumalai, V. C. *Macromolecules*, **1996**, *29*, 730.
60. Hoshino, M.; Takai, M.; Fukuda, K.; Imura, K.; Hayashi, J. *J. Polym. Sci: Pt. A: Polym. Chem.*, **1989**, *27*, 2083.
61. Takai, M.; Fukuda, K.; Hayashi, J. *J. Polym. Sci.: Pt. C: Polym. Lett.*, **1987**, *25*, 121.
62. Doyle, S.; Pethrick, R. A.; Harris, R. K.; Lane, J. M.; Packer, K. J.; Heatley, F. *Polymer*, **1986**, *27*, 19.
63. VanderHart, D. L.; Atalla, R. H. In *The Structures of Cellulose--Characterization of the Solid State*; Atalla, R. H., Ed.; ACS Symposium Series 340; American Chemical Society: Washington, DC, 1987, 88-118.
64. Nunes, T.; Burrows, H. D.; Bastos, M.; Feio, G.; Gil, M. H. *Polymer*, **1995**, *36*, 479.
65. Pines, A.; Chang, J. J.; Griffin, R. G. *J. Chem. Phys.*, **1974**, *61*, 1021.
66. Pawlowski, W. P.; Sankar, S. S.; Gilbert, R. D. *J. Polym. Sci.: Pt. A: Polym. Chem.*, **1987**, *25*, 3355.

Chapter 11

^{13}C NMR Structural Study on Cellulose Derivatives with Carbonyl Groups as a Sensitive Probe

Y. Tezuka

Department of Organic and Polymeric Materials, Faculty of Engineering, Tokyo Institute of Technology, O-okayama, Meguro-ku, Tokyo 152, Japan

A new versatile analytical method for cellulose ester and ether derivatives has been developed, to determine the degree of substitution (DS) at the individual positions (2, 3 and 6) on the glucose residue. Thus unsubstituted hydroxyl groups in the starting cellulose derivatives are first converted to appropriate acyl groups. The obtained peresterified derivatives become soluble in common NMR solvents regardless of the DS of the samples, in contrast to the original cellulose derivatives. The subsequent 13-C NMR analysis with the acyl carbonyl carbon as a structural probe allows to determine the detailed distribution pattern of the unsubstituted hydroxyl groups, and consequently that of the ester or ether substituents in the starting cellulose derivatives. In addition, supramolecular structures of cellulose triacetate in different solutions have been studied by means of 2D-NOESY technique by the detection of the through-space interaction of acetyl protons.

Cellulosic materials have gained a renewed interest along with the recent developments of both "high-tech" and "bio-tech" applications *(1)*. Since most cellulose derivatives are produced through the reaction of the hydroxyl groups either at 2, 3 or 6 positions in an anhydroglucose residue, the precise control and deterimination of the distribution pattern of the substituents are of primary importance not only for the elucidation of structure-property relationships to achieve optimal performance of the final product, but also for the quality control of the product to address the product-liability requirements.

We have so far proposed a new analytical technique applicable for cellulose derivatives, in particular such cellulose ethers as methylcellulose, hydroxyalkylcelluloses and those having both methyl and hydroxyalkyl substituents, in which their peracetylated derivatives are subjected to 13-C NMR measurement *(2-6)*. The following advantages are noted for this procedure *(7,8)*.

1. The acetylation of cellulose ethers is an easy and simple pretreatment to provide products readily soluble in common NMR solvents over a wide DS range of the starting derivative, facilitating the spectral comparison with those of model polymers.

2. The acetyl carbonyl carbon signal can act as a remarkably sensitive structural probe, reflecting its location on the anhydroglucose residue.

3. The acetylation of the hydroxyl groups can eliminate spectral complication arisen from intra- and intermolecular hydrogen bonds.

4. The polymeric form in cellulose derivatives can be maintained during the analytical procedure, to avoid the cumbersome hydrolysis pretreatment inevitable for chromatographic techniques such as GLC and HPLC.

As an extension of the preceding studies, the present paper describes first on the application of this novel technique to industrially important cellulose ester and ether derivatives, namely cellulose acetate (CA) *(9)* and carboxymethylcellulose (CMC) *(10)*, respectively. Secondly, 2D-NOESY results on cellulose triacetate are presented with acetyl protons as structural probe, to detect the through-space interaction of acetyl groups to provide supramolecular structures of cellulose triacetate in different solutions *(11)*.

Results and Discussion

Substituent Distribution in Cellulose Acetate (CA). Cellulose acetate is commercially produced with appropriate degrees of substitutions, and is provided for wide applications such as fibers, plastics, films and coatings. Although the total acetyl content in CA may be determined by a standard titration technique, the DS at the individual positions in the anhydroglucose residue is not readily obtainable by chromatographic techniques since the hydrolysis pretreatment of CA samples is accompanied by the deacetylation. Thus a 1-H NMR technique has been applied by using costly perdeuterioacetyl derivative of CA, since acetyl proton signal in cellulose triacetate appears as a resolved triplet reflecting its location on the glucose residue *(12)*.

An alternative and versatile 13-C NMR technique has been developed by using perpropanoated derivative of CA. Quantitative propanoation of the hydroxyl groups in CA samples without the concurrence of ester exchange reaction is a prerequisite for the precise determination of the DS by use of the propanoated derivative. Such a reaction condition has been realized with propanoic anhydride/pyridine/4-(dimethylamino)pyridine system. Indeed, the possibility of ester exchange reactions was excluded through observing constant total acetyl content during the propanoation treatment, and from the fact that attempted propanoation of cellulose triacetate under the present condition failed to introduce propanoate groups.

The carbonyl region spectrum of a propanoated derivative of CA is shown in Figure 1, together with that of the starting CA. In the latter, carbonyl carbon signal appears as overlapped multiplet peaks, which reflect the substitution patterns of acetyl groups and the hydrogen-bond interactions between acetyl and hydroxyl groups. On the contrary, a remarkably simple spectrum is obtained for the propanoated derivative, where both acetyl and propanoyl carbonyl carbon signals appear separately and are resolved into the two sets of three peaks corresponding to their substitution positions on the anhydroglucose residue. The three peaks are assigned as that on the 2, 3 and 6 positions, respectively, from up field.

Thus a quantitative-mode 13-C NMR measurement for a series of propanoated CA derivatives allows one to determine the individual DS of the starting CA samples. In addition, the propanoyl group distribution, which is complementary to that of acetyl groups, facilitates the determination of the DS of CA samples of low acetyl contents with high precision. Consequently, the present NMR technique is considered as a significantly improved analytical means to determine the distribution pattern of acetyl groups in CA samples of wide range of DS. In the 1-H NMR technique with perdeuterioacetyl derivatives of CA, in contrast, the absolute signal intensity decreases along with the decrease in the acetyl content.

This novel technique has been successfully applied also for a commercially important cellulose ester derivative having two different ester substituents, namely cellulose acetate butyrate (CAB) *(13)*. 13-C NMR carbonyl region spectra listed in Figure 2 demonstrate that both acetyl and butanoyl carbonyl carbon signals appear separately and are resolved into two-sets of three peaks as in the case of propanoated CA samples. In consequence, the relative content of acetyl and butanoyl groups for a series of CAB samples can be determined with a quantitative-mode measurement. The absolute distribution of the two subsutituents is subsequently determined by taking into account the total degree of substitution by the two substituents.

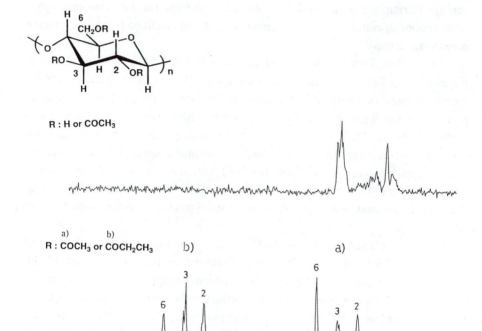

R : H or COCH₃

R : COCH₃ or COCH₂CH₃

Figure 1. 100MHz 13-C NMR carbonyl region spectra of cellulose acetate having the total DS of 1.43 in DMSO-d₆ (top) and of its propanoated derivative in CDCl₃ (bottom). (Reproduced with the permission from ref. 9. Copyright 1995 Elsevier Science Ltd.)

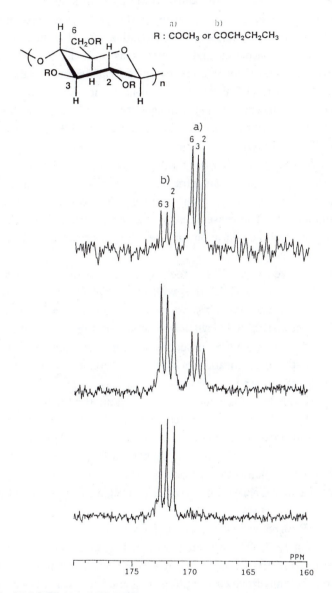

Figure 2. 67.8 MHz 13-C NMR carbonyl region spectra of cellulose acetate butyrate having different substitution patterns in CDCl3. (Reproduced with the permission from ref. 13. Copyright 1993 Elsevier Science Ltd.)

Substituent Distribution in Carboxymethylcellulose (CMC). Sodium carboxymethylcellulose, commonly termed carboxymethylcellulose or CMC, is an important water-soluble cellulose ether derivative applied in such human-contact uses as food, pharmaceutics and cosmetics as well as in such ecologically-sensitive uses as soil treatment, oil recovery and paper sizing processes.

Although the total carboxymethyl content in CMC samples may be determined by a standard titration technique, the individual DS at the 2, 3 and 6 positions on the anhydroglucose residue is obtainable only after the hydrolysis treatment. The hydrolytic pretreatment of CMC, however, requires precautions to avoid the loss of the hydrolysate during both reaction and recovery processes. The side reactions, in particular intra- and intermolecular lactonization between carboxylic acid and hydroxyl groups under acidic condition, should also be carefully eliminated. The presence of anomeric isomers in the hydrolysate may also cause spectral complication in the chromatographic and spectroscopic analysis.

An alternative 13-C NMR technique has been developed by using peresterified derivative of CMC, thus without the hydrolysis pretreatment. Since CMC possesses carboxylate salt substituent, it is soluble only in water. By this characteristic property, the acylation reaction of the unsubstituted hydroxyl groups is circumvented to give a product soluble in any of common NMR solvents. Hence, a two-step derivatization of CMC has been employed, where sodium carboxymethyl groups are first converted into methyl ester groups by the reaction with dimethyl sulfate. Although this methylation treatment was occasionally accompanied by undesirable side reactions to result in partly insoluble product, subsequent propanoation treatment of the residual hydroxyl groups could produce a series of peresterified CMC samples soluble in DMSO-d_6. The acetylation in place of the propanoation of the hydroxyl groups proceeds as well, while the acetyl carbonyl carbon signal appears in the coincided region with that of carboxymethyl carbonyl signal.

13-C NMR carbonyl region spectra of a series of peresterified CMC samples are listed in Figure 3. The propanoyl carbonyl signal appears as three resolved peaks regardless of the DS of the sample, and corresponding to its substitution positions on the anhydroglucose units. In addition, the carboxymethyl carbonyl carbon signal, being resolved also as three peaks in case of high total DS, provides the complementary information on the distribution of the substituents. Minor unassignable signals occasionally visible at 169-170ppm may be arisen from side

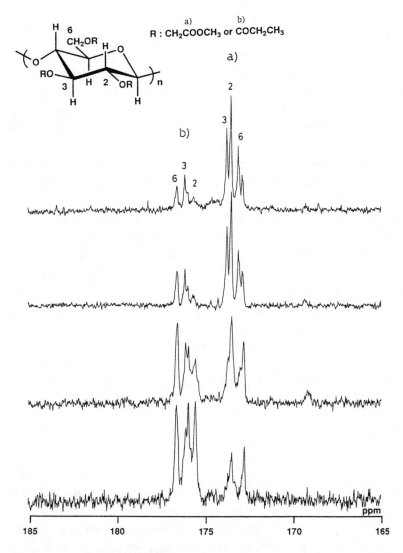

Figure 3. 100MHz 13-C NMR carbonyl region spectra of peresterified carboxymethylcellulose having different DS values in DMSO-d$_6$. (Reproduced with the permission from ref. 10. Copyright 1996 Elsevier Science Ltd.)

reactions during the methylation process. Finally, a quantitative-mode 13-C NMR measurement was performed to determine the individual DS of the starting CMC samples having a wide range of degrees of substitution.

Interaction between Substituents in Cellulose Triacetate (CTA). The control of the substitution pattern in cellulose derivatives may lead to the design of supramolecular structures of cellulosics in bulk and in solution. The detection of the interaction between the substituents is a key to elucidate the chain conformation and the resulting supramolecular structures. In the preceding study *(2)* , the through-space interaction between the substituents at the 3 and 6 positions in the anhydroglucose residue on acetylated methylcellulose has been postulated from the chemical shift change in the methoxy methyl carbon signals.

The detection of the through-space interaction between acetyl groups in cellulose triacetate dissolved in different solvens is now achieved by means of a 2D-NOESY technique with acetyl proton as NMR probe, since acetyl proton signal in cellulose triacetate appears as a resolved triplet reflecting its location on the glucose residue. Figure 4 shows NOESY spectra of CTA in $CDCl_3$ and in DMSO-d_6 solutions, in which off-diagonal cross peaks are generated besides a large diagonal signal. In $CDCl_3$, the cross-peaks correlate the acetyl proton signal at the 3 and 6 positions on the glucose residue. In DMSO-d_6, on the other hand, the off-diagonal cross-peaks appear at the 2 and 6 positions in addition to the 3 and 6 positions. These results demonstrates that the NOESY technique is a powerful means to observe the interaction between substituents, to provide the solution dynamics of three-dimensional structures of cellulose derivatives.

Experimental

Cellulose acetate (CA) samples of different DS values were prepared by acid hydrolysis of cellulose triacetate (Aldrich or Daicel). Propanoation of a series of CA samples was carried out by propanoic anhydride/pyridine/4-(dimethylamino)pyridine at 100°C for 1h. Cellulose acetate butyrate (CAB) samples of different distribution patterns were supplied from Eastman Chemical Japan, Ltd. Carboxymethylcellulose (CMC) samples of different DS values were obtained from Dai-ichi Kogyo Seiyaku Co. Methyl esterification of sodium carboxymethyl groups in a series of CMC samples was performed by dimethyl sulfate in DMSO at 45°C for 24h. Subsequent propanoation of methyl-esterified CMC samples was carried out by propanoic anhydride/pyridine/4-(dimethylamino)pyridine in DMAc/LiCl at 100°C for 6h.

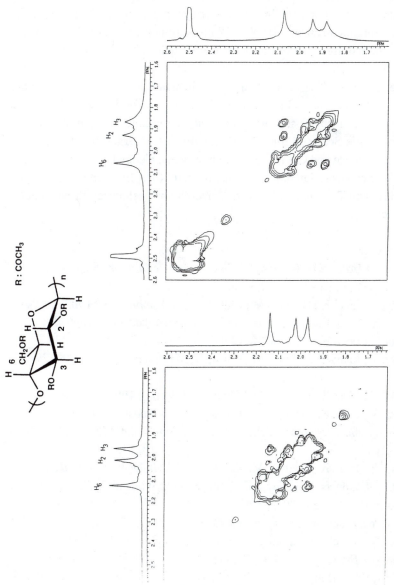

Figure 4. NOESY spectra of acetyl methyl region of cellulose triacetate in CDCl₃ (left) and in DMSO-d₆ (right). (Reproduced with the permission from ref. 11. Copyright 1994 John Wiley & Sons, Inc.)

13-C NMR measurements were performed with either JEOL GX-270 or EX-400 apparatus. Chemical shift values were referenced from the solvent signal of either $CDCl_3$ (77.0) or DMSO-d_6 (43.5). Quantitative-mode 13-C NMR measurements were conducted by a non-NOE gated decoupling technique with a pulse repetition time of 30sec. 2D-NOESY measurements were carried out with a mixing time of 500msec, and 16 transients were acquired for each t_1 value.

Acknowledgments. The author thanks to Eastman Chemical Japan, Ltd., Shin-Etsu Chemical Co., and Dai-ichi Kogyo Seiyaku Co., for the gift of cellulose derivatives. The financial supports from The Agricultural, Chemical Research Foundation, Nestle Science Promotion Committee, and a Grant from the Ministry of Education, Science and Culture are gratefully achnowledged. The author is also grateful to the collaboration of M. Oshima, K. Ito of Shin-Etsu Chemical Co., and of Y. Tsuchiya and T. Shiomi of Nagaoka University of Technology throughout the present study.

Literature Cited

1. *Cellulosics Utilization, Research and Reward in Cellulosics*; Inagaki, H.; Phillips, G.O., Eds.; Elsevier, London, 1989.
2. Tezuka, Y.; Imai, K.; Oshima, M.; Chiba, T. *Macromolecules*, **1987**, *20*, 2413.
3. Tezuka, Y.; Imai, K.; Oshima, M.; Chiba, T. *Carbohydr. Res.*, **1990**, *196*, 1.
4. Tezuka, Y.; Imai, K.; Oshima, M.; Chiba, T. *Polymer*, **1989**, *30*, 2288.
5. Tezuka, Y.; Imai, K.; Oshima, M.; Chiba, T. *Makromol. Chem.*, **1990**, *191*, 681.
6. Tezuka, Y.; Imai, K.; Oshima, M.; Chiba, T. *Polym. J*, **1991**, *23*, 189.
7. Tezuka, Y.; Imai, K.; Oshima, M.; Chiba, T. In *Cellulose and Wood, Chemistry and Technology*, Schuerch, C. Ed.; John Wiley & Sons, New York, 1989, p1011.
8. Tezuka, Y.; Imai, K.; Oshima, M.; Chiba, T. In *Cellulose: Structural and Functional Aspects*, Kennedy, J.F.; Phillips, G.O.; Williams, P.A. Eds.; Ellis Horwood, Chichester, 1990, p251.
9. Tezuka, Y.; Tsuchiya, Y. *Carbohydr. Res.*, **1995**, *273*, 83.
10. Tezuka, Y.; Tsuchiya, Y.; Shiomi, T. *Carbohydr. Res.*, **1996**, *291*, 99.
11. Tezuka, Y. *Biopolymers*, **1994**, *34*, 1477.
12. Goodlett, V.W.; Dougherty, J.T.; Patton, H.W. *J. Polym. Sci. Part-A*, **1971**, *9*, 155.
13. Tezuka, Y. *Carbohydr. Res.*, **1993**, *241*, 2853.

Chapter 12

Novel Approaches Using FTIR Spectroscopy To Study the Structure of Crystalline and Noncrystalline Cellulose

T. Kondo, Y. Kataoka, and Y. Hishikawa

Forestry and Forest Products Research Institute (FFPRI),
P.O. Box 16, Tsukuba Norin Kenkyu, Tsukuba, Ibaraki 305, Japan

This paper deals with introduction of two methods with combination of FTIR which are very powerful to investigate cellulose supermolecular structures. One is using a microscopic accessary and the other is using a special reaction cell for deuteration. The former method depends on interpretation of change of the characteristic IR bands for cellulose Iα and Iβ crystalline phases to determine crystalline structures during coniferous wood cell wall formation, while the latter needs kinetic approaches to characterize amorphous structures by monitoring changes of OH bands during the deuteration process with D_2O as a probe for amorphous 3 film samples, cellulose, 2,3-di-O-methylcellulose (23MC) and 6-O-methylcellulose (6MC).

There are lots of powerful tools to investigate cellulosic structures such as X-ray, electron diffraction and microscope, FTIR, Raman spectroscopy and solid-state NMR. Among them we have been using FTIR to study cellulose structures as well as hydrogen bonding formation (1-3). In particular not only FTIR but also FTIR in combination with suitable attachments could provide us with further information on cellulose supermolecular structure.

In this paper we will introduce the following two IR methods and their applications (Figure 1) to study the supermolecular structures for cellulose and its derivatives. The methods are i) FTIR with a microscopic accessary (4) and ii) FTIR with a special reaction cell for deuteration. In the following, the former was applied to change of crystalline form and crystallinity of wood cell wall cellulose during cell wall formation, whereas the latter was employed for the study of amorphous regions using deuteration process. The advantages of the two combination will be described here briefly. For FTIR with a microscopic attachment, FTIR spectra for such a small amount of sample less than 1 mg can be obtained from small area (minimum : 50 x 50μm^2). Using the special reaction cell for deuteration, FTIR spectra can be obtained throughout a deuteration process of cellulosics and thus this method enables us to perform a kinetic analysis of the reaction.

Experimental
1. FTIR with a microscopic accessary.
Materials.   To investigate each stage of formation for coniferous wood tracheid cell wall, a sectinonal sample was prepared from radial and radial face (perpendicular to radial) direction of *Cryptomeria japonica* D. Don (Japanese cedar) and *Chamaecyparis obtusa* Endl (Japanese cypress) with 30 μm thickness as illustrated in Figure 3. In this section, each stage from cambial zone to mature xylem including the primary wall (P) and secondary walls (S_1, S_2 and S_3) was lined up in order of maturity. Purification was thoroughly performed according to previous manners (*5,6*) to remove pectin, hemicellulose and lignin. After purification it was freeze-dried with *tert*-butyl alcohol and then provided for FTIR measurements.
FTIR measurements.   FTIR attached with a microscopic accessary was Nicolet Magna 550 in combination with a Nicplan microscopic attachment. The spectra from 4000-650 cm^{-1} were the average of 64 scans recorded at a resolution of 4 cm^{-1} with a MCT detector. As shown in Figure 3, IR beam was irradiated from the radial face direction with an area of 50 x 300 μm^2 for each stage of the cell wall formation. When the beam was irradiated from radial direction, we could not obtain any spectra.
2. FTIR with a special reaction cell for deuteration of cellulosics.
Materials.   Three kinds of amorphous cellulosic films were prepared by casting from their N,N-dimethyl acetamide (DMAc) solutions. Namely, they were regioselectively methylated amorphous 2,3-di-*O*-methylcellulose (23MC)(*7*) and amorphous 6-*O*-methylcellulose (6MC)(*8*), and pure amorphous cellulose prepared from a DMAc-LiCl cellulose solution by casting and the subsequent washing with ethanol. The films were sufficiently thin with a thickness from 5 to 10 μm to obey the Beer-Lambert law (*9*). The 3 amorphous samples were confirmed by the very diffused patterns in their wide angle X-ray diffractgrams.
FTIR measurements.   The film samples above mentioned were fixed in a reaction cell as shown in Figure 2. Inside of the cell there was a pool for heavy water and it was filled and saturated with D_2O vapor (*10,11*). IR beam passed through calcium fluoride windows and irradiated the sample and then was detected by a DTGS detector. The IR spectra were obtained every 5 minutes for the first 5 hours and then every 15 minutes for the next 2 hours and finally every an hour until the end of the deuteration using a Nicolet Magna 550 spectrophotometer. The wavenumber range scanned was 4000-400 cm^{-1}; scan rate was 28 scans per minute; 32 scans of 2 cm^{-1} resolution were signal averaged and stored.

Results and Discussion
1. FTIR with a microscopic accessary.   The main advantages of the this method is that FTIR spectra for such a small amount of sample less than 1 mg can be obtained from small area (minimum : 50 x 50μm^2). Therefore this can be applied to the *in situ* biological system. For this application we have to remove H_2O to get completely dried state. Under such condition, we have investigated the crystalline form and crystallinity in both primary and secondary walls during coniferous wood cell wall formation
   The process of biosynthesis and subsequent crystallization of cellulose is the initial stage of plant cell wall formation. Although many *in vitro* investigations on this process have been carried out, the *in vivo* mechanism for cellulose crystallization has still not been clearly resolved. Several years ago, Attala and VanderHart (*12*) found that the cellulose I crystalline structure is really a composite of two allomorphs, thereby providing a new method by which to investigate the mechanism of cellulose crystallization. Why are two cellulose allomorphs, Iα and Iβ, crystallized at the surface of plant cells? An earlier study (*13*) pointed out that cellulose from enlarging

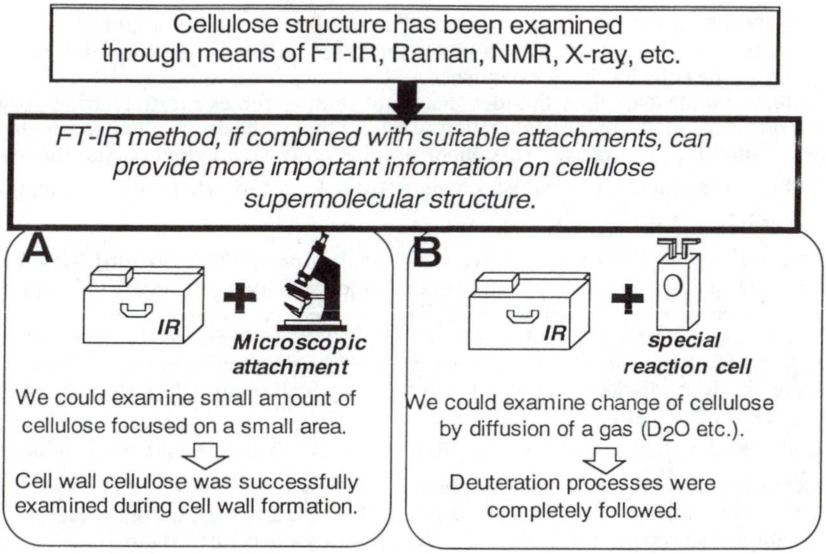

Cellulose structure has been examined
through means of FT-IR, Raman, NMR, X-ray, etc.

*FT-IR method, if combined with suitable attachments, can
provide more important information on cellulose
supermolecular structure.*

A

+ *Microscopic attachment*

We could examine small amount of cellulose focused on a small area.

Cell wall cellulose was successfully examined during cell wall formation.

B

+ *special reaction cell*

We could examine change of cellulose by diffusion of a gas (D_2O etc.).

Deuteration processes were completely followed.

Figure 1. Two IR methods introduced in this present paper.

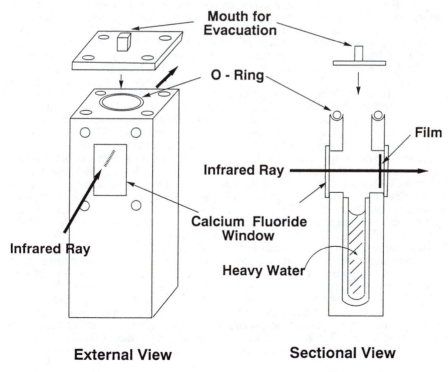

Mouth for Evacuation

O - Ring

Film

Infrared Ray →

Infrared Ray

Calcium Fluoride Window

Heavy Water

External View   **Sectional View**

Figure 2. The IR sample cell used for monitoring the deuteration process.

algae was rich in the Iα phase, whereas the Iβ phase was the dominant cellulose component in higher plants which form thick secondary walls once the primary geometry of the cells has been established.

Here, we want to explore the idea that plant cellular forces exerted during growth have a direct influence on crystallization of newly biosynthesized cellulose which can be gel or liquid crystal phase. This paper provide an *in vivo* evidence that the major crystalline component of cellulose changes from Iα to Iβ when cellular enlarging growth ceases.

The purified sections were analyzed on small areas (300 X 50 μm) by a FTIR spectrometer using a microscopic attachment which was focussed on the radial face of the sections as already described. After FTIR analysis these sections were observed by a scanning electron microscope (JEOL JXA840A) to examine tracheid cell wall formation. This procedure allowed us to examine the developing cellulose crystalline structure in four distinct stages of the forming cell wall (P, P+S1, P+S1+S2, P+S1+S2+S3) as illustrated in Figure 3.

Cellulose Iα shows characteristic IR bands at 3240 and at 750 cm-1, while the characteristic bands for the Iβ phase appear at 3270 and at 710 cm^{-1} (*14*). Significant differences in the FT-IR spectra appeared in 1000 - 650 cm-1 region obtained for both wood samples as shown in Figure 4. The IR absorption band at 750 cm-1 attributed to the Iα cellulose form appeared only in the spectra from the primary wall (P). This band became weaker with the subsequent deposition of the secondary wall layers on the pre-formed wall layers (P+S1, P+S1+S2, P+S1+S2+S3). On the other hand, the band at 710 cm-1 assigned to Iβ cellulose was not detected in the spectra for the primary wall, whereas after the start of the secondary wall deposition, it began to appear as a significant shoulder peak. Because the secondary wall is much thicker than the primary wall, the deposition of the secondary wall containing the Iβ phase results in the baseline intensity increasing and it begins to overlap the Iα absorption. Thus, in the region of the spectra having both walls, the Iα absorption band is not clearly distinguishable. These results show that the cellulose in the primary wall is rich in the Iα phase, whereas the cellulose in the secondary wall is rich in the Iβ phase. The second derivatives of IR spectra obtained from the OH stretching region after deuteration (Figure 5) also supported the above suggestion. Although the absorption due to Iα phase at 3240 cm-1 appeared unclear in the FT-IR spectra even for the deuterated primary wall (a in Figure 5), this absorption was proved clearly in the second derivative spectrum (b in Figure 5). On the other hand, it was not detected in the spectra for the deuterated mature wall (P+S1+S2+S3) in the same Figure. According to Fengel *et.al.* (*15, 16*), the band position due to Iα phase may change after the deuteration while the band due to Iβ phase at 3472 cm-1 is hardly influenced by milling or deuteration. In our case for the comparison between P and P+S1+S2+S3, the second derivative spectra after the deuteration showed the difference significantly as Michel reported previously (*17*).

There are two contradictory reports that wood cellulose is rich in the Iα phase (*18*) while at the same time the Iβ phase is dominant in it (*6*). Our IR results can explain this apparent contradiction because the mixture of Iα-rich primary and Iβ-rich secondary walls was analyzed concurrently in the samples in those reports.

In addition, primary wall cellulose has been described as being disorganized cellulose I (*5*) or totally amorphous (*19*). However, we have obtained a result that crystallinity of tracheid primary wall cellulose can be higher than that of secondary wall in the following. Figure 6 shows the change of crystallinity determined by a IR method for each stage of wood cell cellulose formation. At first stage for the

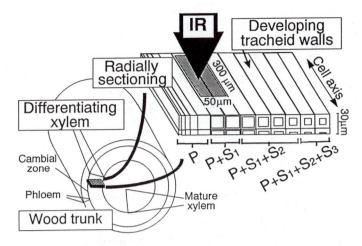

Figure 3. The sectioning method for obtaining IR spectra for each stage of tracheid wall formation. P; primary wall, and S_{1-3}; secondary walls. In the developing tracheid walls S_1 layer deposits on the primary wall from inside (cell side), and in the same manner S_n layer deposits on the S_{n-1}.

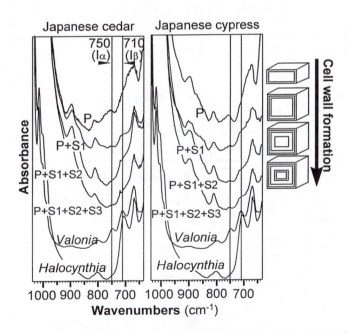

Figure 4. Comparison of IR characteristic absorption bands for the Iα forms of cellulose (at 750 cm^{-1}) with that for Iβ forms (at 710 cm^{-1}) in each stage of two coniferous developing tracheid walls; Japanese ceder (left) and Japanese cypress (right). Spectra for *Valonia* and *Halocynthia* represent standard spectra of cellulose rich in Iα and pure Iβ allomorphs, respectively.

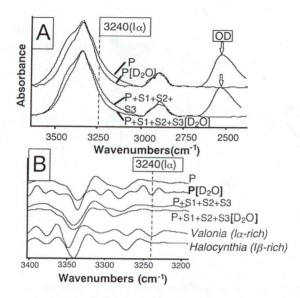

Figure 5. Changes in FT-IR spectra for deuterated cellulose in the region from 3400 to 2800 cm^{-1} (A) and their second derivatives (B) from Japanese cypress tracheid walls composed of P and P+S1+S2+S3, compared with standard second derivative spectra of non-deuterated cellulose rich in Iα (*Valonia*) and pure Iβ (*Halocynthia*) allomorphs. IR absorption bands at 3240 cm^{-1} is characteristic for the Iα form of cellulose.

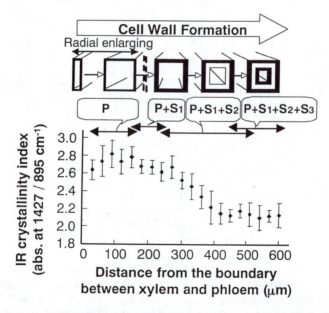

Figure 6. Change of IR crystallinity index (1427 / 895 cm^{-1}) for cellulose during tracheid cell wall formation. Results are averaged of 6 analyses.

formation of primary wall the crystallinity of the cellulose increased, and then with the deposition of secondary wall the crystallinity started to decrease gradually, and finally it became saturated at 2.1 as IR crystallinity index which corresponds to approximately 60 % for the crystallinity index determined by X-ray. The value coincides with the crystallinity for ordinary wood. These results clearly suggest that crystallinity for the primary wall cellulose is higher than that for the secondary wall cellulose. As described above, using FTIR equipped with a microscopic attachment enabled us to obtain *in situ* changes of both cellulose crystalline form and crystallinity during cell wall formation.

2. FTIR with a special reaction cell for dueteration of cellulosics.   In previous papers (*1-3*), we believed that a characterization of the hydrogen bonds found in amorphous cellulose would be of fundamental value. Furthermore, we proposed that a structural study of amorphous cellulose in light of hydrogen bonding might be a first step in uncovering details of how molecules rearrange in going from the liquid to the crystalline state (*3*). For the above objectives, it is important to get the information about "amorphous regions" for cellulose. Currently there is a shortage of structural information about the amorphous regions, perhaps partly because terminology suggests and impression persists that molecular chains in these regions are completely without structure and partly because methodology has been limited to WAXD and CP-MAS ^{13}C-NMR for measuring order in the presence of substantial amount of disorder. It is not uncommon for substrates that are not identifically crystalline by a method such as X-ray diffraction to be labeled "amorphous", but definition of amorphous goes beyond noncrystalline to unorganized and having no pattern of structure (*20*). At present FTIR is one of the best tools available to derive structural information including hydrogen bonding. Therefore, we particularly wanted to examine details of the amorphous regions for the amorphous cellulose film samples, regioselectively methylated 23MC and 6MC amorphous film samples, as model components of amorphous cellulose using FTIR methods (*3*).

Applications of deuteration methods to IR have been so far focussed on the separation of IR spectra for cellulose structure into two parts of crystalline and amorphous regions, respectively, and then only the discriminated crystalline regions have been studied (*21-23*). In this paper we will focus only on the amorphous regions for our cellulose model compounds using the FTIR monitoring of the deuteration process. The main advantages of this method is that the use of our reaction cell has enabled us to obtain not only change of reproducible and correct IR spectra for amorphous regions in cellulose, but also the rate constant for the deuteration reaction which is assumed to correspond to the nature of amorphous regions. Thus we have attempted to characterize the morphology for amorphous regions by analysing the diffusion behavior of D_2O as a probe. Amorphous cellulosic samples 23MC and 6MC we used are shown in Figure 7. As already reported in a previous paper (*2*), each polymer is considered to have different types of inter- and intramolecular hydrogen bonds. Namely, 6MC may have two intramolecular hydrogen bonds, one between the OH at the C-3 position and an adjacent ether oxygen of the glucose ring, and the second between the ether oxygen at the substituted C-6 position and an adjacent OH at the C-2 position while 23MC may have both inter- and intramolecular hydrogen bonds at the C-6 position.

Figure 8 shows change of IR spectra for 6MC accompanied by the deuteration process. When the deuteration procceded, OH stretching vibration around 3470 cm-1 decrease, and instead the bands at 2559 cm^{-1} due to OD at the C-2 and C-3 positions appeared and then increased. The similar behaviors were observed for the OH and OD bands in the amorphous 23MC and cellulose. However, the deuteration was not completed regardless of the amorphous samples, giving unreacted OH groups. The amount of the unexchangable OH was 14.3, 10, and 13.3 % for 6MC, 23MC and amorphous cellulose, respectively. We assume that these unexchangable hydroxyl

(i) 2,3-di-*O*-methylcellulose (23MC):

Intra; OH at C-6 and OCH$_3$ at C-2.
Inter; OH at C-6.

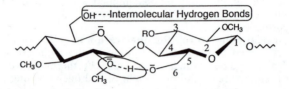

(ii) 6-*O*-methylcellulose (6MC):

Intra; OH at C-3 and O in the neighboring ring,
OH at C-2 and OCH$_3$ at C-6.
Inter; Nothing.

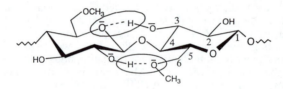

Figure 7. Hydrogen bonding formation proposed for 23MC and 6MC.

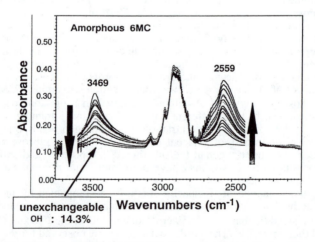

Figure 8. Changes of OH and OD IR absorption bands for monitoring the deuteration of amorphous 6MC.

bands may indicate the presence of domains involving the intermolecular hydrogen bonds which is already proposed as an amorphous model (*3*).

Here we will postulate that the OH-OD exchange reaction obeys pseudo-first order kinetics because of the large amount of D_2O to the amount of OH in each film sample. The general equation, $-d[OH]/dt=k[OH]$, can be employed. Thus the derived relationship, $\ln\{[OH]_t/[OH]_0\}=-kt$, was drawn in Figure 9 for the 3 amorphous samples, 23MC, 6MC and cellulose, respectively. Each of 3 samples showed the similar kinetic behaviors which were devided by three single reactions. In other words, three kinds of the exchange reaction (1, 2, and 3 in Figure 9) can be competitively coexisted. Table I shows the rate constants for each reaction.

Table I   Rate constants (k) from OH to OD for every single reaction in the deuteration processes of the amorphous 3 film samples.

	Time course	$k(h^{-1})$	R^2
Reaction 1			
23MC	0.08 - 5 hr	0.136	0.98
6MC	0.08 - 4 hr	0.247	0.99
Cellulose	0.08 - 2 hr	0.264	1.00
Reaction 2			
23MC	6 - 12 hr	0.085	0.98
6MC	5 - 13 hr	0.057	0.99
Cellulose	3 - 12 hr	0.075	0.98

R: Correlation coefficient.

The rate constant for reaction 1 was relatively different among the 3 samples. On the contrary the rate constant for 2 reaction seems to be similar among them. Of course, this is still speculative. In general adsorption and diffusion in crystalline polymers is considered to occur in amorphous regions. However, it is very complex because the structure for the crystalline regions may have somehow influences on it. In fact the morphological effects has not sufficiently studied yet. In this present case we have only to investigate the adsorption and diffusion of D_2O since the film samples exhibited amorphous. In another paper, details of this interpretation will be reported.

Considering that the samples exhibited amorphous for the X-ray measurements, amorphous regions in these cellulosics may be discriminated into 3 phases i) deuteration is fast, corresponding to reacion 1, ii) deuteration is slower, corresponding to reaction 2, and iii) almost undeuterated, corresponding to reaction 3. Interestingly the second rate constants among the three samples were very similar although the first rate constants for them differed. This indicates that the second one may correlate with the deuteration for the intramolecular hydrogen bonds (see Figure 7) in amorphous phases which are common interactions among the above three samples. Further investigation will be needed because the type of intermolecular hydrogen bonds found at the C-2, C-3 and C-6 hydroxyl positions is participating, to some extent, in determining the structure of amorphous cellulose (*3*).

Acknowledgments
We thank Dr. A. Isogai, Mr. E. Tsushima, Mr. R. Nakata, Dr. K. Takabe and Ms. N. Hayashi for providing the native cellulose samples.

Literature Cited
1. Kondo, T.; Sawatari, C.; Manley, R. St. J.; Gray, D. G. *Macromolecules* **1994**, *27*, 210.

182

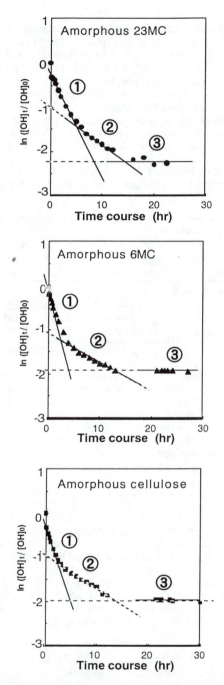

Figure 9. The relationships between decreasing ratio of OH groups, $\ln\{[OH]_t/[OH]_0\}$ and time, t, during the deuteration process for the three amorphous samples, 23MC, 6MC and pure cellulose.

2. Kondo, T. *J. Polym. Sci.B: Polym. Phys.* **1994**, *32*, 1229.
3. Kondo, T.; Sawatari, C. *Polymer* **1996**, *37*, 393.
4. Kataoka, Y.; Kondo, T. *Macromolecules* **1996**, *29*, 6356.
5. Chanzy, H.; Imada, K.; Vuong, R.; Barnoud, F. *Protoplasma* **1979**, *100*, 303.
6. Wada, M.; Sugiyama, J.; Okano, T. *Mokuzai Gakkaishi,* **1994**, *40*, 50.
7. Kondo, T.; Gray, D. G. *Carbohydr. Res.* **1991**, *220*, 173.
8. Kondo, T. *Carbohydr. Res.* **1993**, *238*, 231.
9. Coleman, M. M.; Painter, P. C. *J. Macromol. Sci., Rev. Macromol. Chem.* **1978**, *C16*, 197.
10. Tsuboi, M. *J. Polym. Sci. part C* **1964**, *7*, 125.
11. Smith, J. K. *J. Polym. Sci. part C* **1963**, *2*, 499.
12. Atalla, R. H.; VanderHart, D. L. *Science* **1984**, *223*, 283.
13. Horii, F.; Yamamoto, H.; Kitamaru, R.; Tanahashi, M.; Higuchi, T. *Macromolecules* **1987**, *20*, 2946.; Sugiyama, J.; Okano, T.; Yamamoto, H.; Horii, F. *Macromolecules* **1990**, *23*, 3196.
14. Sugiyama, J.; Persson, J.; Chanzy, H. *Macromolecules* **1991**, *24,* 2461.
15. Fengel, D. *Holzforschung* **1993**, *47*, 103.
16. Fengel, D.; Strobel, C. *Acta Polymer.* **1994**, *45*, 319.
17. Michell, A. J. *Carbohydr. Res.* **1993**, *241*, 47.
18. Tanahashi, M.; Goto, T.; Horii, F.; Hirai, A.; Higuchi, T. *Mokuzai Gakkaishi* **1989**, *35*, 654.
19. Nowak-Ossorio, M.; Gruber, E.; Schurz, J. *Protoplasma* **1976**, *88*, 255.
20. Rowland, S. P.; Howley, P. S. *Text. Res. J.* **1988**, *58*, 96.
21. Marrinan, H. J.; Mann, J. *J. Polym. Sci.* **1958**, *32*, 357.
22. Jeffries, R. *Polymer* **1963**, *4*, 375.
23. Taniguchi, T.; Harada, H.; Nakato, K *Mokuzai Gakkaishi* **1966**, *12*, 215.

Chapter 13

Molecular Weights and Molecular-Weight Distributions of Cellulose and Cellulose Nitrates During Ultrasonic and Mechanical Degradation

Marianne Marx-Figini

Polymer Division, INIFTA, National University, La Plata, Argentina

Cellulose and cellulose nitrate were subjected to depolymerization by ultrasonication and high-speed mechanical agitation (stirring). Both treatments resulted in significant, but different, depolymerization with the resulting degradation products having considerably narrower molecular weight distributions than the parent substrates. In all cases a level-off degree of polymerization was reached which was different for unsubstituted cellulose and the derivatives. The results obtained with unsubstituted cellulose are explained with the existence of "weak links" in cellulose backbones.

The treatment of polymers with ultrasound has received considerable attention in the past (1-6). Cellulose depolymerization by ultrasonification, by contrast, has been largely overlooked (7-8); additionally, this work predates the advent of modern and reliable cellulose molecular weight determination methodology. Since one of the unusual features of polymer degradation with ultrasound is the narrowing of the molecular weight distribution in addition to the apparent existence of a limiting level-off DP, it was of interest to reexamine cellulose (derivative) degradation by ultrasound in view of the potential of generating cellulose preparations with narrow molecular weight distributions. Alternatively, this task requires time consuming fractional precipitation for preparative size exclusion chromatography (SEC) experiments.

Likewise, cellulose degradation studies using high speed mechanical agitation (9-12) were performed at a time when reliable molecular weight methodology did not exist. Since an understanding of the molecular weight behavior of cellulose under mechanical stress can provide insight into the effect of industrial processing on macromolecular properties of cellulose, cellulose depolymerization during mechanical high speed stirring was investigated as well.

Experimental Section

I. Determination of Molecular Weights and Molecular Weight Distributions: Molecular weights were determined viscosimetrically as well as by size exclusion chromatography (SEC). The resulting molecular weight data were interpreted in terms of degrees of polymerization (DPs) so as to reach a universally applicable polymeric size parameter that is dependent only on the number of repeat units and not on the average weight or substitution pattern of repeat units. DP is independent of the nature and degree of substitution of the anhydroglucose repeat unit. Viscosimetric measurements were carried out in ethyl acetate and cupriethylenediamine (cuen) solution for cellulose nitrate and unsubstituted cellulose, respectively, according to Marx-Figini and Schulz (13). All measurements were performed with solutions whose concentration was adjusted so that η_{sp} was between 0.3 and 0.6. All intrinsic viscosity measurements were expressed as DP using Equations 1 to 4.

$$[\eta]_{Ea} = 5.70 \times DP^{0.76} \quad (1) \quad \text{and} \quad [\eta]_{cuen} = 2.29 \times DP^{0.76} \quad (2) \quad \text{for DP>950}$$

$$[\eta]_{Ea} = 1.06 \times DP \quad (3) \quad \text{and} \quad [\eta]_{cuen} = 0.42 \times DP \quad (4) \quad \text{for DP<950}$$

where Ea stands for ethylacetate. These equations were recently established using more than 60 representative light scattering and ultracentrifugation data (14). Adjusting the pertinent values of $[\eta]$ to standard conditions, these equations were found to produce considerable agreement for data from many different authors. In addition, the fact that the ratios of intrinsic viscosities of cellulose nitrate in two different solvents as well as those for cellulose nitrate and cellulose remain constant over the entire DP range serves as an indication that the exponent α in the Mark Houwink equation is invariable for all different solvents investigated (15).

Molecular weight determinations by SEC were performed using a Waters HPLC unit with UV-detector (model 441) and a set of ultrastyragel (10^6 Å) and microstyragel (10^5, 10^4, 10^3 Å) columns. THF served as solvent, and known cellulose nitrate samples were used as calibration standards. DP was derived by

$$\log DP = A - B V_e \quad (5)$$

where V_e is elution volume, and A and B are three different pairs of parameters for DP ranges between 100 and 8,000. Different constants for A and B were used for the DP ranges of 100 to 500; 500 to 1000; and 1000 to 10,000, and their mathematical derivation is described elsewhere (17, 18). The resulting DP values agreed well with DP_η and DP_n determined by viscosimetry and osmometry, respectively. (See reference 16 for more information.)

II. Molecular Degradation: *Ultrasonic Degradation* - Five nitrated cotton cellulose samples with DP_η between 400 and 7,500 and a degree of substitution of 2.90 ± 0.02 were subjected to sonication in ethylacetate solution at a concentration between 0.045 and 0.15%. The sonication apparatus was a Bandlin-Sonopuls HD-60 apparatus with a frequency of 20 kHz and a delivered power of 22 W. Temperature was maintained constant at 20°C. (Concentration was found to play an insignificant role in the molecular degradation). Samples taken after different sonication times were analyzed by viscosimetry and SEC. DP_n, DP_η, and DP_w values were calculated. In each case DP_η-values obtained by viscosimetry agreed well with those obtained by SEC.

High Speed Stirring - Six cotton cellulose samples with DP-values between 300 and 7,500 in N-butanol suspension (150 mg cellulose per 100 mL butanol) were subjected to a high speed rotary homogenizer operating at 13,500 rpm. Degradation experiments were performed at a constant temperature of -30°C, and the molecular weights of the filtered and dried degradation samples were measured in cuen as well as by SEC following nitration.

Results and Discussion

The macromolecular degradation of cellulose nitrate samples during exposure to ultrasound reveals that the degradation rate increases with increasing DP (Figure 1). As DP_0 declines, the effect of sonication time (t) diminishes. All cases seem to result in a limiting DP (DP_∞) below which no further degradation takes place. Similar results were reported for other polymers, principally low dispersity polystyrenes (5). Despite the large differences between the DP_0-values, all DP_∞-values lie in a narrow range, between DP_η 500 and 700. Samples having a DP_0 lower than the lowest DP_∞ do not degrade. In agreement with observations made with synthetic polymers (5), cellulose depolymerization proceeded faster, and to a lower level-off DP_∞, if delivered sonication power was raised from 22 to 37 W.

Corresponding observations were made with the depolymerization of suspended cellulose using high speed mechanical stirring (Figure 2). A limiting, level-off DP is reached following an exponential degradation rate. However, in contrast to degradation by sonication, mechanical degradation results in (a) a higher DP_∞-value; (b) DP_∞ is independent of DP_0; and (c) the decrease of DP is independent of the rate of stirring (8,000 or 13,500 rpm). The results suggest that

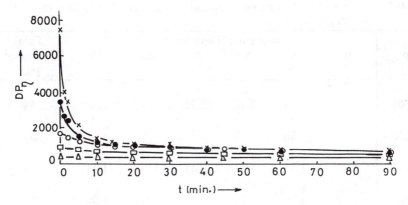

Figure 1. Behaviour of DP_η with sonication time. Key: × is $DP_\eta (0) = 7500$; O is $DP_\eta (0) = 1650$; ◁ is $DP_\eta (0) = 392$; ● is $DP_\eta (0) = 3430$; and ❏ is $DP_\eta (0) = 900$.

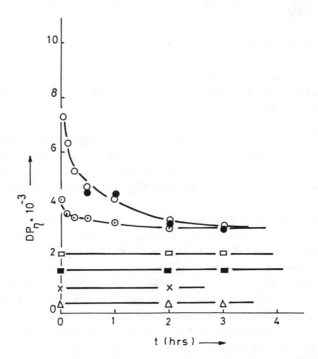

Figure 2. Behaviour of the DP_η with high-speed stirring time (rpm = revolutions per minute). Key: O, ❂, ❏, ■, ×, △ indicates rpm = 13,500; ● is rpm = 8000; O, ● is $DP_\eta (0) = 7300$; ❏ is $DP_\eta (0) = 2207$; × is $DP_\eta (0) = 915$; ❂ is $DP_\eta (0) = 4150$; ■ is $DP_\eta (0) = 1560$; △ is $DP_\eta (0) = 330$.

Table I: Cellulose Nitrate Polydispersity-Values (U)[1] as Related to Sonication Time.

$DP_{\eta(0)}$	Sonication Time (min)						Total Decrease[2] (%)
	0	5	10	20	45	90	
7500	1.66	1.33	1.32	1.32	1.34[3]	1.30	54
3700	1.89	1.86	1.39	1.44	1.43	1.38	57
1700	1.92	1.53	1.53	1.45	1.41	1.41	55
937	1.92	1.75	1.70	1.57	1.56	1.49	47
455	1.96	2.08	1.99	2.11	2.05	2.02	--

[1] $U = DP_w/DP_n$
[2] Of U
[3] Sonication time of 40 min.

cellulose degradation by stirring, like cellulose nitrate degradation by ultrasonication, does not proceed in a random scission process but by selective bond rupture. This is suggested by the behavior of the molecular weight distributions.

Polydispersity, U, (DP_w/DP_n) reveals a significant narrowing of the molecular weight distribution with sonication time (Figure 3). U was found to depend on both sonication and DP_0 (Table I). This suggests that molecular degradation proceeds more rapidly with high DP preparations than with lower ones.

The degradation behavior of macromolecules during ultrasonication is generally assumed to proceed by cleavage of the molecules in their center, or according to a Gaussian distribution on both sides adjacent to the center. For this case, bimodal DP distributions were predicted (on mathematical grounds) and also experimentally observed (2, 4). The molecular weight distribution curves obtained with cellulose in this study, however, are uniform rather than bimodal. Bimodal distributions were obtained only from nearly monodisperse polystyrenes (U <1.2) (2,4). The present cellulose nitrate samples showed initial distributions having a U of 1.6 to 2.0. Macromolecules with an initial polydispersity of U >1.5 are also predicted to produce degraded polymers with uniform distributions (19). This agrees with the current findings. It can therefore be concluded that ultrasonic degradation of cellulose nitrate proceeds via central scission or scission near the

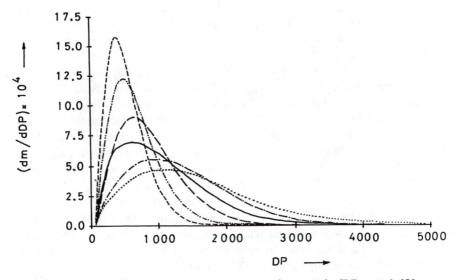

Figure 3. Distribution curves of a representative sample ($DP_{\eta(0)}$ = 1,650, $DP_{n(0)}$ = 943 at different times of sonication.
Starting sample (....); degradation time 5 minutes (-.-.-);
degradation time 10 minutes (——); degradation time 20 minutes (— —);
degradation time 45 minutes (-..-); degradation time 90 minutes (----).

center of the molecule. This is in contrast to synthetic polymers which depolymerize following a Gaussian distribution (2, 3, 19).

Native (with respect to DP) celluloses, derived by biosynthesis are monodisperse (20). Unprocessed cotton cellulose, however, is available only with a polydispersity of about 1.5 or 1.6 due to unavoidable degradation reactions during storage. This U still increases with industrial processing, and it eventually approaches a value of 2.0. Molecular degradation by ultrasonication produces uniform materials with reduced U-values. High speed stirring of cellulose suspensions produces DP_∞ values of nearly 3,200 regardless of DP_0. This chain length corresponds approximately to the number of anhydroglucose units between two "weak links" in a cellulose molecule of high DP (21). Such "weak links" have recently been revealed in studies on the acid catalyzed hydrolysis of cellulose (22). According to Marx-Figini and Coun-Matus (22), each native cellulose molecule (with DP_0 of approximately 13,000 (20)) has three to four "weak links." The distance between two weak bonds would therefore be approximately 3,200 monomer units. Under acid hydrolysis conditions, "weak links" break 10^6 times faster than regular glycosidic bonds. "Weak link" hydrolysis has a lower activation energy than that found for glycoside hydrolysis (22), but it is high enough to permit the assumption that there are native modifications in the chemical structure of the cellulose molecule which cause the rapid splitting. The alterations in the glucose unit may, for example, involve the occurrence of 2-deoxyglucosides or fructosides, or xylan-units, the hydrolysis rate constants of which are known to be 10^3 - 10^5 times larger than the corresponding β-1,4 glucoside. An analytical proof of the chemical nature of the alterations, however, appears impossible because of the very low fraction of the non-glucosidic units. Assuming that mechanical force only breaks "weak links," a narrowing of molecular weight distribution can be expected, and this was experimentally confirmed (21) (Figure 4).

Assuming that only "weak links" were cleaved during cellulose degradation, polydispersities should decline as depolymerization progresses. A comparison of experimental with theoretical data provides compliance with this model (Figure 5). The mathematically derived data (21) were derived in accordance with an equation system previously developed by Schulz and Husemann. (23). The data shown in Figure 5 are based on the starting values of DP and U obtained by SEC on the initial material. The data of Figure 5 also reveal that high speed mechanical stirring also produces reduction of U.

Since there is no definitive evidence as yet in support of a relationship between crystallite size and DP, no attempt was made to explain the mechanical degradation behavior with crystallite dimensions. Since former detailed investigations on the heterogeneous hydrolysis of cellulose (22) have shown that heterogeneous state fails to influence degradation behavior (by, for example, diffusion control), no effect of crystallite size on polydispersity index was expected.

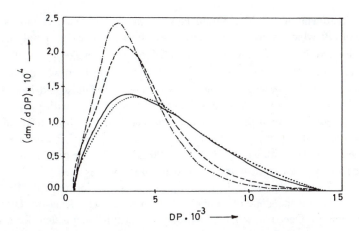

Figure 4. Distribution curves of a representative sample ($DP_{\eta(0)}$ = 7,500, $DP_{n(0)}$ = 4,630) after different times of high-speed-stirring.
Starting sample (...); degradation time 15 minutes (———);
degradation time 30 minutes (---); degradation time 2 hr (-.-.-).

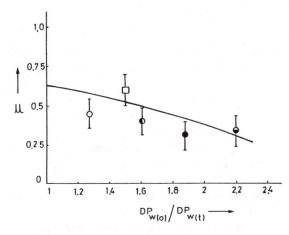

Figure 5. Polydispersity (U) as a function of degree of degradation. Continuous line: mathematical model.
Experimentally determined data points: -O-, -□-, -◐-, -●-, and -◒-
represent degradation times of 5 minutes (-O-), 15 minutes (-□-), 30 minutes
(-◐-), 2 hr (-●-), and 3 hr. (-◒-).

Conclusions

Treatment of cellulose nitrate in solution with ultrasound causes a rapid decline of DP which is accompanied by a significant narrowing in polydispersity. This is limited, however, to cellulose derivatives having a $DP_{\eta(0)}$ $<<700$. Native celluloses with $DP_{\eta\,(0)}$ $>10,000$ and U <1.4 already are narrowly dispersed as a consequence of biosynthesis. Sonication of these materials would indeed further reduce their polydispersity, but according to results obtained with polystyrene (3, 4), the degradation would provoke a bimodal distribution. This means that only polymers with an initial polydispersity of between 1.6 and 2.0, corresponding to cellulose derivatives with a $DP_{\eta(0)}$ between 900 and 8,000, are capable of generating degraded polymer solutions with narrow distribution values.

Suspended cellulose is degraded by mechanical action only if DP_0 $>3,000$. Degradation is independent of applied shear forces. The level-off DP-value is reached rapidly. Once reached, the limiting DP, which is independent from DP_0, no longer declines. Both observations, the existence of a level-off DP and the narrowing of U during degradation, suggest that mechanical shear forces degrade cellulose by cleaving "weak links" present in native cellulose. The results observed in this study are consistent with this hypothesis (21). The results therefore suggest that mechanical treatments of cellulose during industrial processing, including storage, are responsible for significant molecular degradation.

Literature Cited

1.   Jellinek, H. H. G. and G. White. J. Polym. Sci. VI, N 6, (1951) 757.
2.   Glynn, P. A. R. B. M. E. Van der Hoff. J. Macromol. Chem. A7/8(1973) 1695.
3.   Van der Hoff, B. M. E. and P. A. R. Glynn. J. Macromol. Sci.-Chem., A8/2 (1974) 429.
4.   Niezette, J. and A. Linkens. Polymer 19 (1978) 939.
5.   Price, G. J. and P. F. Smith. Polymer International 24 (1991) 159.
6.   Price, G. J. and P. F. Smith. Polymer 34 (1993) 4111.
7.   Thomas, B. B. and W. J. Alexander. J. Polym. Sci. 15 (1955) 361.
8.   Thomas, B. B. and W. J. Alexander. J. Polym. Sci. 25 (1957) 285.
9.   Grohn, H. J. Polym. Sci. 30 (1958) 551.
10.   Grohn, H. and W. Deters. Faserforsch. und Textiltechn. 13 (1962) 544.
11.   Deters, W. and H. Grohn. Faserforsch. und Textiltechn. 14 (1963) 58.
12.   Ott, K. L. J. Polym. Sci. Part. A2 (1964) 973.
13.   Marx-Figini, M. and G. V. Schulz. Makromol. Chem. 31 (1959) 140.
14.   Marx-Figini, M. and G. V. Schulz. Makromol. Chem. 54 (1962) 102.
15.   Marx-Figini, M. Angew. Makromol. Chem. 72 (1978) 161.

16. Soubelet, O., M. A. Presta and M. Marx-Figini Appl. Macromol. Chem. 175 (1990) 117.

17. Vribergen, R. K., A. A. Soeteman, G. A. M. Smit. J. Appl. Polym. Sci. 22 (1978) 1267.

18. Andreetta, A. H. and R. V. Figini. Appl. Macromol. Chem. 93 (1981) 143.

19. Ballauf, M. and B. A. Wolf. Macromolecules 14 (1981) 654.

20. Marx-Figini, M. J. Polym. Sci. C', N 28 (1969) 57.

21. Marx-Figini, M. and R. V. Figini. Appl. Macromolec. Chem. 224 (1995) 179.

22. Marx-Figini, M. M. Coun-Matus, Makromol. Chem. 182 (1981) 3603.

23. Schulz, G. V., E. Husemann, Z. Phys. Chem., Abt. B, 52 (1942) 23.

Chapter 14

Depolymerization of Cellulose and Cellulose Triacetate in Conventional Acetylation System

Shu Shimamoto, Takayuki Kohmoto, and Tohru Shibata

Research Center, Daicel Chemical Industries, Ltd., 1239 Shinzaike, Aboshi-ku, Himeji, Hyogo, 671-12 Japan

In a conventional system for the acetylation of cellulose, acetic acid, acetic anhydride and sulfuric acid act as a diluent, an acetylation reagent and a catalyst, respectively; the sulfuric acid acts as a catalyst not only for acetylation but also for depolymerization. The depolymerization behaviors of cellulose and cellulose triacetate were studied so that the degree of polymerization of the final product could be predicted. Model experiments in which the acetylation reagent was absent revealed that the depolymerization of cellulose in earlier stages of the reaction proceeds considerably faster than that of cellulose triacetate, and depolymerization of cellulose triacetate is random whereas that of cellulose is not. Simulations of the final degree of polymerization of the product were carried out using the activation energies obtained by the model experiments. The estimated degree of polymerizations showed good agreements with the experimentally obtained ones when the depolymerization rate of cellulose was assumed to be proportional to the amount of sorbed sulfuric acid.

Among known systems for the acetylation of cellulose, one comprised of acetic acid, acetic anhydride and sulfuric acid has been widely employed in commercial productions of cellulose triacetate (CTA) and cellulose diacetate because of its efficiency (1). Sulfuric acid catalyzes depolymerization as well as acetylation in the system. Although there are some studies on the depolymerization during the pretreatment (2) and acetylation (3), a quantitative prediction of degree of polymerization (DP) of the final product based on depolymerization kinetics of cellulose and the intermediate product has not been reported as far as we know. In an attempt to accomplish this, depolymerization behaviors of cellulose and CTA were studied in the system in the absence of the acetylation reagent in this paper.

Result and Discussion

Evaluation of DP of cellulose. Although there are several known methods to evaluate DP of cellulose, we established an novel method in which cellulose was converted to CTA without remarkable degradation in order to apply the same Mark-Houwink-Sakurada equation to the evaluation of depolymerization of cellulose and

194

CTA. That was to avoid a systematic error caused by applying different Mark-Houwink-Sakurada equations to cellulose and CTA.

In order to examine the degradation during the acetylation using a lithium chloride / dimethyl acetamide mixture, cellulose regenerated from CTA with hydrazine was converted to CTA again by the method described in the experimental section. The original and the resulting CTA were analyzed by a GPC low angle laser light scattering (GPC-LALLS) technic. The results are shown in figure 1 as molecular weight distribution curves. The distribution curves were almost identical although there was a slight difference at the higher molecular weight region, and the difference in weight average molecular weight was quite small, only 8 %. From these results, it was concluded that there is no remarkable degradation of cellulose during the acetylation using a lithium chloride / dimethyl acetamide mixture.

Depolymerization of cellulose and CTA in acetic acid / sulfuric acid system. Depolymerization of cellulose and CTA in an acetic acid / sulfuric acid system in the absence of an acetylation reagent were studied separately. Cellulose degraded much faster than CTA in earlier stages of the reaction as shown in figure 2.

It is well known that number average DP at reaction time t ($DP_{n,t}$) in a random depolymerization is given by the following kinetic equation $(3,4)$.

$$\frac{1}{DP_{n,t}} = \frac{1}{DP_{n,0}} + k \cdot t \tag{1}$$

$DP_{n,0}$ is the initial DP_n and k is the rate constant of depolymerization. Weight average DP (DP_w) at reaction time t can be expressed by the similar manner by assuming a Schulz - Zimm type distribution of DP and the polydispersity factor of 2 (5).

$$\frac{1}{DP_{w,t}} = \frac{1}{DP_{w,0}} + \frac{1}{2} \cdot k \cdot t \tag{2}$$

Equation 2 means that a plot of DP_w versus time t gives a linear relationship in a random depolymerization. CTA gave an almost straight line in the plots whereas cellulose did not, as shown in figure 3. These results suggest that depolymerization of CTA in the system is random but that of cellulose is not.

In figure 3a, temperature dependence of CTA depolymerization is clearly seen ; the higher the temperature is the faster the depolymerization rate is, as expected. The activation energy of the CTA depolymerization obtained by a plot of the logarithmic rate constant versus the reciprocal temperature was 13.7 kcal/mol. On the other hand, cellulose depolymerization showed much smaller dependence on temperature (figure 3b). It is known that sulfuric acid is sorbed on cellulose in an acetic acid / sulfuric acid system (6). Malm et. al. showed that the depolymerization rate of cellulose correlates with the quantity of sulfuric acid sorbed on cellulose (6). Temperature dependence of sulfuric acid sorption is shown in figure 4. The amount of sulfuric acid sorbed on cellulose decreased remarkably with increasing temperature. The small temperature dependence of cellulose depolymerization could be explained by the temperature dependence of sulfuric acid sorption.

Simulation of DP of the final product in actual acetylation system. Simulations of the DP of the final product in an actual acetylation system comprised of acetic acid, acetic anhydride and sulfuric acid were attempted by using the above mentioned results.

In order to obtain a kinetic equation for the non-random depolymerization of cellulose, it was assumed that cellulose was composed of two regions both of which were depolymerized by random manners. We tentatively call the faster depolymerizing

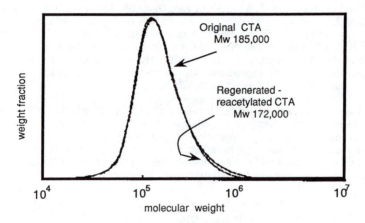

Figure 1. Molecular weight distribution of original and regenerated - reacetylated CTAs.

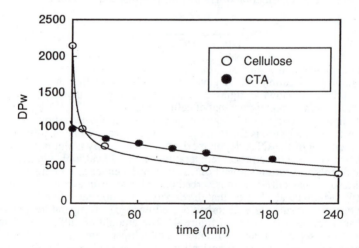

Figure 2. Plots of DP_w of cellulose and CTA vs. time in acetic acid / sulfuric acid system at 50°C.

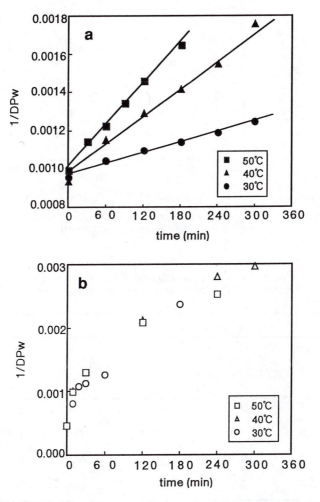

Figure 3. Plots of 1 / DP_w vs. time in acetic acid / sulfuric acid system.
a : CTA. b : cellulose.

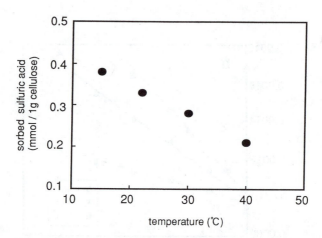

Figure 4. Effect of temperature on amount of sulfuric acid sorbed on cellulose in cellulose / acetic acid / sulfuric acid system (1 g / 43 g / 0.45 mmol).

Table I. **Comparison of experimental DPw and calculated ones for acetylation of cellulose (DPw 2,100) in acetic acid / acetic anhydride / sulfuric acid system.**

No.	Reaction conditions			DPw		
	Liquid to solid ratio	Amount of sulfuric acid, g / 100g cellulose	Temperature, °C	Experimental	Calc. 1	Calc. 2
1	83.8	9.5	40	324	303	326
2	7.0	8.0	40	715	114	780
3	6.0	4.2	60	609	141	631
4	6.0	2.1	60	625	262	757

Calc. 1 : assuming that the rate constants of depolymerization of cellulose were proportional to the amount of total sulfuric acid.
Calc. 2 : assuming that the rate constants of depolymerization of cellulose were proportional to the amount of sorbed sulfuric acid.

region amorphous and the slower one crystalline in this paper. In this case, DP_w at time t is given in the following equation.

$$DP_{w,t} = \frac{\alpha}{DP_{w,0}^{-1} + 0.5 \cdot k_{crys} \cdot t} + \frac{1-\alpha}{DP_{w,0}^{-1} + 0.5 \cdot k_{amo} \cdot t} \tag{3}$$

Where, α is the weight fraction of the crystalline region and k_{crys} and k_{amo} are the rate constants of depolymerization of the crystalline and amorphous regions. The rate constants of depolymerization were determined by a multi-variable fitting method using the experimental results. In the fitting procedure, α was adjusted at 0.3 to obtain the best fitting. The activation energies of depolymerization, obtained by plots of the logarithmic rate constant versus the reciprocal temperature without considering the variation of the sulfuric acid sorption with temperature, were 0.8 kcal/mol and 6.7 kcal/mol for the crystalline and amorphous region, respectively. However, they were calculated to be 8.3 kcal/mol and 14.2 kcal/mol for the crystalline and amorphous regions by assuming that the depolymerization of cellulose was propotional to the amount of sorbed sulfuric acid. While the latter set of activation energies seemed more reasonable, both of the sets of activation energies were used in the following simulation.

The following equation was employed to simulate the DP of the final acetylation product.

$$DP_{w,t} = \frac{\int_0^t F(t,\tau) \cdot G(\tau) d\tau}{\int_0^t G(\tau) d\tau} \tag{4}$$

In equation 4, the function $F(t,\tau)$ represents the $DP_{w,t}$ of the molecules converted from cellulose to CTA at reaction time τ. The function $G(\tau)$ stands for the weight fraction of the molecules converted from cellulose to CTA at reaction time τ.

The simulation results obtained under various reaction conditions are listed in Table I with experimental results. The values in the column of Calc. 1 were obtained by assuming that the rate constants of depolymerization of cellulose and CTA were proportional to the amount of total sulfuric acid. The values in the column of Calc. 2 were obtained by assuming that the depolymerization rate of cellulose was proportional to the amount of sorbed sulfuric acid. The latter simulation gave much closer values to the experimental results than the former did.

Conclusion

Depolymerization of cellulose is much faster than that of CTA in earlier stages of the reaction. Depolymerization of CTA in the acetylation system is random, whereas that of cellulose is not. The estimated degree of polymerizations agreed well with the experimentally obtained ones when the depolymerization rate of cellulose was assumed to be proportional to the amount of sorbed sulfuric acid.

Experimental

A prehydrolyzed kraft pulp and CTA with degree of substitution of 2.9 were used throughout this study. DP_w of these materials evaluated by the method described below were 2,100 and 1,000 for prehydrolyzed kraft pulp (cellulose) and CTA, respectively. The polydispersity factors evaluated by a GPC-LALLS technic were 2.3 for cellulose and 1.9 for CTA.

The general procedure of depolymerization experiment of cellulose was as follows. To a suspension comprised of 0.5 g of cellulose and 21 g of acetic acid was added 0.4 g of acetic acid containing 0.25 mmol of sulfuric acid and it was retained at an isothermal condition for a required period. The reaction was terminated by an addition of sodium acetate dissolved in an acetic acid / water mixture. The degraded cellulose was recovered by filtration and washed with water, saturated aqueous sodium hydrogencarbonate and water, successively. The depolymerization experiment for CTA was carried out at the same conditions with a solution of CTA in acetic acid.

DP of the degraded cellulose was evaluated by means of viscometry after it was acetylated by the following procedure. A degraded cellulose sample was washed with dimethyl acetamide by using a sintered glass filter to remove water. It was dissolved in a lithium chloride / dimethyl acetamide mixture at 100 °C for 3 hours then acetylated at 60 °C for 5 hours by adding pyridine and acetic anhydride to the solution. CTA as the reaction product was recovered by pouring the reaction mixture into an excess amount of water, then it was washed carefully with water several times and dried *in vacuo*. The limiting viscosity number ($[\eta]$) of the resulting CTA in dimethyl acetamide was determined by a usual extrapolation method. The viscosity average DP was calculated from $[\eta]$ by the equation established by Kamide et. al. (7). The DP_w was assumed to be the same as the viscosity average DP.

Literature Cited

1. Malm, C. J.; Hiatt, D. G. *Cellulose and Cellulose Derivatives* ; Interscience Publishers, 1954; 763 - 824.
2. Malm, C. J.; Barkey, K. T.; Lefferts, E. B.; Gielow, R. T. *Ind. Eng. Chem.* **1958**, *50*, 103.
3. Frith, W. C. *Tappi* **1963**, *46*, 739.
4. af Ekenstam, A. *Bericht* **1936**, *69*, 553.
5. Saito, O. *Statistical Properties of Polymers*; Publisher of Chuoh University, 1992; 149 - 158.
6. Malm, C. J.; Barkey, K. T.; May, D. C.; Lefferts, E., B. *Ind. Eng. Chem.* **1952**, *44*, 2904.
7. Kamide, K.; Miyazaki, Y.; Abe, T. *Polym. J.* **1979**, *11*, 523.

Chapter 15

Progress in the Enzymatic Hydrolysis of Cellulose Derivatives

B. Saake, St. Horner, and J. Puls

Institute of Wood Chemistry and Chemical Technology of Wood Federal Research Centre of Forestry and Forest Products, 21031 Hamburg, Germany

Single enzymes of the cellulase complex are efficent tools for understanding the biodegradation of cellulose derivatives. Sensitive size exclusion chromatography makes it possible to detect minor fragmentations. The accessibility of cellulose derivatives is clearly a function of the degree of substitution (DS). In addition, charge and size of the substituents and the substituent distribution play major roles for enzymatic attack. Whereas carboxymethyl cellulose of a certain DS was fully resistant against endoglucanase action, a methylcellulose of the same DS was markedly fragmented. Cellulose acetate was taken as an example for demonstration of the impact of esterases besides cellulases in biodegradation. The presence of acetyl esterase enabled the endoglucanase to degrade cellulose acetate much faster than when this type of enzyme was absent.

There is a continuous interest in the enzymatic hydrolysis of cellulose derivatives, starting with the pioneering work of Husemann (*1*) and Reese (*2*). Whereas Husemann was mainly interested in the location of the hydrolysis, Reese's main focus was to prevent a microbial attack to cellulose. One of the means to prevent an enzymatic degradation was identified to consist of an exchange of the free OH groups of the anhydroglucose units by bulky substituents. In contrast to the early research, today's concern is at least partly the conservation of the biodegradability of cellulose derivatives, i.e., cellulose should be substituted by certain derivatives only to the extent that the biological attack is not impaired. However, early papers as well as recent publications discuss the action of cellulases against cellulose derivatives in order to draw conclusions about the location of the substituents within the anhydroglucose units, as well as about the substituent distribution along the cellulose chain. Unfortunately, the early work was exclusively performed using enzyme cocktails of different compositions and origins; for this reason this overview was supplemented by additional practical experiments conducted in the authors' laboratory. In these experiments a mono-component cellulase, namely a fungal endoglucanase. was used.

Analytical Methods for Measuring the Enzymatic Degradation of Cellulose Derivatives. The analytical methods mainly used to follow the enzymatic hydrolysis of water-soluble cellulose derivatives are based on measuring changes in viscosity, the amounts of glucose, and reducing sugars. From these data only indirect conclusions about the length of cleaved chains and the relative amount of derivatized cellulose fragments can be drawn.

In more recent investigations the enzymatic action was visualized by changes in the molecular weight distribution, as revealed by size-exclusion chromatography (SEC). The main problem with SEC-characterization of polyelectrolytes such as CMC and the separation of their enzymatic degradation products was the selection of an appropriate eluent. The nature of the buffer and the ionic strength were essential factors in avoiding unwanted interaction with the stationary phase material (3). In order to differentiate the DS, ^{13}C-NMR spectroscopy was used (4).

Demeester et al. could demonstrate the usefulness of light scattering for the detection of the enzymatic action (5). The cellulolytic degradation of cellulose was also considered as a sample pretreatment prior to the structural characterization of cellulose derivatives (6). This method, however, could not be applied for high DS samples.

The Cellulose Degrading Enzyme System. The complete hydrolysis of natural cellulose demands the action of exoglucanases (also called cellobiohydrolases, CBH), endoglucanases (EG), and ß-glucosidases. Only a few microorganisms produce the complete set of cellulases for efficient degradation of insoluble cellulose. The general term cellulase usually refers to this complete set of cellulolytic enzymes. According to the current hypothesis, EGs initialize the attack on cellulose by randomly hydrolyzing internal bonds in amorphous regions of the cellulose. The action of EG thus produces new chain ends, which then become available for CBHs, which are believed to hydrolyze the chains from both the reducing (CBH I) and the non-reducing end (CBH II), whereby cellobiose is liberated (7,8). Finally, ß-glucosidases hydrolyze cellobiose into glucose. The different cellulases have been shown to hydrolyze cellulose synergistically.

The proposed degradation pathway, as well as the synergy, has been the subject of extensive discussion during recent years. Thus the present classification of specific cellulases as being either CBH or EG seems dubious. Today cellulases are classified for their amino acid sequence, followed by a classification by means of hydrophobic cluster analysis, HCA, in addition to classification according to their action against specific substrates (9). The HCA considers the different charges of amino acids and indentifies hydrophobic clusters along the two-dimensional sequences. Based on this information a three-dimensional structure can be predicted. This method is especially useful for the detection of similar folds in different enzymes with low sequence identity. Of the 11 cellulase families currently classified (10), fungal cellulases are found in six.

The only true exoglucanase structures solved are the two CBHs of the fungus *Trichoderma reesei*. The three-dimensional structures of the catalytic domain of *T. reesei* CBH II (family 6) and CBH I (family 7) revealed that the active site of both enzymes was identified to be situated in a tunnel ranging through the whole domain and formed by stable surface loops (11,12). It therefore seems that the active site tunnel is a general feature of exoglucanases. Indeed, exoglucanases are believed to be unable to attack cellulose derivatives, and the existence of the active site inside a tunnel could explain why especially bulky cellulose derivatives could get stuck inside the tunnel. Thus, EGs should have similar overall folds, but the active site should be more open, allowing a random hydrolysis of the cellulose (*11*). The three-dimensional structure of the *Thermomonospora fusca* EG E2 (*13*), belonging to the same family 6 as CBH II of *T. reesei*, confirmed this hypothesis. A similar structure was found for EG I (family 7) from *Humicola insolens* (*14*). EG V from *Humicola insolens* belongs to family 45 and cleaves 1,4-ß-glucosidic linkages with inversion of configuration, whereas EG I catalyses cleavages with retention of configuration. As a general feature, the active sites of endoglucanases were found to be located in long open grooves.

The structure of EG V has been solved, and it was found that this protein has 7 subsites (A to G) for binding with its substrate, the cleavage taking place between subsites D and E. On the basis of mutation experiments of the active site of EG V, there was strong evidence that an aspartate near subsite E, sitting in a predominantly hydrophobic environment, acts as the proton donor in the hydrolysis mechanism, whereas another aspartate functions as the base, activating the nucleophile. The active site residues of the *H. insolens* endoglucanase I are two glutamates and an aspartate, the latter functioning as the proton donor (*14*). This enzyme has the negative charge of its active site in common with EG V of *Humicola insolens*. In all cases investigated so far, the major role of the carboxylic amino acids glutamate and aspartate for scisson of the polysaccharide has been identified (*15*).

Accessibility of Cellulose Derivatives to Enzymes. It is well known that the accessibility of cellulose to enzymatic hydrolysis is dependent upon physical and structural features of the polymer. The relevant features are crystallinity, degree of swelling, solubility, and presence of other structural components (lignin and hemicelluloses). Cellulose derivatives are prepared by exchanging hydrogen atoms on primary and secondary hydroxyl groups with various functional groups such as methyl, carboxymethyl, diethylaminoethyl. Modification of cellulose usually makes it non-crystalline and in many cases soluble. Both factors increase the susceptibility of the cellulose to enzymatic hydrolysis. The susceptibility of cellulose derivatives to enzymatic hydrolysis increases as the substrate becomes less crystalline and more water-soluble. Water-solubility is dependent upon the degree of substitution (DS). The DS value at which complete water-solubility is achieved ranges from DS 0.4 to 0.7, depending on the solvation capacity and on the pattern of substitution (*16*). However, when on the average more than one substituent occurs per each anhydroglucose unit, the enzymatic hydrolysis rate decreases, and higher DS values result in a complete inertness of the polymer (*16*).

When the degree of substitution approaches an average of one substituent per glucose unit, steric factors become important, and enzymatic hydrolysis is retarded. The International Union of Pure and Applied Chemistry has published recommendations on the measurement of endoglucanase activities. This commission recommended substrates such as carboxymethyl cellulose (CMC) of DS 0.7 and hydroxyethyl cellulose (HEC) of DS 0.9 - 1.0 (*17*). In connection with the use of CMC as a test substrate for endoglucanase activity measurement, it has been stated that besides the DS, the same origin (i.e., the same method for the preparation of the derivative) is absolutely necessary in order to ensure a reproducible distribution of the substituents (*18*).

Reasons why HEC is preferable to CMC or to other water-soluble substrates have been discussed (*19*). According to earlier reports (*20*) two or more contiguous non-substituted anhydroglucose units in CMC are necessary for an enzymatic scission of the polysaccharide chain. In contrast Ach claims that cellulose acetate of DS 2.5 is biodegradable, although degradation proceeds extremely slowly (*21*). For these reasons it seems advisable to differentiate the accessibility of cellulose ethers and cellulose esters.

Enzymatic Degradation of Insoluble Cellulose Derivatives. Only a few papers deal with the heterogeneous enzymatic hydrolysis of cellulose derivatives. Highly substituted and water-insoluble cellulose derivatives (i.e., cellulose triacetate) are regarded to be completely resistant to enzymatic attack, because of a combination of factors - a lack of hydrophilicity, reduced swelling, and the hindrance caused by spatial substituents (*22*).

Enzymatic Degradation of Cellulose Ethers. Most of the work published on the enzymatic degradation of cellulose derivatives was performed with HEC and CMC (*20, 23-26*). Some results have also been obtained with MC (*1*). From Husemann, it may be assumed that the small methyl substituents inhibit degradation to a lesser extent than do the more bulky carboxymethyl groups (*1*). Intensive studies on the enzymatic degradation of cellulose ethers have established that the degree and the uniformity of substitution are decisive factors for their accessibility (*2, 20, 26, 27*).

According to Reese (*2*), a scission of monosubstituted MC is possible. This finding was confirmed by Schuseil (*28*), whereas CMC requires a sequence of three non-substituted anhydroglucose units (*24*). Wirick (*20, 26*) came to the conclusion that glycosidic bonds within cellulose ethers are only enzymatically cleaved in the case that two neighbouring non-substituted anhydroglucose units or that eventually a non-substituted and a C-6 substituted anhydroglucose unit are present. This result is in accordance with Bhattacharjee and Perlin (*25*), who pointed out that a scission between non-substituted units as well as between a substituted and a non-substituted anhydroglucose unit seems possible depending on the position of the substituent.

According to Kasulke et al. (*29*), three non-substituted glucose units were required in order to achieve the liberation of glucose. In this case, both hydrolysis products carried a non-substituted anhydroglucose unit at the end.

Enzymatic Degradation of Cellulose Esters. In the hydrolysis of water-soluble cellulose acetates Kamide et al. (*30*) found this polymer to be degradable up to DS 1. The authors observed a gradual precipitation of high DS cellulose acetate in the course of enzyme incubation, indicating that short highly-substituted cellulose acetate blocks can exist in water-soluble cellulose acetate chains (*4*). According to Buchanan et al. (*31*), the biodegradable DS range of cellulose acetates is even wider. The authors found cellulose acetates with DS values between 1.7 and 2.5 to biodegrade by mixed culture systems, at least partly following deacetylation. A series of other esters besides acetates was synthesized by Glasser et al. (*32*). The biodegradability of cellulose esters using cellulolytic enzymes was found to depend on two factors: degree of substitution and substituent size. The cellulose esters had acyl substituents ranging in size from propionyl to myristyl and DS values between 0.1 and nearly 3.0. The maximum degree of acylation, that the enzyme could tolerate before the polymer became undegradable and that resulted in degradation in excess of 10 % by weight, ranged from DS 0.5 to at least 1.0, depending on the ester type. It became evident that the larger the substituent of the cellulose ester, the harder it became for the enzyme to recognize the macromolecule and degrade it.

Influence of Substituents and Degree of Substitution. A set of CMC of DS, ranging from 0.6 to 2.1 and different molecular weights were incubated with three cellulase preparations of different origin and purity (*33*). A decrease in enzymatic hydrolysis with increasing DS was observed for all enzyme activities. This decrease could be explained by the difficult access of the enzyme to the 1,4-ß-glycosidic linkage due to the steric hindrance from the carboxyl groups, located mainly at C-2. The action of cellulase enzyme preparations on CMC samples of DS 0.7 and varying MW ranging from 0.55 to 8.71 x 10^6 g/mol was not affected by the large difference in molecular size. Complete hydrolysis was obtained with a purified cellulase at hydrolyzing CMCs with DS up to and including 0.9. Philipp et al. (*18*) reported on two different phases in the enzymatic degradation of CMC, both of them first-order reactions. The first (quick) reaction was explained by the combined action of endo- and exo-enzymes, which was favoured by the presence of longer sequences of non-substituted anhydroglucose units. The second (slower) phase was explained by sporadic EG-mediated "hits", which were made possible by shorter (possibly two) non-substituted units.

Influence of Derivatization Procedure. Cellulose derivatives are obtained mainly by heterogenous derivatization, which is normally considered to yield a statistic distribution of substituents. Cellulase hydrolysis yields large fragments from the more highly substituted regions, whereas small fragments are obtained from those regions which had a lower degree of substitution (*16*). Philipp et al. (*34*) could clearly show that the enzymatic hydrolysis of CMC in the DS range of 0.5 to 0.8 depends not only on the DS, but also on the procedure applied for preparing samples, with different distributions of substituents along and between the polymer chain. A CMC sample, which has been prepared after solubilization of the cellulose, excels by low values for liberated glucose in comparison to a CMC sample prepared in a heterogenous reaction system. Beyond a DS of 0.7 this value seems to be the determining factor for the liberation of reducing sugars viz. glucose (in case sufficient ß-glucosidase activity is present). In this higher DS range factors derived from the procedure applied for the production of the derivatives seem to be of minor importance. Philipp et al. (*34*) draw the conclusion that longer sequences of non-splitable linkages were present in higher DS CMCs. For the DS region 0.5 to 0.7, average sequences of derivatized anhydroglucose units of 4 to 6 were given, whereas in the DS region >1 averages of up to 40 carboxymethyl-glucose blocks were calculated.

Recently Heinze at al. stated that CMC samples, which were synthesized via an induced phase separation, contained a significantly higher amount of both tricarboxymethylated and unsubstituted units than those obtained in a slurry of cellulose in isopropanol/water at comparable DS values (*35*). This finding pointed out that it is not adequate to automatically assume a homogeneous substituent distribution from a homogenous reaction mixture nor a heterogenous distribution from a heterogenous system. Since the substituent distribution along the polymer chain is difficult to access, detailed insight on its impact on enzymatic degradation is still missing.

Influence of the Position of the Substituents. Little is known about a possible toleration of endoglucanase activity for a certain position carrying a substituent. The three possible positions for derivatization in cellulose are the hydroxyl groups at C-2, C-3, and C-6. The experimentally determined molar distribution of carboxymethyl groups in CMC has been reported to be 2:1:2.5 for C-2, C-3, and C-6 hydroxyls (*36*). While the C-6 primary hydroxyl is generally thought to be the most reactive in this particular system, this is not always so. There is considerable evidence that the C-2 hydroxyl group is the most acidic in cellulose (*37*). As a result, many equilibria and rate-controlled reactions that involve cellulosic alkoxide ions appear to favour this site. NMR studies of substituent distribution indicate C-2 > C-6 > C-3 (*6*). The finding that reducing end residues liberated during enzymatic hydrolysis of cellulose were never substituted at the 2-position (*6*) is a strong indication that this position seems to be needed for binding of the enzyme protein to the substrate.

The Influence of the Type of Substituent. The susceptibility of cellulose derivatives to cellulose hydrolysis is also dependent on the type of the substituent (*25*). The derivatization will certainly modify the charge of the polysaccharide and thus influence the interaction of the enzyme's active site with its substrate. Wirick (*20, 26*) investigated the enzymatic hydrolysis of CMC, MC, HEC, and hydroxypropyl cellulose. In a comparison of the above-mentioned cellulose derivatives, and in accordance with Husemann, CMC was the most resistant substrate.

Interestingly, Philipp and Stscherbina (*22*) reported a shift in the pH-optimum from pH 4 (DS 0.7) to pH 7.5 (DS 1.0) in the degradation of CMC using the enzyme system of *Penicillium citrioviride*. The nature of CMC to be a polyelectrolyte was also pointed out by Nicholson and Merritt (*38*). According to Philipp and Stscherbina (*22*) only the reaction velocity was reduced compared to the degradation of non-ionic cellulose derivatives. The extent of hydrolysis was not impaired by the charge of the substrate.

An increase in the negative charge resulted in a reduction of the hydrolysis rate, whereas the presence of positive charges markedly increased the enzyme activity (*39*). This finding could be verified and explained by recent investigations on the reaction mechanism (*15*) and the fine structure of endoglucanases (*14*).

Experimental

Substrates. All CMCs were gifts of Wolff-Walsrode (Walsrode/ Germany). The MC samples came from Kalle, a subsidary of Hoechst AG (Wiesbaden / Germany). The DS 0.7 CA was a gift of Hoechst-Celanese (Charlotte, N.C. / USA). The DS 0.9 to DS 2.9 CAs were gifts of Rhône-Poulenc Rhodia (Freiburg / Germany). The DS-values were determined by the manufacturers.

Enzymes. The mono-component endoglucanase preparation was an experimental product from a genetically modified *Aspergillus* strain, obtained from Novo Nordisk (Bagsværd / Denmark) and used after removal of low-molecular weight components by ultrafiltration. The *Aspergillus* enzyme mix was obtained from Nagase & Co (Tokyo / Japan) and used after ammonium sulfate precipitation and desalting.

Enzyme Treatments. 0.2 % solutions of CMC, MC, and the water-soluble DS 0.7 CA in 0.1 M sodium nitrite were incubated with 10000 nkat per mg substrate. The incubations were performed in an Eppendorf Thermomixer at 45 °C and 1200 rpm for 3 or 6 days. After completion, the samples were boiled for 3 minutes for protein denaturation. The precipitated enzyme protein was removed by centrifugation at 2000 g for 10 minutes. Substrate blanks were incubated for 3 days and treated correspondingly. The incubation of the water-soluble DS 0.7 CA with the enzyme mix was performed as a 1 % solution in bi-destilled water at 40 °C and 1200 rpm. The enzyme dosage was 500 nkat/mg. Enzyme protein was removed according to the procedure mentioned above. For SEC analysis, samples were diluted to a 0.2 % solution using sodium nitrite. Liberated acetic acid was determined using the Boehringer Test Combination Kit (Cat. No. 148 261 Boehringer, Mannheim / Germany).

For endoglucanase treatment of cellulose acetate higher in DS than 0.7, 1 % solutions or suspensions in bi-destilled water were incubated with 10000 nkat *Aspergillus* endoglucanase / mg substrate. The incubation was finalized by ammonium hydroxyde addition and incubation at 30 °C overnight for protein denaturation and acetyl saponification. The solutions were freeze-dried and afterwards stored in vacuum over P_2O_5. Blanks underwent the same procedure.

Carbanilation. 100 mg of saponified samples were suspended in 100 ml of pyridine and derivatized at 80 °C using 7 ml phenylisocyanate. After 48 h the reaction was stopped. A refinement procedure by direct evaporation of the pyridine (*46*) was applied as published previously (*40*).

SEC of Water-Soluble Derivatives. SEC was performed using sample concentrations of 0.2 % and sample volumes of 100 µl injected into three SEC columns, coupled in line (TSK G5000PW$_{XL}$, G4000PW$_{XL}$, G3000PW$_{XL}$, 300x7.8 mm each, and a G2500PW$_{XL}$ guard-column, 40x6 mm, TosoHaas, Stuttgart / Germany). The column temperature was kept constant at 40 °C. The mobile phase (0.4 ml/min) was 0.1 M sodium nitrite in water. The elution profiles were detected by changes in refractive index. The WINGPC 3.0 software (Polymer Standard Service, Mainz / Germany) was used for data aquisition.

SEC of cellulose carbanilates. 0.5 mg cellulose derivative per ml of stabilized tetrahydrofurane (THF) was shaken for 3 days for dissolution. Samples were then centrifuged for 2 h at 20 °C (16000 to 22000 g), and the supernatant was used for further analysis. SEC was performed using a Waters 510 pump, a Kontron 360 autosampler (100 µl), Waters Ultrastyragel 10^3 Å and 10^4 Å columns (300x7.8 mm each), a Spectra Physics SP8400 UV detector, a Shodex RI-71 detector, and WINGPC 3.0 software. The eluent was THF with a flow rate of 1 ml/min at 20 °C. Polystyrene standards were monitored at 254 nm and cellulose tricarbanilates at 235 nm. A calibration curve was obtained by a broad fit of two cellulose tricarbanilate samples with a molar mass of 210000 g/mol and 1000000 g/mol (*40*) to a polystyrene standard curve. The Mark-Houwink constants obtained by this procedure were K: 1.756 +10^{-3} ml/g and α: 0.890.

Results and Discussion

Endoglucanase Degradation of CMC. CMCs in a DS range from 0.6 to 2.4 were incubated for 3 and 6 days with an overdosage of a mono-component *Aspergillus* endoglucanase (10000 nkat/mg substrate), in order to reach the final possible stage of fragmentation. Substrate blanks were incubated for 3 days. Figure 1 demonstrates the hydrodynamic volume of the blank samples to be too high for a perfect separation in the chromatographic system. This is especially the case for the DS 0.6 and DS 0.8 samples, starting at an elution volume of 13.75 ml, with a pronounced steep shoulder. While most blanks had a narrow elution curve, the DS 1.2 and 1.6 samples showed a tailing in the lower molar mass region.

The intensity of endoglucanase degradation was strongly dependent on the DS. DS 0.6 and 0.8 samples gave a strong shift of the elution profile to the low-molecular region and displayed distinct signals for oligomeric compounds. The fragments which were washed from the columns at the elution volume of around 29.6 ml can be attributed to monomeric degradation products. It becomes evident from Figure 1 that the maximal possible fragmentation of the polymer was completed after 3 days of incubation.

DS 0.9 and 1.2 CMCs were also markedly fragmented by the endoglucanase, which is visualized by pronounced shifts of the elution profile to longer elution volumes (Figure 1). However, the degradation of these derivatives was less intense in both regions, the higher molar mass region and the oligomeric range. In the degradation of these samples, a prolongation of the reaction time gave rise to an additional fragmentation. Possibly the prolonged and slower phase can be explained by sporadic cuts between shorter non-substituted or less substituted regions, according to the explanation given by Kasulke et al. (*27*). In addition, tertiary conformation effects of the charged polymer could limit or slow down the accessibility of the enzyme to positions which principally can be cleaved.

The DS 1.6 sample shows only a minor shift in the elution curve and a small bimodality in the lower molar mass range after endoglucanase treatment. For the DS 2.4 CMC sample, the effect of the treatment on the elution curve was even more reduced. However, a small but significant bimodality demonstrates the availability of some positions along the CMC chain to the endoglucanase. This minor change had no effect on the elution curve of the main polymer peak. For both the DS 1.6 and DS 2.4 samples only minor changes occurred even after prolonged incubation times. These results are in accordance with Philipp et al. (*34*), who could demonstrate a reduction in viscosity for DS 1.7 CMC after incubation with a *Gliocladium* culture filtrate. Due to the close involvement of the carboxylates of two amino acids in its active site, the reduced activity of endoglucanase against the negatively charged CMC becomes explainable.

It is noteworthy that the intensity of the elution curve of the enzyme treated DS 2.4 sample was slightly higher than the substrate blank. It could be excluded that this phenomenom was derived from experimental errors. However, it was noted that the number of visible gel particles in enzyme-incubated samples was significantly reduced; this effect explains the higher yield of enzyme-treated samples. It was even more pronounced for the treatment of both MC and water soluble CA, discussed in the following paragraphs.

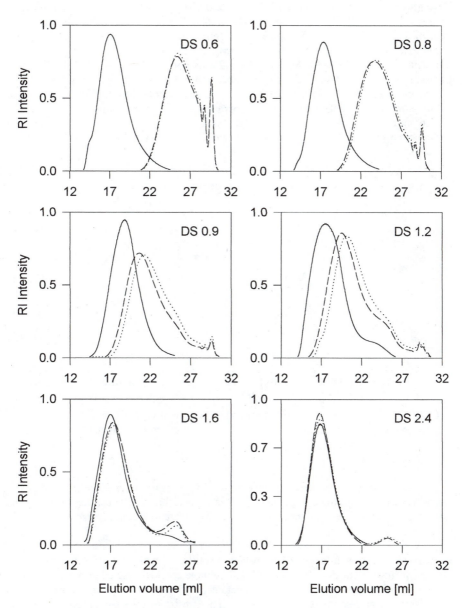

Figure 1. Influence of DS on the fragmentation of carboxymethyl-cellulose by *Aspergillus* endoglucanase, after 3 days and 6 days incubation, monitored by aqueous SEC.

blank: ——— 3 days: _ _ _ _ 6 days: ....

Endoglucanase Degradation of MC. It could be verified that endoglucanase fragmentation of MC in the range from DS 0.5 to 2.1 was much more effective then that of CMC. The DS 0.5 sample was more or less completely degraded into oligomeric products (profile not shown). After 3 days of degradation, the fragmentation of the DS 1.5 MC in Figure 2 was comparable to the DS 0.6 CMC in Figure 1. The MC curve had an even more intensive shift to low molar mass and oligomeric products. At 31.2 to 31.3 ml, the elution curve falls almost rapidly, since monomeric MC products elute into the separation limits of the column.

In another comparison, the DS 2.1 MC was significantly fragmented, whereas the elution curve of the DS 2.0 CMC was only slightly modified, compared to the substrate blank.

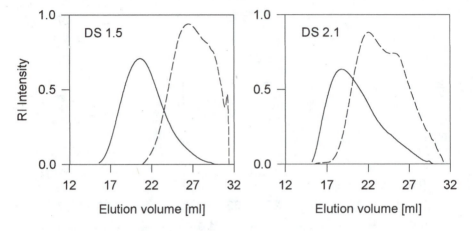

Figure 2. Influence of DS on the fragmentation of methyl cellulose by *Aspergillus* endoglucanase (3 days incubated) monitored by aqueous SEC. blank: ——— enzyme treated: ----

All endoglucanase-treated and centrifuged MC samples have in common that the intensities of the elution curves were by far higher than the substrate blanks. This phenomenon underlines the possibility of endoglucanase-mediated improvements in solubility.

Endoglucanase Degradation of Cellulose Acetate. Cellulose acetate fragmentation was performed with DS 0.7 to DS 2.9 samples, which included heterogeneous hydrolysis of water-insoluble cellulose acetate powder of DS > 1 and homogeneous hydrolysis of water-soluble samples DS < 1. The substrates have been prepared by acid saponification of cellulose triacetate. This is the reason that the molecular weight of the starting material was reduced with decreasing DS (Table I). Due to the heterogeneous hydrolysis of most of the cellulose acetate substrates, the material could not be directly analyzed by gel permeation chromatography, but had to be gently saponified using ammonium hydroxide prior to tricarbanilation and SEC analysis in tetrahydrofurane (*40*).

Table I. Homogeneous and heterogeneous endoglucanase hydrolysis of cellulose acetate samples as revealed by SEC of the corresponding carbanilates

DS	Starting material			Endoglucanase treated material		
	M_W	M_W/M_N	DP	M_W	M_W/M_N	DP
0.9	16 000	1.5	31	1 860	1.1	4
1.2	44 000	1.8	85	2 500	1.2	5
1.6	72 000	1.7	138	14 000	5.1	27
1.7	48 000	1.7	92	26 000	1.2	50
1.9	98 000	2.0	189	52 000	11.0	100
2.5	164 000	3.2	316	159 000	3.0	306
2.9	201 000	4.1	387	204 000	3.3	394

Due to the surplus of endoglucanase activity used in this study, the DS 0.9 sample was extensively degraded. Caused by acid saponification prior to the hydrolysis step, the partly deacetylated material was already considerably reduced in degree of polymerization (DP = 31); the endoglucanase treatment, however, caused a further degradation by homogeneous hydrolysis to DP = 4. A similar result was obtained for the DS 1.2 sample, which was degraded from molecular weight (M_W) 44000 to M_W 2500, equivalent to an average DP of 5.5, indicating a reasonable accessibility of both cellulose acetate samples to endoglucanase action. The accessibility was a clear function of the DS. From DS 1.6 on, the degradation by endoglucanase action was considerably retarded, which was certainly caused by two facts: the water-insolubility of the material and the shielding by increasing amounts of acetyl substituents. In spite of the unfavourable conditions, the material was still markedly degraded up to a DS of 1.9 (from M_W 98000 to M_W 52000). Cellulose acetate of DS 2.5, which is normally used for technical applications, was more or less resistant.

On first glance, these results might be not fully consistent with the results given by Buchanan et al. (31), who found cellulose acetate up to DS 2.5 to be biodegradable. However, these authors used a mixed culture system instead of a single, cell-free enzyme. It was noted in their investigations that cellulose acetate degradation was followed by deacetylation. In our own experiments, no acetyl was released by the *Aspergillus* endoglucanase.

Acetyl Esterase Involvement in Cellulose Acetate Fragmentation

In the search for suitable enzyme preparations, it was found that most commercial cellulase preparations had the capability of releasing varying amounts of acetyl groups. Due to the fact that the same endoglucanase dosage (tested with HEC as substrate) was used, it could be concluded that acetyl release is not a common feature of endoglucanase activity, but must be deduced to a separate enzyme. It is known from earlier studies that acetyl xylan esterase (41) and acetyl mannan esterase (42) are common features in hemicellulolytic enzyme systems. From more detailed work it is known that some esterases are highly specific, whereas others are not (43). Due to the man-made origin of cellulose acetate, nature will not have provided microorganisms with specific cellulose acetate esterases, but will probably have provided non-specific esterases capable of deacetylating cellulose as well as other polymers such as pectin and hemicelluloses.

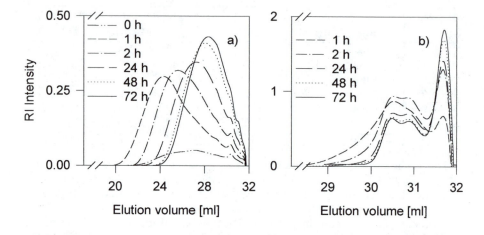

Figure 3. Elution profiles of cellulose-acetate (DS 0.7) monitored by aqueous SEC, fragmented by *Aspergillus* endoglucanase (a) and an *Aspergillus* enzyme mix (b).

In order to learn more about this phenomenon, a DS 0.7 cellulose acetate sample was comparatively incubated with the mono-component *Aspergillus* endoglucanase and another commercial *Aspergillus* enzyme mix (Celluzyme), which had previously been used in the authors' laboratory as a source for the isolation of mannanase activity (*44*) and acetyl mannan esterase activity (*42*). The incubated samples were directly analyzed by aqueous SEC. The time course of the degradation is illustrated by the elution curves from 0 to 72 hours incubation (Figure 3 a and b). The improving RI intensity, in conjunction with the increasing degradation rate, especially as compared to the substrate blank (0 h) with the sample after 1 hour incubation, can be explained by the improved water-solubility of the material. The shift to smaller fragments equivalent with longer elution times is particularly obvious within the first 24 hours of incubation. There was almost no change between 48 hours and 72 hours of incubation. The substrate blank was not included in Figure 3b in order to allow an improved resolution of the low molecular weight fragment (note the different ml-scale in Figures 3a and 3b). Indeed the polymeric material was already drastically reduced in chain length after one hour of incubation. After 24 hours only minor additional changes occurred.

This difference in speed and extent of degradation has been made possible by the presence of an acetyl esterase, in addition to the endoglucanase activity. Unambiguously, the presence of this enzyme was established by the release of acetic acid into solution (Figure 4). The speed of acetic acid liberation was going hand in hand with the fragmentation of the polysaccharide and became slower after 24 hours of incubation. About 50 % of the total acetyl-content of the material was cleaved off after 72 hours. There could be various reasons for the limited release of acetic acid: 1. The acetyl esterase could be a specific enzyme acting exclusively on intact polysaccharides. With increased endoglucanase action, the enzyme could have lost its specificity due to the presence of shorter fragments. 2. The acetyl esterase could be restricted in its catalytic capacity to certain positions within the anhydroglucose unit. 3. The enzyme could be restricted in its action due to the presence of a non-accessible, highly substituted region within the cellulose chain.

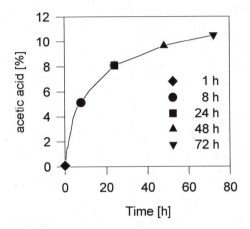

Figure 4. Time course of acetic acid release from cellulose acetate (DS 0.7) during incubation with the *Aspergillus* enzyme mix.

There is some probability for the third possibility, because no acetic acid was released from the DS 2.5 cellulose acetate, incubated with the enzyme mix. However, a similar phenomenon was found for the acetyl xylan esterase, which acted preferentially on acetylated xylan and not on xylan fragments (*45*). The question of the existence of a specific acetyl cellulose esterase or a rather unspecific acetyl esterase could not be solved in this study, but will be the subject of further investigation.

Conclusion

The accessibility of cellulose derivatives clearly was a function of the degree of substitution (DS), in that the material became less degradable with increasing DS. In addition to this factor, charge and size of the substituents play major roles for enzymatic attack. Whereas CMC of DS 1.6 was almost resistant against endoglucanase action, a MC of DS 2.1 was markedly fragmented into shorter chains. Although low DS cellulose ethers were quite markedly degraded by the mono-component endoglucanase, they could not be completely converted into very short fragments. This is a strong indication for the occurrence of non-degradable, highly substituted regions. Consequently, the allocation of the substituents within the anhydroglucose unit and along the cellulose molecule are additional impacts on the availability of the polysaccharide chain for enzymatic fragmentation.

Cellulose acetate was more degradable than could be anticipated from the literature. Cellulose acetate was taken to demonstrate of the impact of esterases, besides endoglucanases, in biodegradation. The presence of acetyl esterase enabled the endoglucanase to degrade cellulose acetate much faster and intensively, and with less enzyme protein. Until now this class of enzymes has been neglected in the consideration of biodegradability of cellulose esters. The specificity of this enzyme is not yet understood and will take our attention in the near future.

The comparison of different authors' work was complicated by the fact that most researchers have drawn their conclusions from viscosity measurements as well as from the determination of nonsubstituted anhydroglucose values and reducing sugars, using culture filtrates containing different ratios of endoglucanases, cellobiohydrolases, ß-glucosidases, and possibly acetyl esterases. It is clear that these catalysts contribute to a different extent of fragmentation, leading to the increase in reducing sugars and liberated anhydroglucose from cellulose derivatives.

Acknowledgment:

This study was supported by DFG-project Pu 81/5-1 and EU-project BIO2-CT94-3030. We thank the suppliers of enzymes (NOVO-NORDISK A/S and NAGASE CORP.) and cellulose derivatives (HOECHST AG, HOECHST-CELANESE CORP., RHONE-POULENC-RHODIA AG, WOLFF-WALSRODE AG) for their generous support.

Literature Cited

(1)   Husemann, E. *Das Papier* **1954**, _8_, 157-162.

(2)   Reese, E. T. *Ind. Eng. Chem.* **1957**, _49_, 89-93.

(3)   Hamacher, K.; Sahm, H. *Carbohydr. Polym.* **1985**, _5_, 319-327.

(4)   Iijima, H.; Kowsaka, K.; Kamide, K. *Polym. J. (Tokyo)* **1992**, _24_, 1077-1097.

(5)   Demeester, J.; Eigner, W.-D.; Huber, A.; Glatter, O. *J. Wood Chem. Technol.* **1988**, _8_, 135-153.

(6)   Parfondry, A.; Perlin A.S. *Carbohydr. Res.* **1977**, _57_, 39-49.

(7)   Ståhlberg, J.; Divne, C.; Koivula A.; Piens K.; Claeyssens, M.; Teeri T. T.; Jones, T. A. *J. Mol. Biol.* **1996**, _264_, 337-349.

(8)   Harjunpää, V.; Teleman A.; Koivula A.; Ruohonen L.; Teeri T.T.; Teleman O.; Drakenberg T. *Eur. J. Biochem.* **1996**, _240_, 584-591.

(9)   Henrissat, B. *Biochem. J.* **1991**, _280_, 309-316.

(10) Tomme, P.; Warren, R. A. J.; Miller Jr., R. C; Kilburn, D. G.; Gilkes, N. R. In: *Enzymatic Degradation of Insoluble Carbohydrates*; Saddler, J. N., Ed.; ACS: Washington, DC, **1995**, Vol. 618; pp. 142-163.

(11) Rouvinen, J.; Bergfors, T.; Teeri, T.; Knowles, J. K. C.; Jones, T. A. *Science* **1990**, _249_, 380-386.

(12) Divne, C.; Ståhlberg, J.; Reinikainen, T.; Ruohonen, L.; Pettersson, G.; Knowles, J. K. C.; Teeri, T. T.; Jones, T. A. *Science* **1994**, _265_, 524-528.

(13)   Spezio, M.; Wilson, D. B.; Karplus, P. A. *Biochemistry* **1993**, *32*, 9906-9916.

(14)   Davies, G. J.; Schülein, M. In: *Carbohydrate Bioengineering (Progress in Biotechnology, Vol. 10)*; Peterson, S. B.; Svensson, B.; Pedersen, S., Eds.; Elsevier: Amsterdam, Netherlands **1995**, pp. 225-237.

(15)   Withers, S. G. In: *Carbohydrate Bioengineering (Progress in Biotechnology, Vol. 10)*; Peterson, S. B.; Svensson, B.; Pedersen, S., Eds.; Elsevier: Amsterdam, Netherlands, **1995**, pp. 97-124.

(16)   Focher, B.; Marzetti, A.; Beltrame, P. L.; Carniti, P. In: *Biosynthesis and Biodegradation of Cellulose*; Haigler, C.H., Ed.; Marcel Dekker Inc.: New York, NY, **1991**, pp. 293-310.

(17)   Ghose, T. K. *Pure & Appl. Chem.* **1987**, *59*, 257-268.

(18)   Philipp, B.; Kasulke U.; Lukanoff B.; Jacopian V.; Polter, E. *Acta Polymerica* **1982**, *33*, 714-718.

(19)   Deemeester, J.; Bracke M.; Lauwers A. In: Hydrolysis of Cellulose: Mechanisms of Enzymatic and Acid Catalysis; Brown Jr., R. D.; Jurasek, L., Eds.; ACS: Washington, DC, **1979**, Vol. 181; pp. 91-125.

(20)   Wirick, M. G. *J. Polymer Sci.: Part A-1* **1968**, *6*, 1965-1974.

(21)   Ach, A. *J. Macromol. Sci., Pure Appl. Chem.* **1993**, *A30*, 733-740.

(22)   Philipp, B.; Stscherbina, D. *Das Papier* **1992**, *46*, 710-722.

(23)   Almin, K. E.; Eriksson K.-E. *Arch. Biochem. Biophys.* **1968**, *124*, 129-134.

(24)   Eriksson, K.-E.; Hollmark B.H. *Arch. Biochem. Biophys.* **1969**, *133*, 233-237.

(25)   Bhattacharjee, S. S.; Perlin A. S. *J. Polymer Sci.: Part C* **1971**, *36*, 509-521.

(26)   Wirick, M. G. *J. Polymer Sci.: Part A-1*, **1968**, *6*, 1705-1718.

(27)   Kasulke, U.; Dautzenberg, H.; Polter, E.; Philipp, B. *Cell. Chem. Technol.* **1983**, *17*, 423-432.

(28)   Schuseil J. Charakterisierung von Methylcellulosen: Verteilung der Substituenten   innerhalb der monomeren Einheiten und entlang der Polymerketten; Thesis; University of Hamburg, Germany, **1988**, 75 pp..

(29)   Kasulke, U.; Linow K.-J.; Philipp B.; Dautzenberg H. *Acta Polymerica* **1988**, *39*, 127-130.

(30)   Kamide, K.; Iijima, H.; Kowsaka, K. In: Cellul. Sources Exploit; Kennedy, J. F.; Phillips, G. O.; Williams, P. A., Eds.; Ellis Horwood, Chichester, UK, **1990**, pp. 365-370.

(31)   Buchanan, C. M.; Gardner, R. M.; Komarek, R.J. *J. Appl. Polym. Sci.* **1993**, *47*, 1709-1719.

216

(32) Glasser, W. G.; McCartney, B. K.; Samaranayake, G. *Biotechnol. Prog.* **1994**, *10*, 214-219.

(33) Melo, E. H. M.; Kennedy, J. F. *Carbohydr. Polym.* **1993**, *22*, 233-237.

(34) Philipp, B.; Kasulke, U.; Dautzenberg, H.; Polter, E.; Hübert, S. *Acta Polymerica* **1983**, *34*, 651-656.

(35) Heinze, T.; Erler, U.; Nehls, I.; Klemm, D. *Angew. Makromol. Chem.* **1994**, *215*, 93-106.

(36) Croon, I.; Purves, C.B. *Svensk Papperstidn.* **1959**, *62*, 876-882.

(37) Lenz, R.W. *J. Amer. Chem. Soc.* **1960**, *82*, 182.

(38) Nicholson, M. D.; Merritt, F.M. In: *Cellulose Chemistry and its Applications;* Zeroninan, S. H.; Nevell, T. P., Eds.; Ellis Horwood: Chichester, UK, **1985**, pp. 363-383.

(39) Boyer, R. F.; Redmond, M. A. *Biotechnol. Bioeng.* **1983**, *25*, 1311-1319.

(40) Saake, B.; Patt, R.; Puls, J.; Linow, K. J.; Philipp, B. *Makromol. Chem., Macromol. Symp.* **1992**, *61*, 219-238.

(41) Biely, P.; Puls, J.; Schneider, H. *FEBS Lett.* **1985**, *186*, 80-84.

(42) Puls, J.; Schorn, B.; Schuseil, J. In: *Biotechnology in Pulp and Paper Manufacture*; Kuwahara, M.; Shimada, M., Eds.; Uni Publishers Co.: Tokyo, Japan, **1992**, Vol. 57; pp. 357-363.

(43) Tenkanen, M.; Schuseil, J.; Puls, J.; Poutanen, K. *J. Biotechnol.* **1991**, *18*, 69-84.

(44) Yamazaki, N.; Dietrichs, H. H. *Holzforschung* **1979**, *33*, 36-42.

(45) Tenkanen, M.; Poutanen, K. In: *Xylans and Xylanases (Progress in Biotechnology Vol. 7)*; Visser J.; Beldman, G.; Kusters-van Someren, M. A.; Voragen, A. G. J., Eds.; Elsevier: Amsterdam, Netherlands, **1992**, pp. 203-212.

(46) Wood, B. F.; Conner, A. H.; Hill Jr., C. G. *J. Appl. Polymer Sci.* **1986**, 32, 3703-3712.

Supramolecular Structures

Chapter 16

Evidence of Supramolecular Structures of Cellulose Derivatives in Solution

Liane Schulz[1], Walther Burchard[1,3], and Reinhard Dönges[2]

[1]Institute of Macromolecular Chemistry, University of Freiburg, D-79104 Freiburg, Germany
[2]Hoechst AG/Kalle-Albert, Forschung Alkylose, D-65174 Wiesbaden, Germany

Recent studies on the solution properties of cellulose derivatives disclosed that two classes have to be distinguished, i.e. the fully and partially substituted chains. Molecularly dispersed solutions are obtained when all OH groups are substituted, while in the other case free OH groups can undergo hydrogen-bonding resulting in association or aggregation. The measurements gave strong evidence for a non random aggregation that might result from uneven derivatization along the chain leading to blocks of low-substituted chain segments. A fringed micellar structure is concluded from the global properties of the aggregates. The model is confirmed and further specified by the angular dependence of the scattered light. The fringed micelle consists of a hard core of laterally aligned segments from which flexible chains emerge. The anisotropic hard core was confirmed by TEM and flow birefringence measurements. The flexibility of the dangling chains could be detected by dynamic light scattering. However, the flexibility became strongly reduced when the concentration was increased beyond the overlap concentration. This behavior results from the fact that the dangling chains cannot penetrate the hard core of the fringed micelle.

Cellulose derivatives are semisynthetic polymers composed of a natural backbone and synthetic side groups. As a consequence of their chemical structure they own a broad range of application comprising chemical, pharmaceutical and food industries, where they are mainly used as thickeners for solutions. Recent measurements have disclosed (1-6) that depending on the degree of substitution, two classes of cellulose derivatives have to be distinguished, e.g. fully and partially substituted chains. The main reason for such differentiation results from non substituted hydroxyl groups that interact in a very specific manner by *intra*- and *inter*molecular hydrogen bonding. In general, specific

[3]Corresponding author.

interaction leads to the formation of aggregates or associates that can be significantly influenced by the solvent. In contrast to these aggregated structures, molecularly dispersed polymer solutions can be obtained if the cellulose backbone is fully derivatized. Typical examples are cellulose tricarbanilate (CTC) (7), cellulose trinitrate (CTN) (8) and cellulose in complexing solvents such as cuoxam (CUOXAM-Cell) (9) which prevents intramolecular hydrogen bonding. In spite of being molecular dispersed these three systems have special disadvantages. These are:

CTC: Difficulties in the preparation of high molecular weight polymers and reduced solubility of derivatives with degree of polymerization DP > 6000 based on hydrogen bonds that have to be broken up cooperatively.

CTN: Tendency to degradation.

CUOXAM-Cellulose: Decreasing solubility for macromolecules with DP > 6000 because of intermolecular crosslinking as a consequence of the complexation process (9).

This contribution deals with the solution structure of partially substituted cellulose derivatives. The aim of this contribution is a presentation of characteristic features of partially substituted methylhydroxypropyl- or methylhydroxyethyl-celluloses (MHPC, MHEC) in water and trifluoroethanol, and of partially substituted carboxymethyl celluloses (CMC) in 0.1 M NaCl solution (Scheme 1). Surprisingly polyelectrolytes as well as hydrophobically and hydrophilically modified chains display very similar behavior. These features are striking and call for an interpretation of the mechanism of structure formation. The reasons for this behavior will be discussed in detail. Most interesting was the question whether these aggregates are random in nature or have already supramolecular characteristics. The present contribution is essentially a report on the main features and is not considered as a detailed communication on all experimental details. Parts have been already submitted (1-4,30) other details will follow.

The paper is divided in three main sections. One is concerned with the global structure, e.g. the hydrodynamic radii R_h and radii of gyration R_g, the intrinsic viscosities [η] and the second virial coefficients A_2. Another one deals with the shape of the particles determined by the angular dependence of the scattered light in static light scattering (SLS) experiments. Finally, in further sections we discuss particularities of the internal or segmental mobility determined by dynamic light scattering measurements (DLS), flow birefringence, rheo-optics and electron microscopy.

Experimental

Materials. The cellulose ethers were laboratory samples from Hoechst-Kalle AG, and the cellulose acetates were from Rhône-Poulenc. The cellulose tricarbanilates were prepared in the Freiburg laboratory by treating cellulose in hot pyridin with phenylisocyanate as described previously (49). The cellulose ethers, ionic and non-ionic samples, were dialysed for one to two weeks against distilled water and the result was controlled by conductivity measurements. This was done to assure equilibrium solution states and to free the materials from included salts, respectively. The samples were then freeze dried and kept under dry atmosphere until use. Only in a few cases also the non dialysed samples were chosen. These are indicated by a suffix *u*. The cellulose 2.5-acetates were purified by centrifugation of 1% solutions in acetone followed by precipitation in water and dried in a vacuum oven at about 50°C. The degree of substitution is given in the legends to the corresponding Figures.

A

B

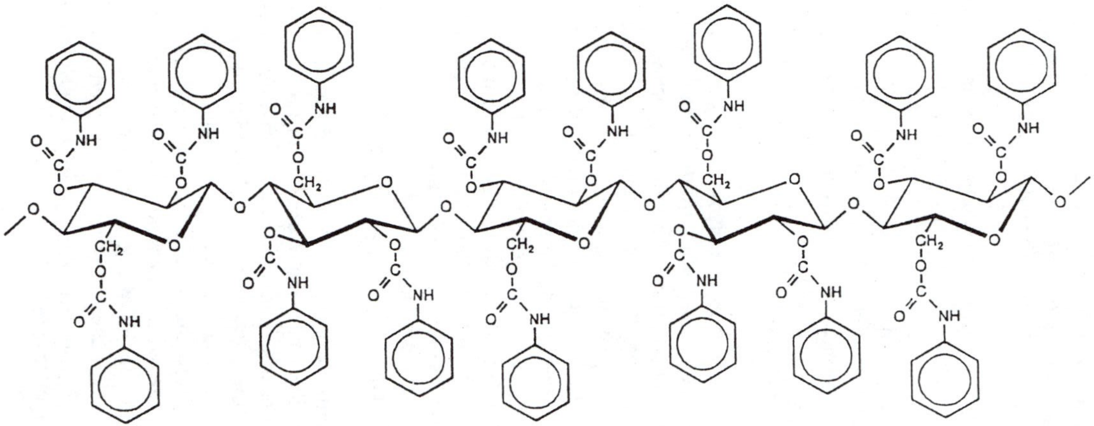

Scheme 1: Schematic representation of the chemical structure of various cellulose derivatives considered in this contribution: **A** - MHPC; **B** - MHEC; **C** - CMC; **D** - CTC.

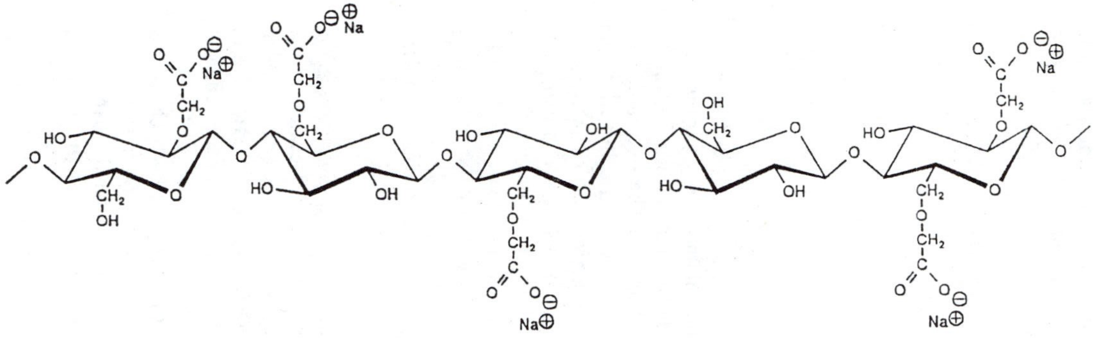

Instrumentation Static light scattering measurements were made with a computer driven modified SOFICA instrument that was equipped with a new light detection system *(50)*. The instrument used for combined static and dynamic LS was an ALV 3000 instrument in combination with the ALV correlator/structurator. The set-up was described previously *(51)*, where also details of the instrument calibration are given. The digital pick-up of data with both instruments allowed an immediate evaluation of molecular parameters by common Zimm-, Berry-, Guinier- and Kratky-plots, respectively.

Rheo-optic A common Bohlin constant stress (CS) rheometer was used for rheological measurements. The instrument was equipped with a home made device for detection of the birefringences induced by the shear. The recently developed set up was made by Richtering et al. *(30)*.

TEM One drop of a 0.2% aqueous MHPC II solution was picked up by a copper grid and negatve-stained with uranyl acetate as usual.

Global Properties

The global properties of polymer solutions are defined by parameters which describe averages over the entire particle including thermodynamic or hydrodynamic interactions among them. All these properties are based on three quantities, i.e. radius, volume and molar mass of the particles. Typical examples are: (i) The two molecular radii R_g and R_h, where R_g, the radius of gyration, is defined geometrically, while the hydrodynamic R_h includes the effect of hydrodynamic interactions which in addition depend on the segment density *(10)*. (ii) The second virial coefficient A_2. It expresses the thermodynamic interactions between two polymer coils in dilute solution. Its value is proportional to the volume of the molecule ($\sim R_g^3$) divided by the square of the molar mass M_w *(11)*. (iii) The intrinsic viscosity $[\eta]$ which is determined by the ratio of the hydrodynamic volumes to their molar mass *(11,12)*. (iv) Another quantity of interest is the ratio of the two radii $R_g/R_h \equiv \rho$; this ρ-parameter allows a rough estimation of the segmental density but depends also on the molecular size non uniformity *(13)*.

Experimental Findings. Partially substituted derivatives have been characterized in different solvent systems over a broad range of concentration *(1,2,6)*. Because of the large number of experimental data we confine ourselves to only a few examples which emphasize the particularities in behavior of the solutions from MHPC, MHEC and CMC. Figures 1 and 2 give a preliminary impression on the deviations of the cellulose derivatives from the behavior of common synthetic polymers. Because of the basic similarity, (the same backbone, fully substituted derivative), CTC has been chosen as reference in the following comparison. According to their global properties CTC dissolved in dioxane behaves like a synthetic polymer *(7)*. In particular the exponents for R_g, R_h, A_2 and $[\eta]$, obtained from the slopes in Figures 1 to 4, obey the predicted scaling properties *(14)*.

$$a_{R_g} = a_{R_h} = \nu$$
$$a_{A_2} = 3\nu - 2 \qquad (1)$$
$$a_{[\eta]} = 3\nu - 1$$

Their values are listed in Table I.

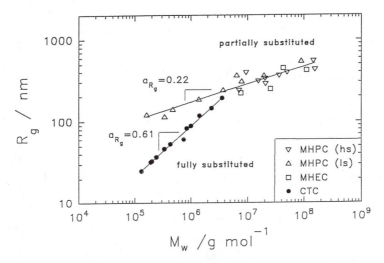

Figure 1 : R_g as a function of particle weight M_w for three cellulose derivatives. (hs): high HP-substituted, (ls): low HP-substituted. (ls): MHPC 1-5; DS = 1.84, MS($-OC_3H_6OH$) = 0.2; η = 19-30 Pa at (2%); (hs): MHPC I-V; DS = 2.07, MS($-OC_3H_6OH$) = 0.78; η = 241-96120 Pa at 2% . MEHC: DS between (ls) and (hs), (estimation). CTC in dioxan; DS = 3.0.

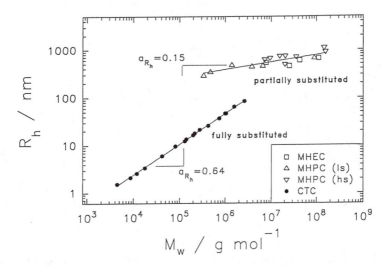

Figure 2 : R_h as a function of particle weight M_w for three cellulose derivatives. Meaning of (hs), (ls) and the other notation as in Figure 1.

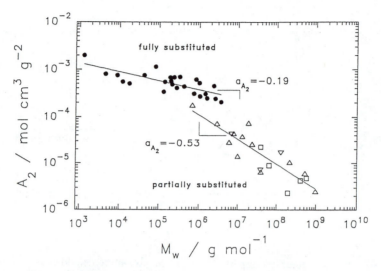

Figure 3 : Molar mass dependence of the second virial coefficient A_2 for CTC in dioxane (fully substituted) and the partially substituted MHPC and MHEC ethers.

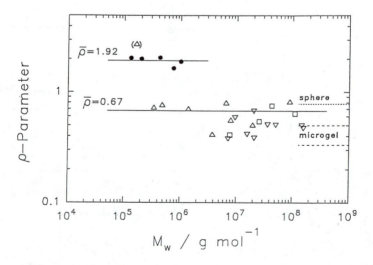

Figure 4: Molar mass dependence of the ρ-parameter for the same derivatives as shown in Figure 3.

Table I

Predicted (columns 2 and 3) and experimentally observed exponents in the molar mass dependencies of the radius of gyration, the hydrodynamic radius, the second virial coefficient and the intrinsic viscosity for three cellulose derivatives

Exponent	good solvent	θ-solvent	CTC dioxan	MHPC/MHEC water	CMC 0.1 M NaCl
$^a R_g = v$	0.60	0.50	$0.61 \pm 0,01$	0.22 ± 0.01	0.22 ± 0.02
$^a R_h$	0.59	0.50	0.64 ± 0.01	0.15 ± 0.02	0.86 ± 0.03
$^a A_2$	-0.20	-	-0.19 ± 0.04	-0.53 ± 0.06	-0.68 ± 0.07
$^a [\eta]$	0.70 - 0.80	0.50	0.89 ± 0.04	not measured	---*

* almost no molar mass dependence, large scatter of data.

Strikingly, the exponents of the partially substituted cellulose derivatives deviate from this behavior while those for the fully substituted CTC fulfil these conditions.

Discussion of the Global Properties. Comparison of the fully and partially substituted derivatives reveals two deviations from the properties of molecularly dispersed flexible chains. A much weaker increase in R_g and R_h is found for the MHPC and MHEC samples than expected and observed for linear chains. In other words, the molar mass increases far more pronounced than the dimensions that only increase weakly. Evidently, the arrangement of the polymer segments becomes more densely packed with growing molar mass. In this context it is relevant to stress that the actually measured molar mass is at least by a factor ten larger than 130 - 200 kmol/g of common cellulose pulp with a DP of 800 - 1200. These two findings are obvious indications for the presence of aggregates.

In order to find an explanation for the uncommon slopes expressed by $a_{R_g} = v$ the concept of fractal dimensions d_f may be consulted according to which $d_f = 1/v$ (*15*). The fractal dimension is a quantity that usually indicates self similarity of the polymer structures. In principal d_f can vary between 1 (rodlike structures) and 3 (compact structures). Values higher than 3 are physically irrelevant and have to be considered as meaningless. The values one would obtain from the slopes observed with the cellulose derivatives are in the range of 4.5 to 5 and cannot be sensible dimensions; they need another explanation. One possibility would be the following: Because of the high molar mass aggregation has occurred, and the explanation may be sought in this fact. Indeed, the observed phenomena are in agreement with a model of fringed micelles (*16,17*) (Figure 6). This model consists of laterally aligned chains which form a rather compact and probably geometrically anisotropic core from which dangling chain sections emerge. Due to the lateral aggregation the radius of gyration changes only little while the mass increases strongly.

The question arises whether there exist other indications for the formation of the above introduced aggregation model. Plots of A_2 and ρ reveal behavior that again is consistent with a dense packing of segments (see Figures 3, 4 and 5). Indeed, the stronger negative exponent a_{A2} is typical for crosslinked or associated structures, and the low value of the ρ-parameter is characteristic of a high structural density that is to be expected with fringed micelles.

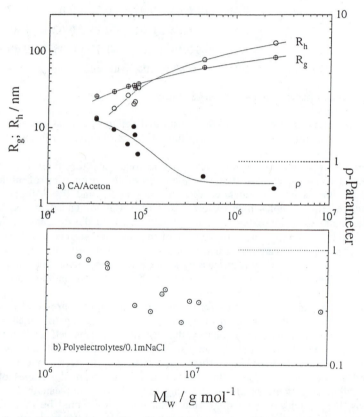

Figure 5 : (a) Cellulose 2.5-acetates in acetone: R_g, R_h and ρ-parameter as a function of molar mass. (b) CMC in 0.1M NaCl solution: ρ-parameter. DS = 0.93; η = 118-66900 Pa at 2%.

In fact, the model reconciles the main, apparently contradicting properties:
- Very high molar masses.
- Increasing segment density with molar mass.
- A weak increase of the dimension.

On the other hand, the global parameters are not sufficient for distinguishing more detailed features of fringed micelles. Various shapes are still compatible with the above findings; a few of them are shown in Figures 6a, b and c together with two other modifications of laterally aligned chain sections.

Shape of Aggregates

General Remarks and Results. For further information on the structure it is useful to measure the angular dependence of the scattered light. There exist several techniques of presenting the data in suitable graphs (*18*). For the present study the Kratky presentation (*19*) proved to be most instructive. In such plots the normalized scattering intensity R_θ is multiplied by q^2 and plotted against $u=qR_g$, where R_θ is the Rayleigh ratio at the scattering angle θ and $q = (4\pi/\lambda)\sin(\theta/2)$ is related to the scattering angle. As usual, the radius of gyration R_g can be determined from the initial slope of the inverse scattered intensity as a function of q^2 (Zimm plot) (*20*). The Kratky representation has the advantage that the asymptotic part of the scattering function at $qR_g > 1$ is amplified by $(qR_g)^2$, which makes even small changes clearly detectable. This asymptotic regime is sensitive to the internal architecture of the particles. Therefore, one can easily distinguish rods from random coils or hard spheres. Branched materials and compact aggregates are expected to develop curves in between of those for hard spheres and random coils. A qualitative estimation can be made already on a first sight. This facilitates the qunatitative evaluation by fits to special models. Figure 7 exhibits some examples from the same sample measured under different conditions. For comparison the curves for random coils and hard spheres are added. Stiff rods exhibit a linear increase with qR_g in such a plot .

Another, commonly used way of plotting the data is the double logarithmic graph of the scattered intensity against q. The plot, that is not presented here, displays a bent curve at low q-values but develops a power law decay at large *q*. The slope gives the ensemble fractal dimension $<d_f>$. Their values deviate strongly from $d_{f,app} = 4 - 5$, that was found with the data from the molar mass dependence of the radius of gyration when using the samples of different degree of polymerization of the individual nacromolecules. The values obtained for the individual samples are all around $<d_f> = 2.0$ and 2.9, and lie fully in the range of physically meaningful values.

Actually the true fractal dimension is not obtained from the asymptotic slope of the scattering curve. It still contains an influence of the molar mass distribution which results from the fact that the mean square radius of gyration is a z-average while the molar mass is a weight average (*21*). However, as long as the ratio M_z/M_w does not significantly change with M_w the correction is small and can be neglected. For the present cellulose derivatives the molar mass distribution is not known and could not be determined by size exclusion chromatography since the materials adhered at the gel matrix. The width of the size distribution is probably not large as will be demonstrated below.

Attempt of Interpretation. The fractal dimensions varied significantly for different samples from the same homologous series and even more when the solvent was changed and other derivatives were compared. These data for d_f correlate well with

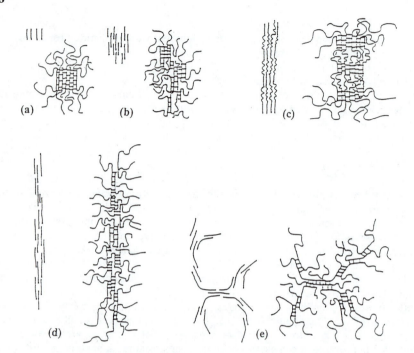

Figure 6 : Three different forms of fringed micelles (above) and two other models of laterally aligned chain sections (below). Only in the first three cases a weak increase of the radius of gyration with the molar mass is obtained. The inserts at the upper left corner indicate the alignment of the chains in the core. A more realistic picture exhibits the electron micrograph by Fink et al. (48).

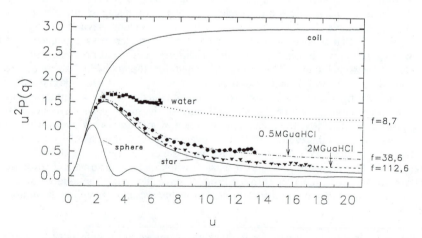

Figure 7 : Kratky plots of a MHPC 2 ($M_w = 7.98 \times 10^6$) in three solvents. The curves are fits to the model of star-branched macromolecules, f denotes the number of arms in a star molecule, see Eq.(2).

the differences in the scattering curves of the Kratky representation (Figure 7 presents only a small selection of measurements).

The observed curves resemble those of star shaped macromolecules with f polydisperse arms. As was shown previously (22,23) the angular dependence of such star molecules with flexible chains could be calculated. It is given in the Kratky representation by the equation

$$u^2P(q) = \frac{u^2(1+u^2/3f)}{\left[1+u^2(f+1)/6f\right]^2} \qquad (2)$$

which for $u = qR_g \gg 1$ approaches an asymptotic plateau of

$$u^2P(q) \rightarrow 12\frac{f}{(f+1)^2} \qquad (3)$$

Thus the plateau height is a sensitive measure of the number of arms f. Most of the experimental scattering curves could be well fitted by the model of star-branched macromolecules, but in a few cases, when $f < 10$, the fit became poor and not applicable at all. Figure 8 demonstrates the effect.

Instead of approaching a plateau a strong increase is now observed at large q-values. Such behavior is characteristic of stiff chains. Now the plot of $qP(q)M_W$ approaches a constant plateau with a height of (25-29)

$$qP(q)M_w \rightarrow \pi M_L \qquad (4)$$

Here M_L is the linear mass density of rod like chain sections of length l_k and molar mass M_k where the subscript k stands for Kuhn segment. The measured linear mass density can be compared with the calculated one for a single strand M_o/l_o, where M_o and l_o are the molar mass and bond length of the repeating unit. From the ratio of the measured to the calculated values the number of laterally aggregated chains can be estimated. Before reaching the asymptotic plateau the curve for semiflexible chains passes through a maximum. Its position should appear at $qR_g = 1.41$ if chains of uniform length are present and at $qR_g = 1.70$ if $M_w/M_n = 2.00$ (26). For even broader size distributions the maximum is shifted towards larger values. In the present examples a value of 1.71 was in no case exceeded.

As a next point we checked whether the fractal dimension d_f is correlated to the number of arms f. Figure 9 shows the result. A strong correlation between the two quantities can be clearly stated, though not a strict functional dependence. These results from the angular dependence of the scattered light confirm the conclusion drawn previously from the global properties and specify further details of the model. The fact that d_f increases with the number of aggregated chains makes clear that the aggregates are not randomly constructed because in this case the fractal dimension would change between 2.0 and 2.5 (15,21). Hence the two models in Figure 6d and 6e can be excluded. The increase beyond a value of 2.5 up to approximately 3.0 indicates an increasingly dense packing. The objects must still have a large number of dangling chains since otherwise the curves could not be fitted by the model of star-branched molecules.

Moreover, for $f < 10$ we observed low fractal dimensions ($d_f < 1.7$) and clear stiff chain behavior. Hence the densely packed core must have a remarkable rigidity

230

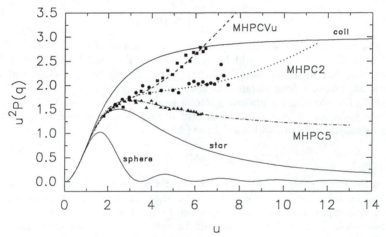

Figure 8 :   Kratky plots of some MHP-celluloses which cannot be fitted to the model
of star molecules. The strong increase at large q-values indicates chain
stiffness. These curves could be fitted with the model of wormlike chains
(*24-27*).

which could be quantified by the fit of the scattering curves with the model of worm-like chains. Applying Koyama's theory of worm-like chains (24) the Kuhn segment length could be determined (27) which ranged from $l_k = 65$ nm up to about 300 nm. (For comparison the Kuhn segment length of polystyrene is about 2 nm and that of DNA molecules around 80 nm.)

We are aware of the approximate character when describing fringed micelles by star shaped objects which include dense and geometrically anisotropic cores, but nonetheless this approach will enable refinements in the derivation of mathematical models which can further be examined by small angle X-ray (SAXS) or small angle neutron scattering (SANS) measurements.

Evidence for Fringed Micelles from other Experiments

The geometric anisotropy could be expected to be sufficiently large such that the core might be visualized by electron microscopy. The geometric anisotropy will be very likely connected with an optical anisotropy which should develop a noticeable birefringence when the material is sheared. Both conjectures could indeed be realized.

Figure 10 shows the shear rate dependence of the viscosity and the development of flow birefringence with increasing shear rate (30). The decrease in viscosity is a common feature of entangled polymer chains, but in the present example it is also accompanied by orientation of the rigid core. This conclusion becomes apparent by the observation that the birefringence just started when the shear thinning exhibited a significant effect.

Figure 11 gives a picture obtained by transmission electron microscopy (TEM). Only the dense, slightly anisotropic cores are seen; the individual dangling chain sections are too small in their cross-section. The cores appear to be isolated from each other. This effect is caused by the dangling chains which keep the cores apart and sterically stabilize the micelles.

Local Dynamics

The dynamics of polymers has been studied so far mostly by oscillatory rheology. Although in the past much progress has been achieved in the development of sensitive fluid rheometers, these instruments work satisfactorily only in semidilute solutions where the viscosity is sufficiently high. Rheology is a mechanical technique that probes the dynamics in a macroscopic manner, and conclusions on the local dynamics can be drawn only via appropriate theories. For about 15 years the chain dynamics can be studied also by dynamic light scattering, which gives a molecular response without disturbing the systems by external forces. Information on the local dynamics can be obtained when the particles have large dimensions, i.e. in the order of the wave length of the light used. The method has recently been applied with success to branched macromolecules (31-33) with the following result:

Much smaller distances than the radius of gyration are examined if an angular region of $qR_g > 3$ is chosen. The influence of the translational diffusion of the center of mass (that dominates at $qR_g < 2$) has there already fully decayed (34), and only the segmental motions are seen. According to Zimm (35) and the refined theories on dynamic light scattering from flexible *linear* chains by Pecora (36), de Gennes (37) and Akcasu (38-40) the first cumulant $\Gamma(q)$ of the time correlation function (TCF) reaches a q^3 dependence (the first cumulant represents the initial slope of the logarithmic TCF against the delay time).

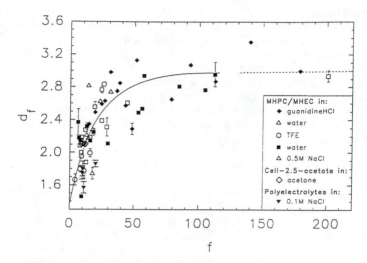

Figure 9 : Plot of the fractal dimensions d_f obtained with various fractions and derivatives of cellulose against the number of arms f. The systems are indicated by the insert. CA: cellulose 2.5-acetate in acetone; Polyelectrolytes: CMC and alginates (not discussed here).

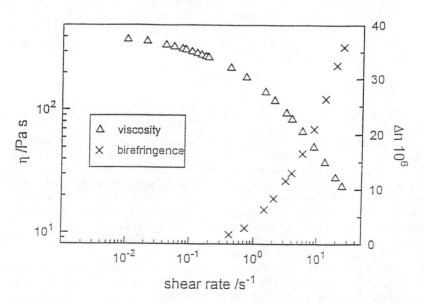

Figure 10: Shear rate dependencies of flow birefringence and shear viscosity of the sample MHPC II in water (30).

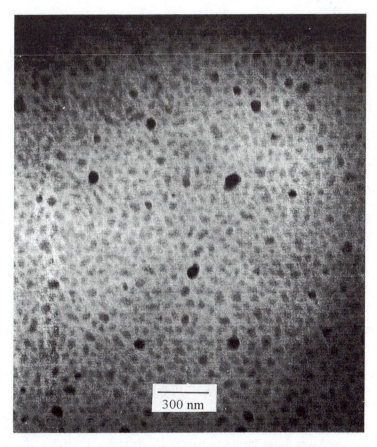

Figure 11 : TEM picture from a 0.2% aqueous solution of the sample MHPC II ($M_w = 2.07 \times 10^7$). Negative staining with uranyl acetate.

It is convenient to define a reduced (dimensionless) first cumulant as

$$\Gamma^*(q) = \frac{\Gamma(q)}{q^3}\left(\frac{\eta_0}{kT}\right) \tag{5}$$

that should approach a constant plateau. The plateau height has been calculated for *linear* chains in good and Θ-solvents, repectively (*40*). In the pre-average approximation the result is

$$\Gamma^*(\infty) = 0.071 \qquad \text{good solvent}$$
$$\Gamma^*(\infty) = 0.053 \qquad \Theta\text{-solvent}$$

For *branched* materials the $\Gamma^*(q)$ values reach no constant plateau; instead, it continuously decreases. This behavior is approximately described by a power law (*31,33,41*)

$$\Gamma^*(q) \sim q^{-0.17\pm0.03} \tag{6}$$

The curves of the different branched materials lie in the region between random coil and hard sphere behavior.

At a value of $qR_g = 3$ the $\Gamma^*(q)$ reached values of 0.040 and 0.030 for randomly crosslinked polyesters and degraded amylopectin molecules, respectively. These observations led us to the conclusion that the height of $\Gamma^*(q)$ at large $qR_g > 3$ is a measure of the internal flexibility. This conclusion is supported by the value for hard spheres that for large qR_g approaches $\Gamma^*(q)\rightarrow0$. The decrease in the flexibility is evidently due to branching and the resulting high segment density (number of segments per volume). Theories on the dynamics of branched macromolecules are presently still missing.

So far, all theories and experiments were confined to infinitely dilute solutions. In the present study we now extended the region of measurement to semidilute solutions and performed measurements with the MHPC1 sample in the three solvents (i) water, (ii) 2M guanidine HCL/water and (iii) trifluoroethanol (TFE). The $\Gamma^*(q)$ curves exhibited in some cases a minimum and finally increased again, but most of the curves assymptotically approached a constant value. We compared the values found at $qR_g = 3$ which are shown in Figure 12.

A very pronounced decay of the $\Gamma^*(qR_g=3)$ values with increasing concentration was found. For a better display of the data we thus plotted $\sqrt{\Gamma^*(qR_g = 3)}$ against the concentration. Even at $c = 0$ the highest value (0.033) was much lower than 0.071 for linear chains, and this is considered being a result of the aggregated state. The reduction of the $\Gamma^*(qR_g=3)$ values depended slightly on the nature of the solvent used, and the lowest value was found for the sample in 2M guanidin HCl/water where also the highest aggregation number was observed. The drastic decrease of the $\Gamma^*(qR_g=3)$ values with increasing concentration, (that is *not* found with polystyrene), gives evidence for a remarkable inhibition of the segmental motion. In fact, the dangling chains of the fringed micelles can interpenetrate only partly to form a transient network of entangled chains. Beyond a certain concentration, however, the chains hit the dense core where they cannot penetrate any further. Hence the chains become more and more compressed, and this finally causes a complete loss of mobility.

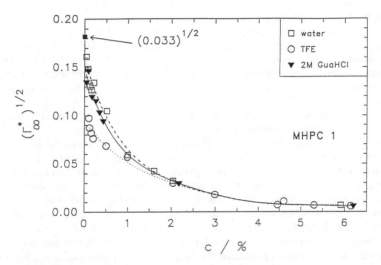

Figure 12 : $\Gamma^*(qR_g=3))$ as a function of polymer concentration for measurements of MHPC 1 ($M_w = 6.47 \times 10^6$) in water, 2M Guanidine HCl/water and Trifluoroethanol (TFE). The overlap concentration $c^* = A_2 M_w c$ lies in all cases around 0.3%.

Conclusions

The following 5 facts may be summarized: (i) The very weak increase of the dimensions that is accompanied by a very pronounced increase in the molar mass indicates the formation of compact aggregates similar to that in micelle formation. (ii) The analysis of the angular dependence of static light scattering measurements revealed similarities to star branched macromolecules which posses a remarkable chain stiffness (averaged over the whole structure). These two observations give strong evidence for a fringed micellar structure as depicted in Figure 6a. (iii) The anisotropy of the hard core that is expected from laterally aligned chain segments could be visualized by TEM micrographs and was confirmed by flow birefringence. (iv) The flexibility of dangling chains was proven by dynamic light scattering. (v) The loss of internal mobility on increasing the polymer concentration is in agreement with the existence of a hard core that cannot be penetrated by the dangling chains. All these facts are in close agreement with the fairly regular fringed micelle structure of Figure 6a. Since this structure is observed essentially with all not fully substituted cellulose chains it can be considered as a rather universal feature of a special super molecular structure. In searching for reasons of this uncommon aggregation structure two explanations can be offered. Both are based on an uneven substitution along the chains.

(i) The uneven derivatization is very likely caused by the morphology of the native or regenerated cellulose in the solid state. These fibers show a structure consisting of crystalline laterally aligned chain sections interrupted by amorphous but still highly oriented domains (*42,48*). This particular structure was given the name fringed micellar crystals. Most of the derivatives are prepared heterogenously, and thus the easier accessible amorphous chain sections will be preferably derivatised leaving the crystal regions widely unaffected. The highly substituted chain sections dissolve in the various solvents while the crystalline part remains stabilized by the many hydrogen-bonds. Solubility of the colloidal particles (based on macromolecules) results from the entropy of mixing of dangling chains with the solvent (*43*). The mechanism is the same as predicted by Wegner (*44,45*) and Ballauff (*46*) for their hairy rod systems. Thus the first explanation is based on an incomplete dissolution process.

(ii) The chains in the crystalline domains may be less derivatized than in the amorphous domains but still may contain a number of substituents. In this case special solvents may be found which can break up the hydrogen-bonds in the core. In fact the aggregation number could be changed by different solvents. The guaninidin HCl/water solvent was expected preferably to break the hydrogen bonds, but opposite behavior was found. This might be the result of complexities in the hydrophobic interactions. TFE decreased the number down to 4-10 laterally aligned chains. Application of large shear rates may possibly allow full disruption of the structure, but the aggregated fringed micellar structure is expected to reconstitute under equilibrium conditions. Observations with an arabinoxylan (*47*) gave clear indications for a strong increase of chain stiffening as chains start to align laterally, which is probably the result of a cooperative formation of an hydrogen bond system. Once 6-7 chains have been aligned further lateral aggregation took place in which almost no change in the radius of gyration was found.

The final result in both explanations is very similar, and further investigations with other derivatives are needed for distinguishing between the two indicated possibilities of dissolution. Depart of this uncertainty on the mechanism, it may be noted that similarities in the morphology of different cellulose sources and particularities in the chain topology of the individual chains are the mean reason for the surprisingly universal supermolecular structure. This fact offers new chemical routes to transforming these structures into functional elements.

Acknowledgement   The work was partially supported by the Ministry of Research and Development of the Federal Republic of Germany in Bonn.

Literature Cited

1   Burchard, W.; Schulz, L. *Das Papier* **1989**, *43*, 665.
2   Schulz, L.; Burchard, W. *Das Papier* **1993**, *47*, 1.
3   Burchard, W.; Lang, P.; Schulz, L.; Coviello, T. *Makromol. Chem., Macromol. Symp.* **1992**, *58*, 21.
4   Burchard, W.; Schulz, L. *Macromol. Symp.* **1995**, *99*, 57.
5   Lang, P.; Burchard, W. *Makromol. Chem.* **1993**, *194*, 3157.
6   Schulz, L. *Ph.D. Thesis,* University of Freiburg **1996**.
7   Wenzel, M.; Burchard, W. *Polymer* **1986**, *27*, 195.
8   Benoit, H.; Holtzer, A.M.; Doty, P. *J. Phys. Chem.* **1954**, *58*, 635.
9   Seger, B.; Burchard, W. *Macromol. Symp.* **1994**, *83*, 291.
10   Kirkwood, J.G. *J. Polym. Sci.* **1954**, *12*, 1.
11   Yamakawa, H. *Modern Theory of Polymer Solutions*, Harper & Roe, New York 1971.
12   Flory, P.J.; Fox, T.G. *J. Am. Chem. Soc.* **1951**, *73*, 1904.
13   Burchard, W.; Schmidt, M.; Stockmayer, W.H. *Macromolecules* **1980**, *13*, 1265.
14   De Gennes, P.-G. *Scaling Concepts in Polymer Physics*, Cornell University Press, Ithaca, NY, 1979.
15   Daoud, M.; Martin, J.E. In: *The Fractal Approach to Heterogeneous Chemistry*, Ed. Avnir, D., Wiley & Sons, New York 1992, p 109-130.
16   Hermann, K.; Gerngross, O. *Kautschuk* **1932**, *8*, 565.
17   Burchard, W. *TRIP* **1993**, *1*, 192.
18   Burchard, W. *Adv. Polym. Sci.* **1983**, *48*, 1.
19   Kratky, O.; Porod, G. *J. Collod Sci.* **1949**, *4*, 35.
20   Zimm, B.H. *J. Chem. Phys.* **1948**, *16*, 1093.
21   Stauffer, D. *Introduction to Percolation Theory*, Taylor & Francis, San Francisco 1985.
22   Burchard, W. *Macromolecules* **1974**, *7*, 841.
23   Burchard, W. *Macromolecules* **1977**, *10*, 919.
24   Koyama, R. *J. Phys. Soc. Japan* **1973**, *34*, 1029.
25   Schmidt, M.; Paradossi, G.; Burchard, W. *Makromol. Chem., Rapid Commun.* **1985**, *6*, 767.
26   Denkinger, P.; Burchard, W. *J. Polym. Sci.* **1991**, *20*, 589.
27   Dolega, R. *Fitprogram KOYFIT*, Freiburg 1992.
28   Casassa, E.F. *J. Chem. Phys.* **1955**, *23*, 596.
29   Holtzer, A.M. *J. Polym. Sci.* **1955**, *17*, 432.
30   Schmidt, J.; Richtering, W.; Weigel, R.; Burchard, W. *Macromol. Symp.* **1997**, submitted.
31   Trappe, V.; Bauer, J.; Weissmueller, M.; Burchard, W. *Macromolecules*, submitted

32   Galinsky, G.; Burchard, W. *Macromolecules*, submitted.
33   Trappe, V.; Burchard, W. *Proceedings of the Krakow Symposium* **1996**, *Light Scattering and Photon Correlation Spectroscopy*, submitted.
34   Berne, B.J.; Pecora, R. *Dynamic Light Scattering*, Wiley & Sons, New York 1976.
35   Zimm, B.H. *J. Chem. Phys.* **1956**, *24*, 269.
36   Pecora, R. *J. Chem. Phys.* **1968**, *49*, 1038.
37   Dubois-Violette, E.; de Gennes, P.-G. *Physics* **1967**, *3*, 181.
38   Akcasu, A.Z.; Benmouna, M.; Han, C.C. *Polymer* **1980**, *21*, 866.
39   Benmouna, M.; Akcasu, A.Z. *Macromolecules* **1978**, *11*, 1187.
40   Benmouna, M.; Akcasu, A.Z. *Macromolecules* **1980**, *13*, 409.
41   Burchard, W. *Adv. Colloid Interface Sci.* **1996**, *64*, 45.
42   Fengel, D.; Wegener, G. *Wood*, de Gruyter, Berlin 1989.
43   Flory, P.J. *Principles of Polymer Chemistry*, Cornell University Press, Ithaca NY, 1953.
44   Seufert, M.; Fakirov, C.;Wegner, G. *Adv. Mater.* **1995**, *7*, 52.
45   Seufert, M.; Schaub, M.; Wenz, G. Wegner, G. *Angew. Chem. Int. Ed. Engl.* **1995**, *34*, 340.
46   Ballauff, M. *Fluid Phase Equilibria* **1993**, *83*, 349.
47   Ebringerova, A.; Hromadkova, Z.; Burchard, W.; Dolega, R.; Vorwerg, W. *Carbohydr. Polym.* **1994**, *24*, 161.
48   Fink, H.-P.; Purz, H.J.; Bohn, A.; Kunze, J. *Macromol. Sympos.* submitted
49   Wenzel, M.; Burchard, W.; Schätzel, K. *Polymer.* **1986**, *27*, 195.
50   Baur, G., **1988,** Baur Instrumentenbau, Hausen, Germany.
51   Bantle. S.; Schmidt, M.; Burchard, W. *Macromolecules* **1982**, *15*, 1604.

Chapter 17

Phase Behavior, Structure, and Properties of Regioselectively Substituted Cellulose Derivatives in the Liquid-Crystalline State

Peter Zugenmaier and Christina Derleth

Institute of Physical Chemistry, Technical University of Clausthal, D-38678 Clausthal, Zellerfeld, Germany

The helical twisting power, value and handedness, of the super-molecular helicoidal structure of liquid crystalline (lc) cellulose derivative / solvent systems strongly depends on the substituents introduced. Investigations on lc cellulosetrisphenylcarbamate and cellulosetris-3-chlorophenylcarbamate reveal different sign of the twisting power in the same solvent triethylene glycol monomethylether. The parameters which influence such a behavior have been studied. These are the site of phenylcarbamate substitution at the anhydroglucose unit (2, 3, 6), the site of substitution at the phenyl ring (3 or 4) including different substituents (hydrogen, chloro, methyl, fluoro) on regio-selectively substituted chains in ethylene glycol monomethylether acetate and, for one case, a random substitution of two groups along the cellulose chain. A strong polymer solvent effect has been detected which is predominantly influenced by the substitution site at the anhydroglucose unit and at the phenyl ring. A study of the phase diagram supports the idea that clusters with bound solvent are formed. These have to be regarded as the structural units for the lyotropic cholesteric phase.

Lyotropic liquid crystalline cellulose derivatives formed by highly concentrated solutions belong to a special class of chiral materials (*1*). Right- and left-handed super-molecular helicoidal structures of chiral nematic, also termed cholesteric mesophases, are observed with positive or negative temperature and concentration gradients for the twisting power. Studies also show that not all cellulose derivatives exhibit lyotropic liquid crystals, rather some lead from the semi-dilute state, where already microgels appear, directly to the gel state, although the chain backbone of these structures has similar stiffness as compared to those derivatives which produce liquid crystalline (lc) phases. This behavior might depend on the chain length that is the molecular mass. Cellulose derivatives with high molecular mass normally omit the lc state and form gels only. This behavior might also occur with short chain molecules depending on the substituents. Considering these observations, it seems that for the lyotropic lc cellulose derivative systems, the polymer solvent interaction plays an important role. Little is known about this interaction in cellulosic systems. From solvent built-in crystals, which are border line cases of liquid crystals, a fiber structure analysis shows that, depending on the solvent, different conformations of the cellulose backbone may occur, or only the side group conformation changes. In most cellulosics the molecular structure does not change at all, rather the packing of the chains accommodates for the solvent. The other border line case, the semi-dilute state, clearly exhibits for the fully substituted cellulose molecule (degree of substitution DS=3 for the anhydroglucose monomer unit) that reversible aggregates rather than molecular dispersed single molecules form the basic building blocks. This statement should also hold for the liquid crystalline state, since no dissolution of these blocks is observed when going to the liquid crystalline state. It has also been found that the solvent in these lyotropic systems, semi-dilute or liquid crystalline, is rather tightly bound to the polymer chain and not freely available for a crystallization process of the solvent by lowering the temperature. In the dilute state molecularly dispersed single molecules are observed for fully substituted cellulose derivatives not capable of hydrogen bonding.

The driving force of the transition from the isotropic to the anisotropic phase of lyotropic systems for stiff chains is believed to lie in the structuring of the solvent. Studying semi-dilute solutions of cellulose derivatives, we have been able to show that most of the solvent is bound to the polymer that means little free solvent is left and that a structuring of the cellulose chains has to be considered as well forming reversible clusters (*2, 3*).

Lyotropic liquid crystalline cellulose derivatives exhibit chirality at three levels. The chirality caused by the configuration of the molecule (here chiral centers), by the conformation of the macromolecules (here normally left-handed helices) and by super-molecular structures as the cholesteric helicoidal structure of various handedness. Chirality at these different levels should be reflected, e.g., by the twisting power of the cholesteric phase. At the present little is known about the correlation of chirality and twisting power except for the rare case of a thermotropic liquid crystal where two chiral centers are placed far apart in a molecule, and the conformation of the molecules with different configurations seems to be very similar. For this example additivity of the twisting power for the two centers was proven. The twisting power by mixing various configurations of these compounds has been described by the weighed twisting power of the different configurations with the molar fractions as weights (*4*).

The ultimate goal of our investigations is to establish a relationship between chirality and the twisting power of cellulose derivatives knowing the twisting power of the various substituents at the different position of the anhydroglucose unit and the conformational helix. This would enable to predict the helix conformation in the lyotropic liquid crystalline state with the experimentally determined twisting power. Little is known about this correlation at the present time. There are two pathways to start such an investigation: Firstly, to substitute statistically with two different groups and changing the overall composition along the chain, preferential at the different possible sites forming a copolymer. Secondly, to substitute regio-selectively at the various positions 2, 3, 6 at the anhydroglucose unit uniformly along the chain. In this paper we will report results for various but very similar cellulose trisphenylcarbamates also termed cellulosetricarbanilates (CTC, cf. Scheme 1) belonging to both groups that will, on the other hand, establish equally well the importance of the polymer solvent interaction.

Experimental

The cellulosetrisphenylcarbamate and cellulosetris(3-chlorophenylcarbamate) have been synthesized by well-established procedures as was the statistical copolymer with random distributed phenylcarbamate and 3-chlorophenylcarbamate side groups. These derivatives were characterized to establish their chemical constitution by elemental analysis, IR, NMR, etc. (5). The synthesis and characterization of regio-selective derivatives, always trisubstituted, have been described elsewhere (6). An overview of regio-selective materials available for this investigation is summarized in Table I with abbreviations and chemical constitution of the molecules given in Scheme 2.

Table I. Overview of Cellulose Derivatives Synthesized and their Corresponding Polymer Code for the Position of Substitution 2, 3, 6 at the Anhydroglucose Unit. (C: Chloro, M: Methyl, H: Hydrogen, F: Fluoro).

Cellulose Derivative	Polymer Code 2, 3, 6
cellulosetris(3-chlorocarbanilate)	CCC
cellulose-2,3-bis-(3-chlorocarbanilate)-6-(carbanilate)	CCH
cellulose-2,6-bis-(3-chlorocarbanilate)-3-(carbanilate)	CHC
cellulose-2,3-bis-(carbanilate)-6-(3-chlorocarbanilate)	HHC
cellulosetricarbanilate	HHH
cellulose-2,3-bis-(carbanilate)-6-(4-fluorocarbanilate)	HHF
cellulose-2,3-bis-(4-fluorocarbanilate)-6-(carbanilate)	FFH
cellulose-2,3-bis-(3-chlorocarbanilate)-6-(3-methylcarbanilate)	CCM
cellulose-2,3-bis-(3-methylcarbanilate)-6-(3-chlorocarbanilate)	MMC
cellulosetris(3-methylcarbanilate)	MMM

The highly concentrated solutions with triethylene glycol monomethyl ether (TRIMM) for the copolymer and with ethylene glycol monomethylether acetate (EMMAc) for the regio-selectively substituted ones were prepared by mixing an appropriate amount of the derivative with the necessary volume of solvent in an Eppendorf vessel under stirring. Homogeneous textures were obtained by placing the highly concentrated solution between two cover glass plates kept 50 μm apart by a spacer cut out of a commercially available foil of this thickness.

Most of the samples showed selective reflections of one color in the visible light range, proving that a cholesteric helicoidal structure with the helix axis perpendicular to the glass plates was present as depicted in Figure 1. The pitch of the super-molecular structure was then determined by spectroscopic means, by ORD and UV-VIS, taking the zero optical rotation in the anomalous dispersion range or the peak of the transmission as selective reflection λ_o and calculating the pitch P according to de Vries (7) with $P = \lambda_o / \bar{n}$, $\bar{n}$ being a mean refractive index. If the selective reflection was not situated

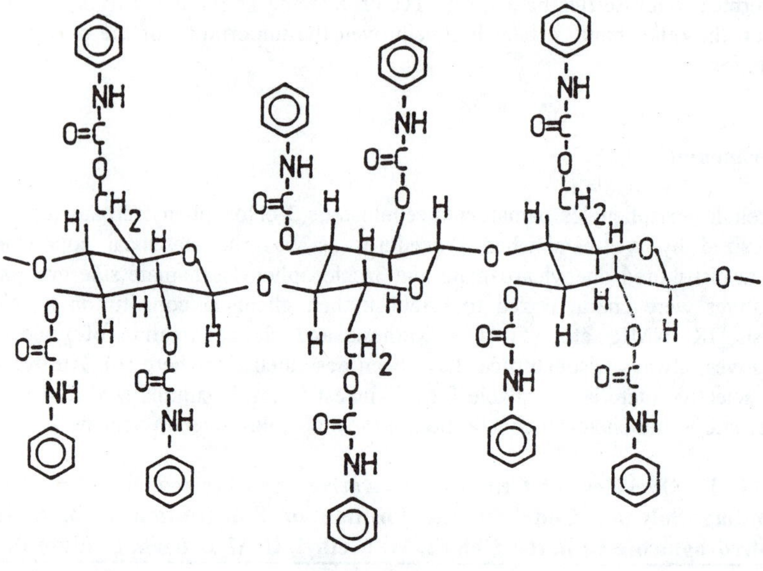

242

Scheme 1. Schematic representation of cellulosetricarbanilate (CTC).

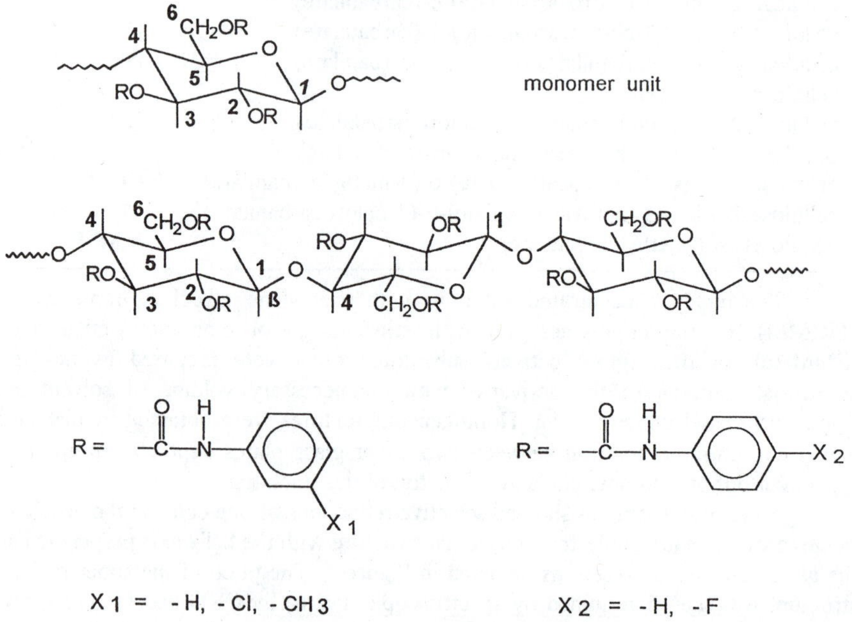

monomer unit

$R =$

$R =$

$X_1 = -H, -Cl, -CH_3$ $X_2 = -H, -F$

Scheme 2. Schematic representation of substituted CTC.

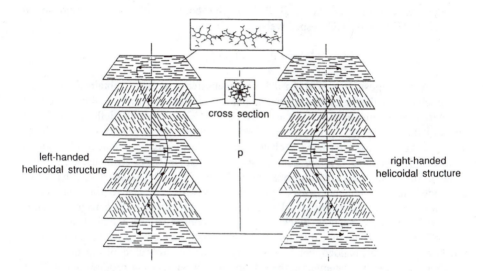

Figure 1. Schematic representation of cholesteric liquid crystalline structures. Right-handed helicoidal structures are assigned a positive, left-handed ones a negative value of the pitch P or twisting power P^{-1}, respectively.

in the visible spectrum range, the Cano-Grandjean (8) method was applied with additional measurement of the handedness of the helicoidal structure by ORD.

Results and Discussion

Statistical Cellulose Copolymer. The twisting power P^{-1} of the lc state as a function of composition (percentage of 3-chlorophenyl versus phenyl groups) in TRIMM is plotted in Figure 2. Pure cellulosetris(3-chlorotricarbamate) (CCC) exhibits a right-handed helicoidal structure, pure cellulosetricarbanilate (HHH) a left-handed one. It is clear from this graph that the twisting power P^{-1} of a known composition x of 3-chlorophenyl side groups may be represented as additive in a first approximation of average twisting powers P_H^{-1} and P_C^{-1} of the corresponding pure trisubstituted compounds HHH and CCC, respectively:

$$P^{-1} = x\ P_C^{-1} + (1\text{-}x)\ P_H^{-1} \tag{1}$$

At a composition x of about 0.5 a compensated structure is obtained with $P^{-1} = 0$ or pitch infinity that resembles a nematic phase. It is also known from solid state fiber structure analysis that both derivatives, HHH and CCC, form similar left-handed 3-fold helices with a repeat of 15 Å as preferred conformations of the cellulosic chains. If the idea is correct that a right-handed chain conformation leads to a right-handed super-molecular lc structure (9), the conformation of a single CCC chain must be right-handed. Since the preferred conformation in the solid state is left-handed, it has to be concluded that the polymer solvent interaction causes a reversal of the twist sense. Such a reversal can be easily visualized, if the solvent causes two adjacent monomer units to become the building block of the chain conformation. A 3-fold left-handed helix of a polymer chain may then turn to a 3-fold right-handed one with the pitch twice the size of the left-handed helix. Assuming a similar overall conformation of the backbone, a different placement in the side group of two adjacent residues in e.g. the 6 position or in the solvent attached to the two adjacent units leads to a dimer as conformational unit and a twist inversion of the conformation.

It is amazing how much a small amount of a second side group with a statistical distribution of both side groups along the chain causes a change in the twisting power although a full helix pitch of one component, right- or left-handed, can not be established. Also the different substitution sites 2, 3, 6 at the anhydroglucose unit differently influences the twisting power. From our investigations on semi-dilute solutions cluster formation was deduced. These clusters have to be considered in the lc state to exist and make an interpretation of the results on the twisting power of a copolymer, as depicted in Figure 2, even more difficult.

Regio-Selectively Substituted Cellulose Derivatives. The optical rotatory dispersion and the transmission curves for right-handed super-molecular lc structures are depicted in Figures 3 and 4 for the system CCH / EMMAc at various temperatures and a fixed concentration of 0.70 g/ml, respectively. A decrease of the peak height at elevated temperatures is observed in the anomalous dispersion range of the ORD curves. This effect is mirrored in the UV-VIS spectra. The transmission T should amount to 50 %

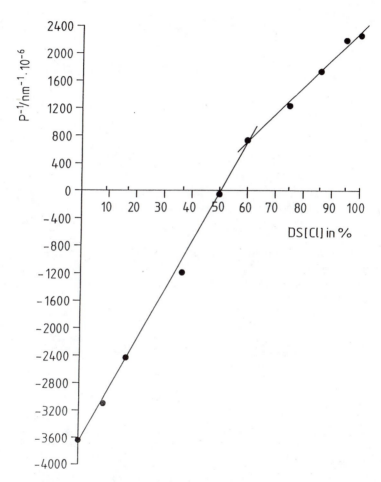

Figure 2. Twisting power P^{-1} as a function of composition for a copolymeric phenylcarbamate- / 3-chlorophenylcarbamate cellulose; random distribution along the chain in triethylene glycol monomethylether (TRIMM); c = 0.8 g/ml, room temperature. Percentage for the degree of substitution for chlorophenyl groups DS(Cl) corresponds to x=DS/100. (Adapted from ref. 5).

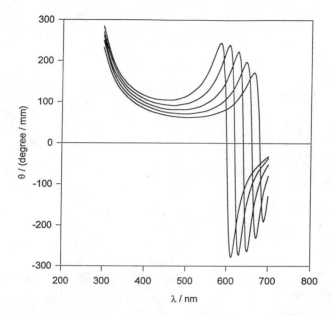

Figure 3.   Optical rotatory dispersion (ORD) curves for the right-handed cholesteric mesophase CCH / EMMAc; c = 0.70 g/ml; sample thickness 50μm. The temperature for the various curves rises from left to right: 301 K, 305 K, 309 K, 313 K, 317 K.

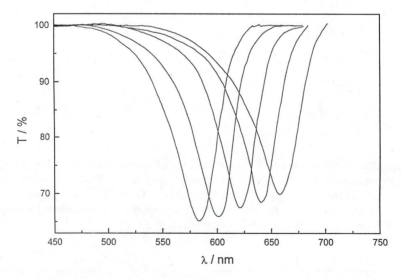

Figure 4. Transmission curves in the visible spectral range (UV-VIS) for lc CCH / EMMAc; c = 0.70 g/ml; sample thickness 50 μm. The temperature for the various curves rises from left to right: 297 K, 301 K, 305 K, 309 K, 313 K.

for linearly polarized light according to theoretical considerations but is found to be only about 35 % or less at higher temperatures. The ORD curves can only be described with the general de Vries's equation when two additional factors are introduced. A dispersion term for the chromophores of the side groups that accounts for the rapid increase of the optical rotation at smaller wavelengths and a damping factor that reduces the optical rotation in the dispersion range. A perfect helicoidal structure should depict a singularity at the selective reflection wavelength λ_o. For a real structure the pitch P or the twisting power P^{-1} are deduced from zero optical rotation in the anomalous dispersion region or the peak in the UV-VIS curves, which represent the selective reflection λ_o. With the knowledge of the mean refractive index $\bar{n}$ of a nematic sheet of the cholesteric structure, the pitch is calculated by $P = \lambda_o/\bar{n}$. The mean refractive index $\bar{n}$ was obtained by measuring the ordinary and extraordinary refractive index of the samples with an Abbé refractometer at the desired temperature.

The temperature dependent measurements of the twisting power P^{-1} are represented for regio-selective lc cellulose derivative systems in Figure 5, and the pitch P listed at T= 303 K in Table II. For some of the derivatives the selective reflection lies outside the instrumental spectral range. The pitch was then determined by the Grandjean-Cano technique.

Table II. Pitch P and Handedness of Various Cellulose Urethane / Solvent Systems at T = 303 K, c = 0.7 g/ml and Slightly Varying Degree of Polymerization DP; Sample Thickness 50 μm.

Polymer/Solvent System	Pitch P / nm	DP[a]
CCC / EMMAc	+ 318	280
CCM / EMMAc[b]	+ 280	250
MMC / EMMAc	+ 315	290
MMM / EMMAc	+ 302	285
CCH / EMMAc	+ 413	210
CHC / EMMAc	+1140	270
HHC / EMMAc	-1015	280
HHH / EMMAc	- 670	245
HHH / EMMAc[c]	- 522	245
HHF / EMMAc[c]	- 488	110
FFH / EMMAc[c]	- 517	100

[a] cf. ref. 6
[b] c = 0.8 g/ml
[c] c = 0.9 g/ml

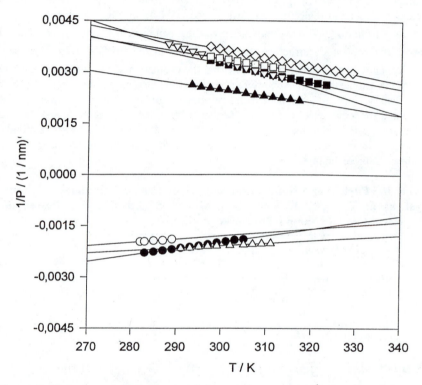

Figure 5. Temperature dependence of the twisting power P^{-1} of various left- and right-handed helicoidal structures of lc cellulose derivatives in EMMAc. CCM ◊; MMM ☐, MMC ∇, CCC ■, CCH ▲, HHH ●, HHF Δ, HHF O (c = 0.80 g/ml); the concentrations for the other lc states are listed in Table II.

A negative helical twisting power, signifying a left-handed super-molecular helical structure of the lyotropic cholesteric phase, is found for cellulosetricarbanilate (HHH) and the 4-fluoro derivatives of the phenyl residue, HHF and FFH, as well as for HHC, the derivative of the tricarbanilate for which a 3-chloro substituted phenyl ring is placed at the 6 position of the anhydroglucose unit. All the other derivatives exhibit right-handed helicoidal super-molecular structures. The temperature gradients of the twisting power for all lc systems shown in Figure 5 are very similar in their absolute values. Nevertheless, they are positive for left-handed structures and negative for right-handed ones.

From a broad study on lc behavior in ref. 3, it can be concluded that all meta substituted cellulosetricarbanilates (CTC) with F, Cl, CH_3O, CF_3 in a variety of solvents led to right-handed super-molecular structures as did all bis-substituted in 3, 4 position at the phenyl ring. All para substituted CTC (Cl, Br, F, CH_3O) exhibit left-handed ones. Only pure lc CTC changes handedness depending on solvent.

Discussing the helicoidal structure in more detail, it is clear from Table II that replacing a 3-chlorophenyl by a 3-methylphenyl group at all sites of the anhydroglucose unit does not change the pitch. However, exchanging a 3-chloro or 3-methyl group by a hydrogen at the phenyl residue and placing this substituent in 2 or 3 position of the anhydroglucose unit drastically alters the pitch in size for CHC and in sign for HHC and HHH. Replacing a 3-chloro group by hydrogen at the phenyl substituent in 6 position slightly increases the pitch only. A compensated chiral nematic phase may be obtained for CHC and HHC by adjusting either the external parameters as temperature and concentration and/or by mixing the two compounds physically or statistically distributing the various groups along the molecular chain. These investigations of the helical twisting power reveal the most sensitive substitution sites for the super-molecular structure of lyotropic lc cellulose derivatives to be the 2 and 3 position at the anhydroglucose unit. Similar conclusions can be drawn for the helix conformation and packing of a single chain in the solid state for which the structure is predominantly determined by the substitution in 2 and 3 position at the anhydroglucose unit (10). The electronegativity of the substituent at the phenyl residue is of minor importance as compared with the substitution sites 3 or 4. The 4-fluoro derivatives HHF and FFH all lead to left-handed structures with similar pitch as HHH in contrast to the 3-chloro derivative CCH with a right-handed one. Since the pitch was found to depend on the molecular mass (11), the degree of polymerization DP is also listed in Table II. From former studies it can be concluded that a DP of >150 is beyond any influence on the size of the pitch, but a small effect may be expected for the fluoro derivatives in comparison with all the other regio-selective cellulose derivatives.

The temperature dependence of the twisting power P^{-1} of chiral nematic phases was investigated by Kimura et al. (12) and Equation 2 derived:

$$P^{-1} = Q (T_n/T - 1) \qquad (2)$$

Q is a factor that includes geometry and volume fraction of the polymer in solution; Q > 0 for right-handed super-molecular structures and Q < 0 for left-handed ones.
T_n represents the inversion temperature for which the super-molecular structure changes the sign of the twisting power. T_n lies above the clearing temperature T_c for the systems considered.

Two plots of the pitch for positive and negative Q values as a function of temperature are shown in Figure 6. Comparing the results for the regio-selectively substituted cellulose derivatives with those in Figure 6 leads to the following conclusions: Curve *b* (left plot) with an increasing pitch or decreasing twisting power best describes the experiments for right-handed helicoidal super-molecular structures (Q > 0). For such a behavior predominant polar and steric effects are responsible. The same argument holds for the left-handed structures. The decreasing size of twisting power with temperature for both types of structures also supports the idea that left-handed conformational helices produce a left-handed super-molecular structure and a right-handed conformational helix a right handed cholesteric structure.

Phase Behavior. The study of the phase behavior and properties at the phase transition represents a crucial test for theoretical models that have been developed for various kinds of liquid crystals. For lyotropic lc CTC with predominant polar interactions, none of the existing models can explain the phase diagram exactly (*2*). Although the volume fraction for which the anisotropic-isotropic transition occurs might be predicted to some accuracy, the actual small biphasic region for a broad molecular mass distribution and the bending of the curves at higher temperatures cannot be described by any model. Figure 7 depicts the phase diagram for CCC / EMMAc taken from texture observations in the polarization microscope. A small biphasic region is detected and a bending of the curves occurs at higher temperatures, the same features as described above, although only steric mixed with polar interactions are detected, and instead of a left-handed super-molecular structure as above, a right-handed one was established. The same difficulties arise for the description of the CCC / EMMAc system adjusting to the theoretical models as for lc CTC, and only fair agreement is obtained with the model of Warner and Flory (*13*), although anisotropic interactions are taken into account. The bending of the curves may be explained by a variation of persistence length with temperature but the small biphasic region cannot be accounted for. It is questionable from the current knowledge of the structure of these systems that the models discussed may be suitable for a description of the isotropic-anisotropic transition in cellulosics. Clearly cluster formation occurs and bound solvent is present in these systems. These effects may play an important role in considering phase transitions. Also a phase separation between lower and higher molecular masses may occur, since different kinds of clusters are observed depending on the size of the cellulosic molecules.

At higher concentrations below the cholesteric phase, a columnar phase appears as in the CTC / diethylene glycol monoethylether system (*14*). In this case an oriented fiber could be produced, and the X-ray analysis revealed hexagonal packing of the molecules.

Acknowledgments. Part of the work reported was supported by a grant from Deutsche Forschungsgemeinschaft.

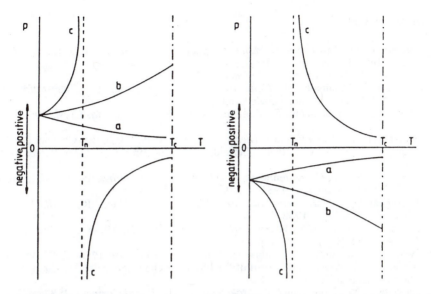

Figure 6. Pitch P versus temperature T according to Equation 2 for right-handed conformational helices Q > 0 (left plot) and left-handed ones Q < 0 (right plot): (a) predominant polar interactions; (b) polar and steric effects; (c) predominant steric effects. (Adapted from ref. 2).

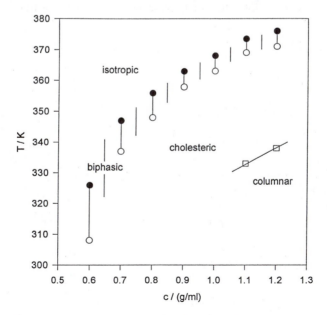

Figure 7. Phase diagram for the system CCC / EMMAc evaluated by texture observations.

Literature Cited

1. Gilbert, R. D. In *Polymeric Materials Encyclopedia;* Salamone, J. C., Ed.; CRC Press: Boca Raton, Florida, 1996, pp 118.
2. Haurand, P.; Zugenmaier, P. *Polymer* **1991**, *32*, 3026.
3. Klohr, E.; Zugenmaier, P. *Cellulose* **1995**, *1*, 259; Klohr, E. *Dissertation*, TU Clausthal, D-38678 Clausthal-Zellerfeld, 1995.
4. Dierking, I.; Gießelmann, F.; Zugenmaier, P. *Mol. Cryst. Liq. Cryst.* **1996**, *281*, 79.
5. San-Torcuato, A. *Diploma Thesis*, Institut für Physikalische Chemie der TU Clausthal, D-38678 Clausthal-Zellerfeld, 1989; Zugenmaier, P. *Das Papier* 1989, **43**, 658.
6. Aust, N.; Derleth, C.; Zugenmaier, P. *Macromol. Chem. Phys.* **1997**, *196*, in press.
7. De Vries, H. *Acta Cryst.* **1951**, *4*, 219.
8. Grandjean, F. *C. R. Acad. Sci. Fr.* **1921**, *172*, 91; Cano, R. *Bull. Soc. Fr. Minéral. Cristallogr.* **1968**, *91*, 20.
9. Hartshorne, N. H. *The Microscopy of Liquid Crystals*, Microscope Publications Ltd., London, England, 1974, p 80.
10. Iwata, T.; Okamura, K.; Azuma, J.; Tanaka, F. *Cellulose* **1996**, *3*, 91 and 107; Möller, R. *Diploma Thesis*, Institut für Physikalische Chemie der TU Clausthal, D-38678 Clausthal-Zellerfeld, 1982.
11. Siekmeyer, M.; Zugenmaier, P. *Makromol. Chem. Rapid Commun.* **1987**, *8*, 511.
12. Kimura, H.; Hosino, M.; Nakano, H. *J. Phys. (France)* **1979**, *40*, C3-174 and *J. Phys. Jpn.* **1982**, *51*, 1584.
13. Warner, M.; Flory, P. J. *J. Chem. Phys.* **1980**, *73*, 6327.
14. Hildebrandt, F.-I. *Diploma Thesis*, Institut für Physikalische Chemie der TU Clausthal, D-38678 Clausthal-Zellerfeld, 1991; Zugenmaier, P. In *Cellulosics: Chemical, Biochemical and Material Aspects;* Kennedy, J. F.; Phillips, G. O.; Williams, P. A., Eds.; Ellis Horwood: New York, NY, 1993, pp 105.

Chapter 18

Blends of Cellulose and Synthetic Polymers

R. St. J. Manley

Department of Chemistry, McGill University, Montreal, Quebec H3A 2A7, Canada

This article reviews and summarizes a series of experiments designed to explore the possibility of forming miscible blends of cellulose with synthetic polymers. It is emphasized that for this purpose it is important to choose synthetic polymers containing functional groups that can interact strongly with the hydroxyl groups of the cellulose chains. The results support the view that cellulose/synthetic polymer blends are not as intractable as they were once thought to be, and that a very intimate level of mixing can be attained in these systems.

During the past decade intense interest has been focussed on polymer blends, partly because of their intrinsic scientific interest and partly because of practical considerations, because it can be anticipated that such materials will show new desirable physical and/or physico-chemical properties not to be expected in conventional homopolymers. Binary blends in which the two components are synthetic polymers have been extensively investigated (1,2). Somewhat surprisingly, however, very little work has been done on blends in which one component is unmodified cellulose. This is primarily because of certain difficulties inherent in the preparation of blends involving cellulose. There are basically two ways in which a blend can be made. One is by mixing the components in the softened or molten state and the other is to blend them in solution. But cellulose cannot be melted (the thermal decomposition temperature lies well below the melting point) and until relatively recently no convenient organic solvent was known. In recent years, however, a variety of new solvent systems for the dissolution of cellulose have been described (3-6), and as it happens these systems will also dissolve many of the synthetic polymers that are of interest. Thus the way is now open for the systematic study of cellulose/synthetic polymer blends and the field has become a subject of increasing interest (7-10).

Polymer blends can be subdivided into different categories. The most important distinction is between the so-called incompatible and compatible blends. Incompatible blends in which the two components consist of separate well-defined phases or domains represent the large majority of all polymer blends. Compatible or miscible blends which consist of a single phase are in the minority. One of the main objectives of our work was to characterize the state of miscibility of a variety of cellulose/synthetic polymer blends. The state of miscibility is an important property because immiscible blends generally have a coarse structure which is reflected in poor mechanical properties. On the other hand miscible blends may combine the properties of the miscible components and hence the mechanical properties may be superior to those of the component polymers.

For blending with cellulose, it is important to choose synthetic polymers containing functional groups that can interact strongly with the hydroxyl groups of the cellulose chains. Such intermolecular interaction is recognized as providing the driving force for the attainment of thermodynamic miscibility (i.e., miscibility down to the molecular level) in polymer blend system_ Many important synthetic polymers satisfy this condition, for example, polyamides, polyesters and vinyl polymers such as poly(vinyl alcohol), poly(acrylonitrile), poly(vinyl pyrrolidone), and poly(4-vinyl pyridine). Figure 1 shows examples of the kind of interactions that are involved.

The purpose of this article is to discuss the principles associated with the formation of miscible cellulose/synthetic polymer blends using four examples drawn from recent research in our laboratory, namely (cellulose/poly(acrylonitrile) (11), cellulose/poly (vinyl alcohol) (12), cellulose/poly (vinyl pyrrolidone) (13), and cellulose/poly(4-vinyl pyridine) (14).

Examples of the Results Obtained with Various Blends

It is appropriate to begin by making brief reference to the methods of sample preparation. The cellulose sample was a wood pulp with a degree of polymerization of ~ 930, corresponding to a molecular weight of about 160,000. The solvents were N,N-dimethylacetamide/lithium chloride or dimethyl sulfoxide/paraformaldehyde. For the preparation of the blends two solutions were made at a concentration of about 1.5% in the same solvent. One was the solution of cellulose and the other of the synthetic polymer. The two solutions thus separately prepared were mixed at room temperature in the desired proportions, so that the relative composition of the two polymers in the mixed solutions ranged from 10/90 to 90/10, in a ratio of weight percent, the first numeral referring to cellulose content. Each blend solution was then coagulated with a non-solvent to form a film or else solution cast directly to form a film.

A commonly used method for investigating miscibility in polymer blends is to measure the glass transition temperature (T_g) of blends of various compositions. It is well known that a miscible blend must show a single glass transition temperature intermediate between the values for the pure components, over the whole range of blend compositions. Figure 2 shows an example of the results of such measurements for the blend system cellulose/poly(acrylonitrile). Here T_g was measured by DSC and dynamic mechanical analysis. The lower curve is the result from DSC measurements, while the two upper curves correspond to the T_g obtained from tan δ and from the loss modulus E'' measurements. The DSC method is much less sensitive than the dynamic mechanical analysis and is unable to detect T_g for compositions higher than 50%. The data show a single T_g for the whole range of compositions as is expected for a miscible system.

CELLULOSE AND POLY(VINYL PYRROLIDONE)

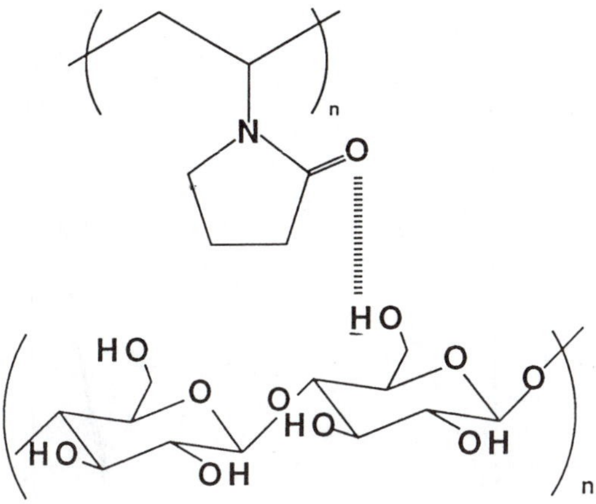

CELLULOSE AND POLYVINYL ALCOHOL

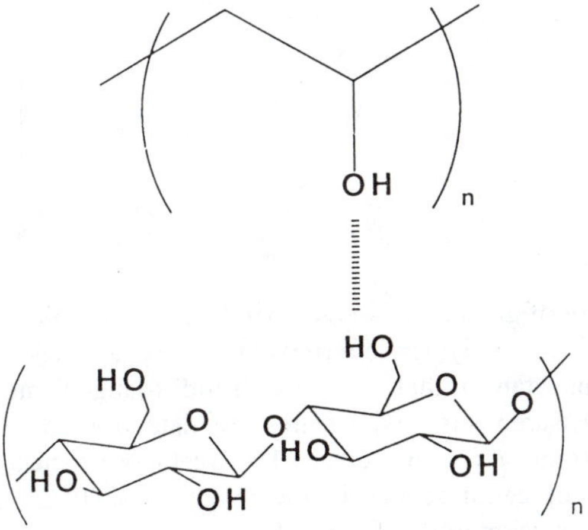

Figure 1.   Examples of hydrogen bonding interactions in cellulose/synthetic polymer blends.

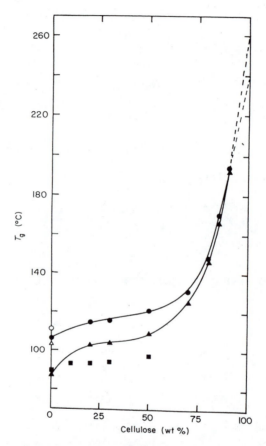

Figure 2.   Plot of glass transition temperature against cellulose content in the
cellulose/poly(acrylonitrile) blends. Filled circles correspond to data
from tan δ measurements, solid triangles are data from E''
measurements, and squares are data from d.s.c. measurements.
Extrapolation of the data to 100% cellulose content indicates that the
T_g of cellulose lies in the range 240-260°C. Reproduced with
permission from reference 11.

Below 50 wt. % cellulose there is only a modest change in T_g with composition indicating a relatively low degree of miscibility, but above 50 wt. % cellulose, T_g increases dramatically indicating a good state of miscibility. By extrapolating the data to 100% cellulose we obtain a T_g of pure cellulose of 240 - 260°C, which corresponds very well with the T_g of cellulose obtained by other methods. It is quite likely that the good state of miscibility at higher cellulose contents is driven by hydrogen bond formation between the CN functionality of the PAN and the OH groups of the cellulose chains. As the cellulose content decreases, the total number of nitrile groups eventually exceeds the number of hydroxyl groups (available for hydrogen bonding) and dipole-dipole association between pairs of nitrile groups becomes progressively dominant.

Next we consider the T_g composition behavior for blends of cellulose with poly (4-vinyl pyridine). In this blend system we expect that there should be hydrogen bonding interaction between the cellulose hydroxyls and the nitrogen of the pyridine rings. The blends were prepared by dissolving the two polymers separately in the DMSO/paraformaldehyde solvent. The two solutions thus separately prepared were mixed in appropriate ratios to give blends with different compositions. Blend films were cast from the blend solutions at room temperature over a period of about 4 hrs. In a first series of blends, the cast films were dried at 125°C overnight in vacuum. In the second series of samples the cast films were steeped in ammonium hydroxide solution, washed in water, and dried in vacuum at 125°C. (Both series of blend films were optically clear, showing no sign of phase separation.) It should be noted that the cast heat treated film is really a methylol cellulose/DMSO complex with a low degree of methylol substitution (0.07), while the other is a true cellulose/poly (4-vinyl pyridine) blend.

Figure 3 shows the T_g/composition data for the two sets of blends as determined from dynamic mechanical measurements of E''. The upper data points (squares) correspond to the cellulose/poly(4-vinyl pyridine) blends (CELL/P_4VPy) while the lower points are from the methylol cellulose/poly(4-vinyl pyridine) (MC/P_4VPy) blends. The solid lines are the T_g/composition dependence predicted by certain semi-empirical equations from a knowledge of the T_g's of the components and their weight fractions. It can be seen that the theoretically predicted T_g's fit the data very well. As seen in the figure, the T_g results for the MC/P_4VPy blend pair are different from those of the CELL/P_4VPy pair. For MC/P_4VPy, the T_g of the blends falls below the calculated weight average values of the T_g's of the components (negative deviation), while for CELL/P_4VPy the T_g of the blends is higher than the corresponding weight average values (positive deviation). Several authors have shown that deviations of the T_g/composition curve can be related to the strength of the interaction between the blend components. Large negative deviations are associated with weak interactions, while a positive deviation has been interpreted as an indication of very strong interactions. Thus the results in the present case suggest that the components of the CELL/P_4VPy pair interact more strongly than they do in the MC/P_4VPy pair.

The next case of interest is that of blends of cellulose with poly(vinyl alcohol) PVA. This is an example of a crystalline/amorphous system because the blend films were regenerated in non aqueous media, with the result that the cellulose component in the blend is predominantly amorphous (12). It is expected that hydrogen bonding between the OH groups of the two polymers should drive the system to thermodynamic

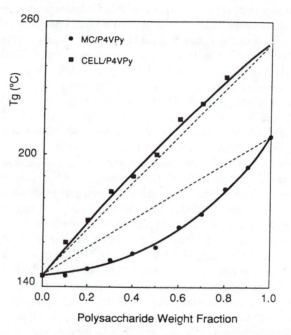

Figure 3.   Theoretical T_g of MC/P$_4$VPy and CELL/P$_4$VPy blends as a function of composition (full line), calculated to give the best fit to the E'' data points. The broken line is the tie line representing the weight-average values. Note that the T_g of CELL was not actually measured; the estimated value of 250°C was used for the calculations. Reproduced with permission from reference 14.

miscibility. If such interaction occurs it is anticipated that it should be reflected in the melting temperature of the PVA in the blends. To this end, the fusion and crystallization behavior of the PVA in the blends was studied by DSC as a function of blend composition. Figure 4 shows the fusion behavior of the specimens. The pure PVA sample gives a large and sharp melting endotherm with a peak maximum at around 230°C. As cellulose is blended with PVA up to 60 wt. %, the endothermic peak of PVA tends to lose its prominence with an accompanying depression in the T_m values. In blends containing more than 70 wt. % of cellulose, it becomes difficult to detect the melting endotherm of PVA in the DSC curves. Table I shows the actual melting temperature T_m of the PVA in the blends, and the heats of fusion per gm of sample as quantitatively assessed by measuring the area under the peaks. It is seen that in going from 0 to 60 wt. % cellulose there is a melting point depression of more than 30°C, which is in fact very large. Simultaneously there is a rapid decrease in the heat of fusion of the PVA, which implies that there is a strong decrease in the degree of crystallinity of the PVA component due to blending with cellulose.

Table I.   Melting Temperature, T_m, and Heat of Fusion , ΔH_f, of Cellulose/PVA Blends Measured by DSC. Reproduced with permission from ref. 12.

Cellulose/PVA w/w	T_m, °C	ΔH_f, cal/g
0/100	230.1	18.7
10/90	226.8	14.7
20/80	224.5	12.4
30/70	220.3	9.4
40/60	212.9	6.2
50/50	205.6	4.0
60/40	197.0	2.0
70/30	~189	-
80/20	-	~0

The phenomenon of the strong depression of the melting temperature of PVA in blends with cellulose can be explained in terms of thermodynamic mixing accompanied by an exothermic interaction between a crystalline polymer and an amorphous polymer. The well-known Flory-Huggins equation for the melting point depression in crystalline/amorphous polymer blends can be written as

$$\Delta T = T_m^° - T_m = - T_m^° (V_{2u}/\Delta H_{2u}) Bv_1^2$$
$$B = RT_m^° (\chi_{12}/V_{1u})$$

where R is the gas constant. Here the subscripts 1 and 2 are used to designate the amorphous and crystalline components respectively, $T_m^°$ is the melting point of pure crystalline polymer 2, T_m is the melting point of the mixture, V is the volume fraction of the amorphous component, V_u is the molar volume of the repeating units, ΔH_u is the enthalpy of fusion per mole of repeating unit, B is the interaction energy density of the

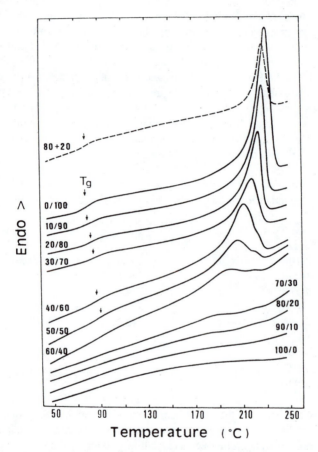

Figure 4.   DSC thermograms for a series of cellulose/PVA blends. The broken-
line curve is a thermogram obtained for a mechanical blend (80+20)
of fine powders of both polymers. The sensitivity of the scans for the
samples containing more than 40 wt % cellulose is twice that for the
others. Reproduced with permission from reference 12.

two polymers, which is related to the thermodynamic interaction parameter χ_{12}. This equation indicates that a plot of ΔT_m versus v_1^2 should be linear with a zero intercept. As seen in Figure 5, the straight line fits the observed values well and yields a slope of 98.2°C and an intercept of 2.9°C. The deviation of 2.9°C is attributed to a residual entropic effect which was neglected in deriving the equation. The interaction parameter χ_{12} derived from these results assumes the large negative value of -0.985 (at 513 K) which is 2 to 5 times larger than values specified in the literature for other polymer pairs. This leads to the conclusion that there is a high degree of interaction between the PVA and the cellulose chains.

Finally, reference is made to some results with the blend system cellulose/poly(vinyl pyrrolidone) (CELL/PVP). In this case a single T_g was observed at every composition, strongly suggesting that the system is miscible. Furthermore, FTIR analysis indicated that the interaction between the two polymers involves the carbonyl functionality of PVP. NMR measurements have provided two types of information. Firstly, as shown in Figure 6, solid state ^{13}C NMR spectra were obtained for the pure homopolymers and for blends of various compositions. The spectra for the blends are not a superposition of the spectra of the constituent homopolymers, because shifts in resonance frequencies are observed. This indicates that the two polymers are interacting with one another. The largest spectral shift is observed for the carbon of the carbonyl functionality of the PVP which occurs at about 175 ppm. This immediately indicates that it is this functionality that is involved in the intermolecular interaction, probably from hydrogen bond formation with the hydroxyl groups of cellulose. The second piece of information provided by the NMR measurements relates to the scale of mixing. When the state of miscibility of a blend is discussed, the question of the scale of mixing has to be taken into consideration. A particular blend may be characterized as miscible with one technique and immiscible with another. For example, visual determination of optical clarity establishes the absence of domains exceeding 200-300 nm. The T_g measured by DSC is sensitive to domain sizes of 35-40 nm, while dynamic mechanical analysis is more sensitive with an upper limit of about 15 nm. From NMR measurements it is possible to estimate domain sizes of a few angstroms to a few tens of nanometers. Thus, using the solid state NMR apparatus it has been possible to measure the so-called spin lattice relaxation time in the rotating frame $(T_{1\rho})$ for the pure homopolymers and for the homopolymer components in the blends. The results are shown in Table II. It is seen that the two components have identical relaxation times which are significantly different from those of the pure unblended homopolymers. This also means that there is extensive mixing of the two polymers. The scale of mixing can be estimated from the relaxation times. It turns out that in the blends the size of the domains is about 2.5 nm. This small domain size means that a very intimate level of mixing has been achieved in these blends. Similar NMR studies on cellulose/poly(acrylonitirile) and cellulose/poly(4-vinylpyridine) indicate that they are also mixed on a very fine scale (i.e., with average minimum domain dimensions in the 3-15 nm range) (19). In a sense, this is somewhat surprising because cellulose is a semirigid polymer while the synthetic polymers with which it has been blended so far are flexible. Flory has predicted that in such a mixture the ordering of the semirigid chains would reject the flexible coil molecules rendering the mixture incompatible. The attainment of a very intimate level of mixing in the

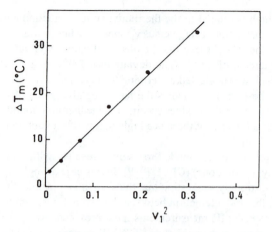

Figure 5. Depression of melting point of PVA in cellulose/PVA blends as a function of volume fraction of cellulose, plotted according to eq. 1. Reproduced with permission from reference 12.

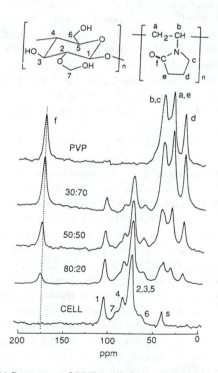

Figure 6. CP-MAS spectra of PVP, cellulose, and three blends. The labelled peaks correspond to the carbons identified in the chemical structure. The peak identified as S in the cellulose spectrum corresponds to bound DMSO. Note that the methylol adduct (C_7OH) can also be on O_2H or O_3H. Reproduced with permission from reference 13.

cellulose/synthetic polymer blends is probably due to the introduction of strong specific interactions between the components (21,11).

Table II.   Proton $T_{1\rho}$ for Solid Films of Cellulose and PVP in their Blended and Unblended States. Reproduced with permission from reference 13.

	$T_{1\rho}{}^{a}$ms	
Cellulose/PVP Blend	Cellulose	PVP
0/100		10.8
30/70	7.3	7.3
50/50	6.2	6.3
80/20	4.2	4.2
100/0	4.2	

$^{a}\pm 5\%$

Conclusions

Cellulose/synthetic polymer blends are not as intractable as they were once thought to be. Perfectly respectable blends can be obtained, and their behavior is entirely analogous to that of the purely synthetic polymer blends. The presence of cellulose in the blends does not seem to lead to any unusual or unexpected behavior. To date several synthetic polymers that form good miscible blends with cellulose have been found, and there seems little doubt that others will be found if we look for them. However, it now seems important to focus on commercially interesting systems. For this purpose we have to recognize that the method of blending used so far is totally unsuitable for industrial applications. The solvent systems (DMSO/paraformaldehyde) or (DMAc/LiCl) are far too expensive and not easily recycled. For commercial applications a more suitable solvent for mixing would be NMMO. Possible applications could be the areas of textile fibers and packaging materials. For example garments made from polyesters have poor comfort properties but good wrinkle resistance, whereas the opposite is true for cellulose. By using a blend it might be possible to obtain the best characteristics of each polymer in a single fiber. This is where the challenge now lies.

Literature Cited

1.   Paul, D.R. and Newman, S. (Eds.) "Polymer Blends", Academic press, New York, 1978.
2.   Olabisi, O., Robeson, L.M., and Shaw, M.T., "Polymer-Polymer Miscibility", Academic Press, New York, 1979.
3.   Hudson, S.M., and Cuculo, J.A., *J. Macromol. Sci. - Rev. Macromol. Chem.* 1980, C18, 1.
4.   Gagnaire, D., Mancier, D., and Vincendon, M., *J. Polym. Sci. Polym. Chem. Edn.* 1980, 18, 13.
5.   Turbak, A.F., Hammer, R.B., Davies, R.E., and Hergert, H.L., *Chemtech.* 1980, 51.

6. McCormick, C.L., and Shen, T.S., "Macromolecular Solutions" (Eds. R.B. Seymour and G.A. Stahl), Pergamon Press, New York, 1982, pp. 101-107.

7. Jolan, A.H., and Prud'homme, R.E., *J. Appl. Polym. Sci.* 1978, 22, 2533.

8. Seymour, R.B., Johnson, E.L., and Stahl, G.A., "Macromolecular Solutions", (Eds. R.B. Seymour and G.A. Stahl), Pergamon Press, New York, 1982, pp. 90-100.

9. Field, N.D., and Song, S.S., *J. Polym. Sci., Polym. Phys. Edn.* 1984, 22, 101.

10. Field, N.D., and Chien, M.-C., *J. Appl. Polym. Sci.* 1985, 30, 2105.

11. Nishio, Y., Roy, S.K., and Manley, R.St.J., *Polymer* 1987, 28, 1385.

12. Nishio, Y., and Manley, R.St.J., *Macromolecules* 1988, 21, 1270.

13. Masson, J-F., and Manley, R.St.J., *Macromolecules* 1991, 24, 6670.

14. Masson, J-F., and Manley, R.St.J., *Macromolecules* 1991, 24, 5914.

15. Bélorgey, G., Prud'homme, R.E., *J. Polym. Sci., Polym. Phys. Ed.* 1982, 20, 191.

16. Bélorgey, G., Aubin, M., and Prud'homme, R.E., *Polymer* 1982, 23, 1053.

17. Kwei, T.K., *J. Polym. Sci., Polym. Lett.* 1984, 22, 307.

18. Pennacchia, J.R., Pearce, E.L., Kwei,T.K., Bulkin, B.J., and Chen, J.-P., *Macromolecules* 1986, 19, 973.

19. Flory, P.J., *Macromolecules* 1978, 11, 1138.

20. Wang, L.F., Pearce, E.M., and Kwei, T.K., *Polymer* 1991, 32 (2), 249.

21. Painter, P.C., Tang, W.-L., Graf, J..F., Thomson, B., and Coleman, M.M., *Macromolecules* 1991, 24, 3929.

Chapter 19

Studies of the Molecular Interaction Between Cellulose and Lignin as a Model for the Hierarchical Structure of Wood

Wolfgang G. Glasser, Timothy G. Rials[1], Stephen S. Kelley[2], and Vipul Davé[3]

Biobased Materials/Recycling Center and Department of Wood Science and Forest Products, Virginia Polytechnic Institute and State University, Blacksburg, VA 24061

Wood and dietary fiber products all belong to a class of biomolecular composites that are rich in cellulose and lignin. The interaction between cellulose and lignin determines such properties as mechanical strength (wood); creep, durability and aging; cellulose purity (pulp); and digestibility (nutrients). The understanding of the interaction between cellulose and lignin can be approached from various types of analyses involving the natural biocomposites, or it can be explored by studying the physical mixtures of the two types of macromolecules. The latter can be prepared by mixing the respective polymers in solid, solution or melt form within the constraints of solubility and melt-flowability. Such mixtures have been examined, and the results suggest that cellulose and its derivatives form two distinct phases with lignin and its derivatives; a crystalline polysaccharide-phase and a continuous amorphous phase that provides evidence for strong intermolecular interaction between the two components. In addition, results suggest that lignin and/or its derivatives are capable of contributing to the supermolecular organization of cellulose (derivatives). The interaction between lignin and cellulose varies in relation to chemical differences as well as molecular parameters. The results are consistent with the view that the hierarchical structure of the natural biocomposite wood is not only the consequence of a sequence of biochemical events, but

[1]Southern Forest Experiment Station, U.S. Forest Service, Pineville, LA 71359
[2]National Renewable Energy Laboratory, Golden, CO 80401
[3]Johnson and Johnson, Skillman, NJ 08558

that it is the result of various thermodynamic driving forces that are independent of the biosynthetic origin.

Hierarchical structures "are assemblages of molecular units or their aggregates that are embedded or intertwined with other phases, which in turn are similarly organized at increasing size levels" (1). It is the multimolecular combination of virtually all biological materials that is responsible for the multilevel architectures that confer the unique properties to the composites of nature. Wood (or more generically, "lignocellulose") is a complex material on all dimensional levels of the structural hierarchy, from the nano- to the millimeter-scale. Lignocelluloses are mixtures of crystalline and non-crystalline polysaccharides with lignin that are assembled into a structural architecture in which the interaction between the biomacromolecules is dictated by the specific sequence of biochemical events that take place during biosynthesis (i.e., plant growth). The resulting multilevel architecture determines all properties, regardless of whether these are mechanical, chemical, sorptive, nutritive, rheological, or degradative in nature. The understanding of hierarchical molecular structures in biological systems is beginning to be taken as a guide for the development of new, man-made materials (1). Several models have been advanced that describe wood (and lignocellulose) as a multiphase material that achieves its remarkable fracture toughness on the basis of the need to create an almost infinitesimal new surface area during fracture (2). It is the creation of interfibrillar cracks during mechanical failure which prevents fiber pullout at all levels of moisture sorption or temperature (2).

Single-phase materials, uniform polymers, often suffer from low impact strength and low dimensional stability when heated. New material properties are achieved when two or more types of molecules are blended or mixed. The resulting morphology of these mixtures is a direct result of the method of blending and the specific chemical and molecular interactions (3). Impact strength and resistance to deformation at elevated temperature rise when mixtures of macromolecules with distinct phases remain molecularly intertwined.

This paper reviews and summarizes a series of experiments designed to explore the specific molecular interactions between cellulose and lignin and their respective derivatives. Man-made blends of these biopolymers are to be compared with the natural biocomposite with a view towards determining the nature of the interaction between the two components. While it is evident that these interactions are also operative during the process of wood formation (i.e., lignification), between lignin precursors and the polysaccharidic matrix, this paper is limited to polymer-polymer interaction arguments, exclusively. However, the reader is referred to the vast body of literature dealing with the biosynthetic aspects of the creation of molecular interactions between polysaccharides and lignin, such as the recent book by Jung et al. (4).

Experimental Section

Materials: All cellulose and cellulose derivatives were obtained as chemically pure, commercially available materials. Lignin and lignin derivatives were obtained from Aldrich Chemical Company except for derivatives described in the primary literature as indicated. Solvents were used as provided from chemical suppliers.

Methods: 1. Blends: Blends of cellulose with lignin were prepared by mixing cellulose solutions in DMAc/LiCl with lignin dissolved in DMAc in accordance with earlier work (5-10). Cellulose derivatives, hydroxypropyl cellulose, ethyl cellulose and cellulose mixed esters (CAB), were blended with lignin by using both melt and solution mixing. Common solvents were pyridine, dioxan and acetone. This work has been described in detail elsewhere (5-7). Melt-blended specimens were produced by injection molding.

2. Thermal Analysis: Thermal analysis was conducted using differential scanning calorimetry (DSC) and/or dynamic mechanical thermal analysis (DMTA) with either thin films (from solvent casting) or melt processed test specimens (dog bones from injection molding).

3. Other Characterization Methods: Ultimate strength was determined by tensile tests using an Instron tensile tester. Dynamic viscosity was determined on a Rheometrics Mechanical Spectrometer (RMS 800) using 20% (w/w) solutions in a parallel-disk geometry (10). Transmission electron microscopy was conducted on a Jeol SEM-100CX-II electron microscope. All methods have been described before where indicated.

Results and Discussion

I.   Wood

The non-crystalline component of wood, which is responsible for the viscoelastic nature of this natural material, gives rise to a variety of responses during heating (11-13). Whereas only a single, broad transition can be detected for native wood at low moisture content, a distinctly trimodal distribution of damping transitions (tan δ-peaks) can be detected at elevated moisture content (Figure 1). Whereas one transition (β) has no discernible impact on the storage modulus, and can safely be attributed to local site exchange of moisture, the occurrence of two distinct glass transition temperatures at which mechanical damping occurs suggests the existence of two different non-crystalline molecular entities. These undergo separate and independent glass to rubber transitions. Specifically, the existence of distinct tan δ-transitions in moist wood, at 30% moisture content, at -10°C (α_2) and 60°C (α_1) (11), suggests the presence of two different molecular components that each reside in their independent phase on the molecular (nano-) level. By investigating the moisture response of these two glass transitions, and by

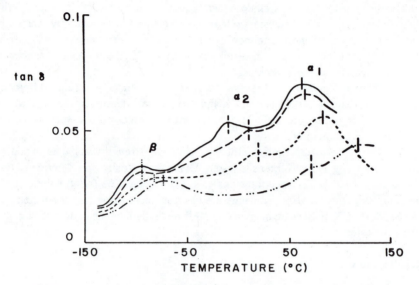

Fig. 1. DMTA spectra (tan δ-transitions) of a section of solid spruce wood recorded at moisture contents rising from 5% (bottom) to 10, 20, and 30% (top). Peaks α_1, α_2, and β reflect large-scale segmental motion and a secondary dispersion (β-transition) characteristic of glass transitions and site exchange of water, respectively. The T_gs were assigned to hemicelluloses (α_2) and lignin (α_1). According to ref. 11.

comparing them to isolated lignin preparations using the model of Kwei (14), α_1 and α_2 were attributed to the presence of the molecular phases representing lignin and hemicelluloses, respectively (11). However, it needs to be pointed out that the existence of separate phases as indicated by DMTA is not inconsistent with the existence of primary or secondary bonds between those phases. Block copolymers between thermoplastic cellulose derivatives and lignin were also found to exhibit phase distinctions with the individual blocks having molecular weights as low as 10^3 daltons (15, 16). There is strong evidence suggesting that the two non-crystalline components of wood, hemicellulose and lignin, are covalently linked in block copolymer fashion (17) and each of these two polymeric phases undergo an independent glass to rubber transition at a different temperature. Due to the hydrophilic nature of these phases, their transition temperatures depend on moisture content; and so does our ability to observe these transitions.

II. Cellulose Derivative/Lignin Blends

Any attempt at recreating wood's native structure by solvent or melt processes is complicated by the variability of the chemical structure of the matrix and the intractability of the cellulose in terms of solubility and melt properties. This limitation may be overcome in part by chemical modification in the form of derivatives. While it is recognized that this severely constrains the realism of the model, it does provide insight into the chemical and physical interaction that can be formed in a binary blend of lignin and cellulose derivatives.

The state of miscibility of a polymer pair is commonly evaluated by studying T_g-behavior in relation to the volume fraction of the respective polymers. Phase-separated mixtures exhibit the T_gs of the individual parent homopolymers while a single transition intermediate to the values of the individual homopolymers indicates miscibility. Partial miscibility is indicated by the T_gs migrating towards a single, common transition in relation to fractional mixing. The DSC-thermograms of a series of hydroxypropyl cellulose (HPC)/lignin (L) blends (prepared by injection molding) provide evidence for strong intermolecular interaction (Fig. 2) (5). A single T_g is observed for blends with a lignin content of up to 55%, and this T_g rises with lignin content.

Dynamic mechanical thermal analysis (DMTA) of this same series of HPC/L blends (Fig. 3) reveals a similar elevation in temperature of damping transitions with lignin content rising. The two tan δ-transitions of the pure HPC-spectrum (at 30 and 85°C) have been assigned to the T_gs of an amorphous phase and that of an organized, liquid crystalline (LC) mesophase, respectively (18). The effect of lignin causes a significant reduction of the temperature range over which the tan δ-transition occurs, with an apparent L-association with the LC mesophase to the exclusion of the lower-temperature amorphous phase (5). This behavior is in conflict with the normal effect of multiphase materials on thermal transitions (19). Normally, the glass to rubber transitions of a mixture of two non-crystalline

270

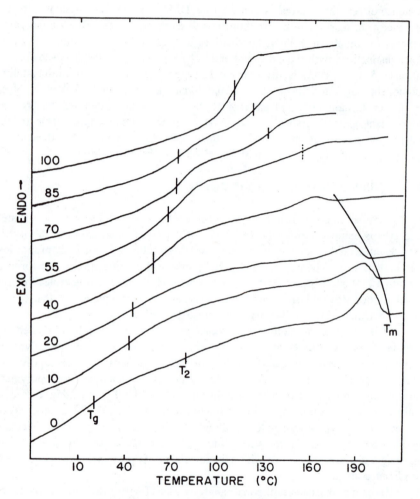

Fig. 2. DSC thermograms of solvent-cast films consisting of mixtures of hydroxypropyl cellulose (HPC) and organosolv lignin (L). Lignin content as indicated by the numbers of each tracing (0 to 100%). Both T_g and T_m show variation in relation to blend composition. T_2 has been attributed to a liquid crystalline mesophase present in HPC (18). According to ref. 5.

polymers broaden on the temperature-scale, and they may stretch over the entire region in which the constitutive components undergo thermal transitions in pure state if they are partially miscible. In the 55% L-content blend, a material with more highly ordered morphology is indicated (Fig. 3). The addition of lignin apparently contributes to the enhancement of HPC's LC mesophase at the expense of an amorphous phase.

A similar rise in relaxation intensity and simultaneous decrease in breadth of the tan δ-transition was observed in blends of HPC with a partially ethylated lignin (EL) (Fig. 4) (6). However, in this case the improved uniformity of phase response results in the complete disruption of the supermolecular structure of the HPC and the formation of a seemingly continuous, miscible, amorphous blend of HPC and EL.

Different observations are made with blends of ethyl cellulose (EC) and L where the addition of a second, immiscible molecular component produces the expected broadening of the tan δ-transition (Fig. 5) (7). Since the resulting two T_g-transitions, however, are found at temperatures below and above those of the respective parent (pure) components, the appearance of two separate phases is explained with the formation of a supermolecularly ordered (discrete) phase with a T_g above that of either parent constituent in addition to a lower-T_g (uniform) amorphous phase. The creation of an LC mesophase architecture by the addition of lignin is also supported by an increase in storage modulus upon passage through T_g (not shown, refer to ref. 7).

An examination of the melting point of the HPC component in relation to lignin content revealed (5) that T_m migrates in relation to the volume fraction of lignin. The degree of melting point depression was approximately dependent on lignin volume fraction and a polymer-polymer interaction parameter, B (20). For blends with components whose molecular weight is >2,000, the melting point depression is approximately related to B according to

$$T^0_{m2} - T_{m2} - \frac{-BV_{2u}}{\Delta H_{2u}} T^0_{m2} \phi^2_1$$

(1)

where the subscript 2 refers to the crystallizable component (i.e., HPC), T^0_{m2} is its equilibrium melting temperature, $\Delta H_{2u}/V_{2u}$ is its heat of fusion per unit volume of repeat unit for 100% crystalline material, V_2 is its molar volume, and ϕ_2 is its volume fraction in the blend. B can be directly evaluated from the slope of a plot of $(T^0_{m2} - T_{m2})$ vs. ϕ^2_1 (Fig. 6) (21). Working with lignin derivatives in which hydroxyl groups were selectively removed (by acetylation or ethylation), a relationship between B and the phenolic hydroxyl content was established (6). The interaction parameter was lowest (and intermolecular interaction most favorable) when the phenolic hydroxyl content was approximately 0.25 per phenylpropane repeat unit

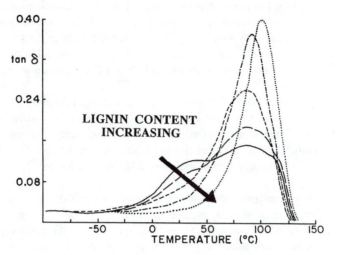

Fig. 3. DMTA Spectra (tan δ-transitions) of melt-processed HPC/L blends containing, 0 (–), 5 (– –), 20 (– – –), 40 (–.–.–), and 55% (.....) L-content. The narrowing of the tan δ-peak with rising L-content suggests an increasing degree of molecular interaction (According to ref. 5).

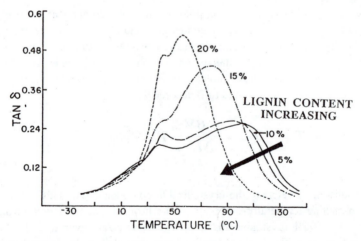

Fig. 4. DMTA spectra (tan δ-transitions) of solvent-cast blend films (dioxan) of HPC with ethyl lignin (EL). EL-content is indicated by the numbers of each tracing (5 to 20%). The progressive narrowing of the tan δ-peak with rising EL-content suggests enhanced molecular interaction with L-content rising. According to ref. 6.

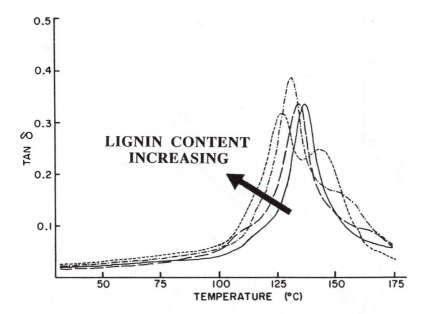

Fig. 5. DMTA spectra (tan δ-transitions) of solvent-cast blend films (dioxan) of ethyl cellulose (EC) with organosolv lignin (L). L-content rises from 0 (--) to 10 (--), 20 (-.-.), and 40% (---). The separation of the tan δ-peak into two distinct transitions is as expected for a molecularly immiscible blend; however, since the higher-temperature transition (at ca. 147°C) is higher than the two parent polymer components, this is attributed to a supermolecularly-ordered phase that is separated from a uniform (continuous) amorphous phase. (Lignin-T_g is at 95°C). According to ref. 7.

274

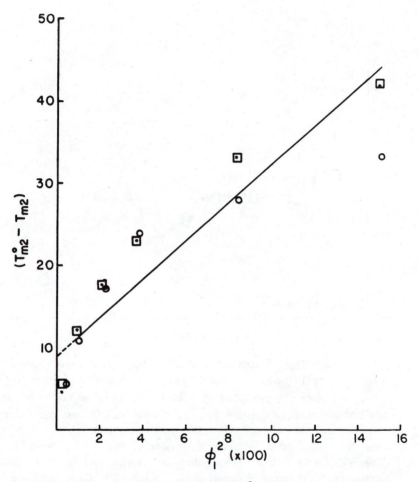

Fig. 6. Relationship between $(T^{\circ}_{m2} - T_{m2})$ and ϕ^2 from eq. 1 for HPC/L blends prepared from dioxan (O), pyridine (●), and from melt (□). According to ref. 5.

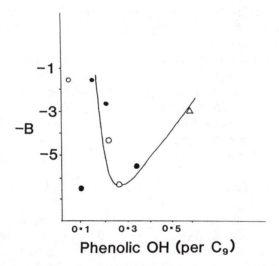

Fig. 7. Relationship between polymer-polymer interaction parameter, B, and a lignin structural feature (i.e., phenolic OH-content per C_9-repeat unit). The lowest value for B denotes greatest interaction between polymers. According to ref. 6.

in lignin (Fig. 7). This indicates that the lignin which has a structure (in terms of functionality) similar to the one in native wood, is the one that provides for conditions most favorable to strong intermolecular interaction between cellulose (derivatives) and lignin, and this favors component miscibility (5-7).

In an attempt to further increase the phase compatibility between lignin and cellulose (derivatives), block copolymers were synthesized which consisted of covalently-linked lignin and cellulose ester segments (or "blocks") (15-16). It was revealed that copolymer architecture, which normally enhances phase compatibility, was unable to provide further improvements in lignin/cellulose (derivative) blend compatibility (16). Glass transition temperatures were shifted towards an intermediate temperature for low molecular weight copolymers, but they did not migrate for high molecular weight components, when blended with cellulose propionate (CP) regardless of whether the lignin was the component of a lignin-CP block copolymer or not (not shown, refer to ref. 16). No improvement in miscibility resulted from the modification of lignin by copolymerization with CP segments.

The blend experiments involving cellulose derivatives and lignin suggest that lignin disrupts both the ordered and the non-ordered forms of cellulose derivative morphology by favoring the formation of an amorphous or liquid crystalline mesophase structure through strong interactive association between lignin and the polysaccharide component.

III. Regenerated Fibers

The spinning of cellulose from an ordered solution state has become commercial practice with the introduction of an N-methyl morpholine-N-oxide-based solvent process (22). Cellulose ester derivatives can also be converted into fibers by spinning from an anisotropic, liquid crystalline solution-state in a variety of solvents (10, 23-26). The formation of anisotropic solutions of cellulose esters in various solvents has been studied in detail (27). Continuous cellulose ester fibers were found to exhibit the expected behavior in terms of mechanical (tensile) properties (i.e., increased strength with increasing orientation) when spun from biphasic or anisotropic solutions (24). The addition of lignin to cellulose and cellulose ester solutions was found to impact the dynamic elastic modulus of concentrated solutions differently: whereas the addition of lignin reduced the dynamic elastic modulus of cellulose ester solutions at all levels of lignin content, cellulose solutions (in DMAc/LiCl) became more viscous (Fig. 8). Considering that isolated lignin has a molecular weight of only ca. 1/100th that of cellulose and cellulose esters, and lignin is usually considered to be a highly compact or spherical molecule, its impact was expected to be one of viscosity-reduction. The fact that the dynamic elastic modulus of cellulose, but not of cellulose ester, solutions increased instead to decline at all shear frequencies is explained with strong secondary interactions of lignin with cellulose in the DMAc/LiCl solvent system.

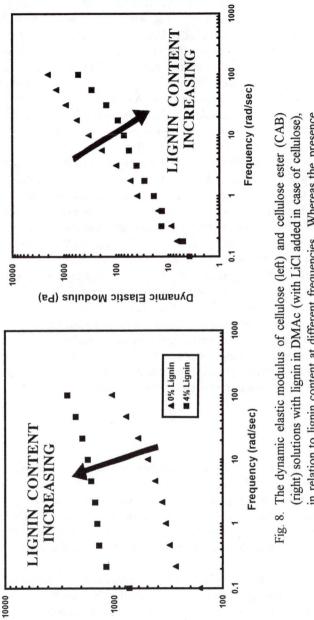

Fig. 8. The dynamic elastic modulus of cellulose (left) and cellulose ester (CAB) (right) solutions with lignin in DMAc (with LiCl added in case of cellulose), in relation to lignin content at different frequencies. Whereas the presence of lignin in the solution raises the modulus of the cellulose/lignin mixture at all frequencies, lignin contributes to a reduction of solution modulus at all frequencies in case of cellulose esters. According to ref. 10.

However, even when cellulose ester (cellulose acetate butyrate, CAB)/L mixtures were spun into continuous fibers from DMAc solution, both fiber tensile strength and modulus increased significantly (10). The strength and modulus (stiffness)-enhancing effect of lignin on cellulose ester fibers was limited to the initial 4%; beyond 4% lignin content, no further positive effect of lignin on fiber strength was noted (Figure 9). It is surprising that the addition of small amounts of (low molecular weight, isolated) lignin neither interfered with the formation of anisotropic solutions nor with the ultimate strength of the resulting fibers (10). A tenacity-increasing effect of small amounts of lignin on cellulose ester fibers can be explained only with a positive effect by lignin on the molecular order of the cellulose derivative in solution and in solid state (10).

The propensity of cellulose esters to form liquid crystalline morphologies has been observed previously (27). A recent study has indicated that the addition of lignin enhances the formation of ordered structures in cellulose acetate butyrate (CAB) (28). Experimental evidence suggests that, as solvent evaporates and both constituents solidify, the surface of phase-separated lignin particles serves to create cholesteric liquid crystalline order by nucleation (Figure 10) (28). The resulting structure provides evidence that lignin phase-separates from cellulose ester derivatives and becomes an integral part of a two-phase architecture in which the degree of organization in the polysaccharidic matrix is substantially increased at the apparent expense of an amorphous phase. The addition of lignin was consistently found to enhance the liquid crystalline mesophase order in non-crystalline cellulose derivatives, and this order is often responsible for increased strength properties. A similar phase-separated morphology also was found in blends of cellulose with lignin (Fig. 11). This morphology reveals heterogeneity at the nano-level, and this provides the basis for a structural hierarchy that has become the trademark of biological materials, such as wood (1).

CONCLUSIONS

Results with blends of cellulose and cellulose derivatives with lignin suggest that the two biopolymers are immiscible.

Experimental evidence supports the hypothesis that lignin enhances the organization of non-crystalline cellulosic structures and gels, and that liquid crystalline mesophase, ordered structures are created that result in multiphase architectures. This enhanced heterogeneity often produces materials with higher modulus and higher strength.

The effect of lignin in blends with cellulose esters is found to disrupt both crystalline order and non-crystallinity by contributing to the formation of a mesophase liquid crystalline morphology that produces an architecture on the dimension of nanometers. This secondary order is held responsible for observed strength gains in lignin/cellulose derivative blends.

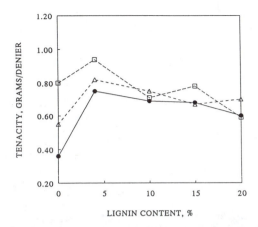

Fig. 9 Relationship between CAB-fiber tensile strength and lignin content at different draw ratios. (Draw ratios increase from 0.8, -●-, to 1.5, -□-, for all lignin contents except 20%, where they ranged between 0.26 and 0.5, respectively.) Tensile strength is seen to increase with lignin content rising to 20%. This is unexpected since lignin has a molecular weight of only ca. 1/100th of that of cellulose derivative and is expected to reduce tensile strength in relation to degree of dilution. According to ref. 10.

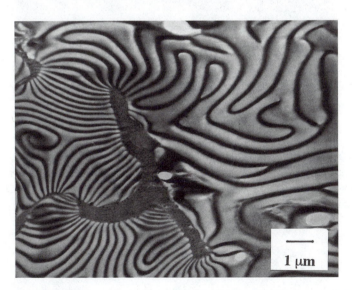

Fig. 10. Transmission electron micrograph (TEM) of a cellulose ester (CAB) film containing 20% lignin. The unstained film shows a phase-separated particle that provides a nucleating surface for cellulose ester liquid crystals. The well-ordered cholesteric arrangement was found to be distinctly more pronounced in the presence of lignin. The periodicity between striation lines was smallest on the surface of the lignin particle indicating lignin's contribution to the cellulose ester's organization. According to ref. 28.

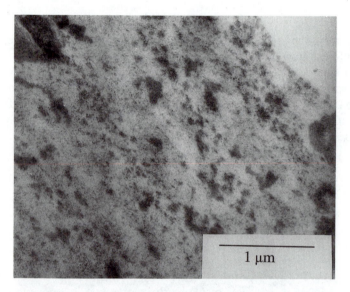

Fig. 11. Transmission electron micrograph (TEM) of the stained cross-section of a cellulose/lignin blend fiber containing 4% (w/w) lignin. There is evidence for an even dispersion of lignin particles in the size-range of 10-20 nm in addition to much larger aggregates.

The results suggest that the natural composite structure of lignified plant materials is not only a consequence of a sequence of well-defined biochemical events, but that it is also a consequence of the thermodynamic driving forces that regulate the interaction between the most prevalent polymer constituents present in lignocellulose, lignin and (cellulosic) polysaccharides.

ACKNOWLEDGMENTS

This report is based on studies financially supported by the National Science Foundation (Washington, D.C.), the Center for Innovative Technology (Herndon, VA), and the USDA (Washington, DC).

LITERATURE CITED

1. National Research Council, "Hierarchical Structures in Biology as a Guide for New Materials Technology," National Materials Advisory Board, ed., National Academy Press, Washington, DC, 1994; pg. 1.

2. G. Jeronimidis, "Wood, One of Nature's Challenging Composites," in "The Mechanical Properties of Biological Materials," SEB Symposium No. 34, Cambridge Univ. Press, Cambridge, England, 1980, 169-182.

3. T. A. Oswald and G. Menges, "Materials Science of Polymers for Engineers," Hanser Publishers, Munich Vienna New York, 1995, 475 pg.

4. H. G. Jung, D. R. Buxton, R. D. Hatfield, and J. Ralph, editors. "Forage Cell Wall Structure and Digestibility," American Society of Agronomy, Inc., Madison, Wisc., 1993; 794 pg.

5. T. G. Rials, W. G. Glasser, J. Appl. Polym. Sci. 37, 2399-2415 (1989).

6. T. G. Rials, W. G. Glasser. Polymer 31, 1333-1338 (1990).

7. T. G. Rials, W. G. Glasser, Wood and Fiber Sci. 21(1),80-90 (1989).

8. V. J. H. Sewalt, W. de Oliveira, W. G. Glasser. J. Sci. Food Agric. 71, 204-208 (1996).

9. G. Garnier, W. G. Glasser, Polymer Eng. Sci. 36, 885-894 (1996).

10. V. Davé, W. G. Glasser, Polymer 38, 2121-2126 (1997).

11. S. S. Kelley, T. G. Rials, W. G. Glasser. J. Mater. Sci. 22, 617 (1987).

12. L. Salmen, J. Mater. Sci. 19, 3090 (1984).

13. A.-M. Olsson, L-Salmén, Chapter 9 in "Viscoelasticity of Biomaterials," W. Glasser and H. Hatakeyama, eds., ACS Symp. Ser. 489, 133-143 (1992).

14. T. K. Kwei, J. Polym. Sci.: Polym. Lett. 22, 307 (1984).

15. W. de Oliveira, W. G. Glasser. Macromolecules 27, 5 (1994).

16. W. de Oliveira, W. G. Glasser. Polymer 35(9), 1977-1985 (1994).

17. N. Terashima, K. Fukushima, L.-F. He, K. Takabe. Chapter 10 "Comprehensive Model of the Lignified Plant Cell Wall," in "Forage Cell Wall Structure and Digestibility," H. G. Sung, D. R. Buxton, R. D. Hatfield,

J. Ralph, Eds., Amer. Soc. Agronomy, Inc., Madison, Wisc., 247-270 (1993).

18. T. G. Rials, W. G. Glasser, J. Appl. Polym. Sci. 36, 749-758 (1988).

19. E. A. Turi, ed. "Thermal Characterization of Polymeric Materials," Academic Press, New York, 1981, 972 pg.

20. P. J. Flory. J. Chem. Phys. 17, 223 (1949).

21. J. E. Harris, D. R. Paul, J. W. Barlow. In "Polymer Blends and Composites in Multiphase Systems," ACS Adv. Chem. Ser., No. 206, C. D. Han, ed., 1984, 17.

22. S. A. Mortimer, A. A. Peguy, Cellulose Chem. Technol. 30, 117-132 (1996).

23. V. Davé, W. G. Glasser. In "Viscoelasticity of Biomaterials," W. G. Glasser and Hatakeyama, eds., ACS Symp. Ser. No. 489, 144-165 (1992).

24. V. Davé, W. G. Glasser, G. L. Wilkes. J. Polymer Sci., Pt. B: Physics, 31, 1145 (1993).

25. V. Davé, W. G. Glasser. J. Appl. Polym. Sci. 48, 683 (1993).

26. V. Davé, J. Wang, W. G. Glasser, D. Dillard. J. Polym. Sci., Pt. B: Physics 32, 1105 (1994).

27. P. Zugenmaier, J. Appl. Polym. Sci.: Appl. Polym. Symp. 37, 223-238 (1983).

28. V. Davé, W. G. Glasser, G. L. Wilkes. Polymer Bulletin 29, 565-570 (1992).

29. J. R. Penacchia, E. M. Pearce, T. K. Kwei; B. J. Bulkia, J.-P. Chen, Macromolecules 19, 973 (1986).

Chapter 20

A Study of the Molecular Interactions Occurring in Blends of Cellulose Esters and Phenolic Polymers

M. F. Davis[1], X. M. Wang[1], M. D. Myers[1], J. H. Iwamiya[2], and S. S. Kelley[1]

[1]Center for Renewable Chemical Technologies and Materials, National Renewable Energy Laboratory, Golden, CO 80401
[2]Advanced Technology Center, Lockheed Martin Missiles and Space, Palo Alto, CA 94304

Relaxation and spin diffusion measurements were performed on a series of blends containing cellulose esters and phenolic polymers such as poly(vinyl phenol) (PVP) and Novolac resins. These experiments were used to determine the extent of molecular-level mixing between the cellulose esters and phenolic polymers. Changes in the peak positions of the phenolic hydroxyl carbon of PVP or Novolac and the carbonyl peak of the cellulose ester were used to study the formation of hydrogen bonds between the two different polymers. ^{13}C CP/MAS and ^{1}H CRAMPS NMR spectroscopy was used to measure the degree of polymer mixing on the molecular scale. Blends of cellulose acetate and phenolic polymers were determined to be mixed on the molecular level. ^{13}C CP/MAS NMR spectroscopy suggested that in the case of cellulose acetate butyrate/phenolic polymers blends that one (or more) butyryl side group attached to a specific position along the anhydroglucose ring was more efficient at forming hydrogen bonds with the phenolic polymers.

Cellulose esters are a class of commercially-important bio-based polymers used for production of fibers, plastics and films. In many of these applications it would be desirable to increase the mechanical strength and improve the melt processibility of the cellulose ester while maintaining optical clarity. There are a number of reports that suggest that the mechanical properties of polymers can be improved and clarity maintained through the preparation of miscible or homogeneous polymer blends (*1*).

Blends of cellulose polymers with synthetic polymers have been studied by several researchers. This work includes mixtures of unmodified cellulose with synthetic polymers (*2*) and blends of synthetic polymers with various cellulose derivatives, including cellulose esters (*3-5*) and cellulose ethers (*6*). Several of these blends appear to be uniformly mixed at the molecular level (2,6), while others appear to be somewhat heterogeneous on a molecular scale (3-5). In the case of the uniformly mixed or miscible blends the mixing seems to be driven by hydrogen-bonding interactions between the cellulosic component and the synthetic polymer.

Miscible polymer blends can be created by selecting polymers that can exploit specific interactions between the different polymers. In order for two or more polymers to form a miscible blend, the free energy of mixing, ΔG_m, must be negative and the second derivative of ΔG_m with respect to the composition must be positive (1). A reduction in the ΔG_m for specific combination of polymers is generally driven by secondary interactions, such as a dipole-dipole interaction or a hydrogen-bonding interaction, which can cause the enthalpy of mixing to be negative or exothermic. The entropic contribution to the ΔG_m will also be slightly negative resulting in the formation of a miscible blend. In the case of blends of cellulose and synthetic polymers the secondary interaction appears to be between the hydroxyl groups on the cellulose and hydroxyl or amide groups on the synthetic polymer. The hydroxyl groups of the cellulose acts as a hydrogen-bonding donor in these blends. However, in the case of cellulose ethers, the hydrogen bonding appears to be between the ether linkages within or between the anhydroglucose rings and hydroxyl groups on the synthetic polymer. The cellulose ether then acts as a hydrogen-bonding acceptor.

Phenolic polymers such as polyvinylphenol (PVP) or Novolac resins are also known to form miscible blends with several polymers, including ester-containing polymers, e.g., acrylates and methacrylates (7-11), and ether-containing polymers, e.g., poly(vinyl methyl ether) and polyethylene glycols (12,13). The miscibility of PVP/methacrylate blends are very sensitive to changes in the chemical structure of the methacrylate polymers. For example, poly(methyl methacrylate) and poly(methyl ethylacrylate are miscible with PVP while poly(methyl butylacrylate) may or may not be miscible, depending on its tacticity. Based on these results, it appears likely that miscible blends of a variety of cellulose esters and phenolic polymers can be created by exploiting the hydrogen-bonding interaction between the phenolic hydroxyl group and either the hydroxyl groups of the cellulose backbone, the ring oxygen, the glycosidic oxygen, or the carbonyl of the substituted ester group. These possible interactions are shown below in Scheme 1.

Recently, in our laboratory, blends of cellulose esters and phenolic polymers have been examined using Differential Scanning Calorimetry (DSC) and Fourier Transform Infrared Spectroscopy (FTIR). Cellulose acetates and cellulose mixed esters blended with either PVP or Novolac resins show a single glass transition temperature (Tg) intermediate between the two pure components. The single Tg varies with the composition of the blend. This observation is consistent with a well-mixed blend, although inhomogeneities on the 10-20 nanometer scale are not easily detected with DSC. Analysis with FTIR spectroscopy shows changes in the both the hydroxyl stretch (3000-3800 cm^{-1}) and the carbonyl stretch (1700-1800 cm^{-1}). These results are consistent with secondary interactions between the phenolic hydroxyl hydrogen and the carbonyl oxygen. Smaller, more subtle changes were also observed in the ether stretching region. As the acetate side groups were substituted with butyrate side groups, it was observed that the hydrogen bonding interaction became weaker as evidenced by smaller shifts observed in the FTIR spectra.

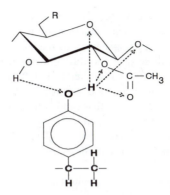

Scheme 1. Possible hydrogen bonding interactions (represented by dashed lines) that could occur in cellulose acetate/poly(vinyl phenol) blends.

Experimental

All of the cellulose esters were commercial samples obtained from Eastman Chemical Co., Kingsport, TN. The cellulose acetate (CA) was CA 398-30 which has a degree of substitution (D.S.) of acetyl groups of 2.45. The cellulose acetate butyrate (CAB) was CAB-381-20 with a acetyl D.S. of 0.4 and a butyryl D.S. of 2.2. The PVP and polystyrene were obtained from Polymer Scientific and had nominal molecular weights of 15,000 daltons. The Novolac polymer was obtained from Plastics Engineering Co., Sheborgan, WI. Solution NMR was used to confirm the chemical composition of these materials.

The blends were created by dissolving both components in acetone at 5% solids. All of the solutions were transparent. The solutions were cast onto glass plates and the solvent allowed to evaporate. All of the resulting films were clear except the CA/poly(styrene) blend. Residual acetone was removed by heating under vacuum at 60 C for 1 hour.

^{13}C cross-polarization/magic-angle spinning (CP/MAS) NMR spectra were obtained using a Bruker DSX-200 spectrometer (4.7T field strength), 1 ms contact time, and a ^{1}H $\pi/2$ pulse width of 4μs. Chemical shifts are externally referenced to the aromatic resonance of hexamethylbenzene (132.3 ppm). Proton combined rotation and multiple pulse (^{1}H CRAMPS) spectra were obtained on a Bruker MSL-200 spectrometer (4.7T field strength) using a tau value of 3.0 μs and $\pi/2$ pulse width of 1.5 μs. The ^{1}H spectra were externally referenced to tetramethylsilane (TMS).

NMR Experiments Exploring the Extent of Polymer Mixing

Nuclear magnetic resonance can be used to study the interactions that are important when different polymers form miscible blends and, by observing the rate of proton spin diffusion, the intimacy of mixing over a range of tens of nanometers (14-20). Spin diffusion is a process in which magnetization is transferred throughout the proton spin reservoir using the dipolar couplings of the nuclear spins. The measurement of spin diffusion can be made directly if the different polymers within the sample can be distinguished by either differences in chemical shifts or can be resolved because of differences in relaxation times. These experiments rely on the application of a suitable filter to create a non-equilibrium state that can be monitored as it returns to equilibrium. Spin diffusion can also be monitored indirectly by measuring relaxation times, such as T_{1H}, the proton spin-lattice relaxation time, or $T_{1\rho H}$, the proton spin-lattice relaxation time in the rotating frame, and observing the effect of the polymer blending on these relaxation times. T_{1H} values have been used to determine a upper limit for the domain size in polymer blends. Generally, T_{1H} values are at least an order of magnitude larger than values for $T_{1\rho H}$ and as such can be used to probe larger distances.

A series of $T_{1\rho H}$ experiments were used to probe the extent of mixing in CA/PVP and CA/PS blends and the results are shown in Table I. The $T_{1\rho H}$ values were measured at 2 locations in the spectrum; 74 ppm, assigned to the C2, C3, and C5 carbons of the CA anhydroglucose unit, and 40 ppm, assigned to the methylene carbons of the PVP backbone. The $T_{1\rho H}$ values measured for the blends were the same (within experimental error) for both components and were intermediate when compared to the $T_{1\rho H}$ values of either the pure CA or PVP. A lower limit to the domain size can be estimated from the $T_{1\rho H}$ data using the relationship (19)

$$x^2 = \frac{4}{3}Dt \qquad (1)$$

where x is the distance that spin diffusion can occur over during a time, t. The diffusion constant, D, is estimated to be $\sim 10^{-12}$ cm^{2}s^{-1}. Using a value of ~ 10 msec (an average $T_{1\rho H}$ value) for t, the domains are estimated to be less than 3 nm.

Table I. $T_{1\rho H}$ values measured for cellulose acetate, poly(vinyl phenol), poly(styrene) and various blends.

Sample	PVP, PS $T_{1\rho H}$ (msec)	CA $T_{1\rho H}$ (msec)
cellulose acetate	----	17.1
poly(vinyl phenol)	7.7	----
25/75 (w/w) CA/PVP	8.8	9.3
50/50 (w/w) CA/PVP	10.1	11.3
75/25 (w/w) CA/PVP	12.8	13.0
50/50 (w/w) CA/PS	6.1	11.4

To contrast to the CA/PVP results, $T_{1\rho H}$ results from a 50/50 mixture of CA/PS are shown in Table I. The $T_{1\rho H}$ value measured for CA is different (and longer) than the $T_{1\rho H}$ value measured for the PS for the 50/50 blend. This indicates that the CA is not intimately mixed with the PS suggesting the importance of the hydrogen-bonding interactions between the phenolic hydroxyls and the cellulose esters in the formation of a well-mixed polymer blend.

T_{1H} values measured for pure CA and PVP were the same (within experimental error) and, as such, were not useful for determining the extent of polymer mixing. It is interesting to note that the T_{1H} of the 50/50 CA/PVP blend (5 seconds) was longer than the T_{1H} of either the CA or PVP polymers (~1.5 seconds). The relaxation time data could be fit to a single exponential indicating that the polymer blend consisted of a single phase. The increase in the T_{1H} for the blend with respect to the constituents suggests that the molecular-level processes that are responsible for the T_{1H} in the blend are occurring on a different time scale. This indicates that the mobility of one or both of the polymers that make up the polymer blend has been effected by the mixing of the polymers. Proton spin-lattice relaxation times, T_{1H}, were also measured for the 50/50 (w/w) CA/PS blend. Different T_{1H} rates were measured for each of the different polymers in the PS/CA blend; 3.0 s for the CA polymer and 6.4 s for the polystyrene. Using equation 1 and the longer PS T_{1H} value, the maximum domain sizes for the 50/50 CA/PS blend are estimated to be larger than 70 nm.

The relaxation results agree with the DSC results (data not shown) that show a single T_g for the CA/PVP blends as opposed to two T_g's, one corresponding to CA and the other corresponding to PS, for the CA/PS blend. The DSC results indicate the polymer separation in the CA/PVP blends to be less than 20 nm (the limit of the DSC sensitivity) and the NMR $T_{1\rho H}$ results extend this limit to less than 3 nm.

Spin diffusion between the CA/PVP and CA/PS blends was also probed directly by using a chemical shift filter to select only the proton magnetization from aromatic protons. The extent of polymer mixing is determined by observing how quickly the magnetization is transferred between the different polymers comprising the polymer blend. The chemical shift filter used was a proton analog of the SELDOM (selectivity by destruction of magnetization) sequence used to select magnetization from specific resonances that are observed using a dilute spin such as ^{13}C (*21*). The SELDOM strategy uses a train of $\pi/2$ pulses spaced with an appropriate delay to select the resonance of interest. In proton spin systems, the proposed selection scheme based on a SELDOM-type strategy can be diagrammed as follows:

$$\{prep_y - [multipulse\ cycle]_n - prep_{-y} - \tau\}_k - \tau_{ddk} - multipulse\ detection \qquad (2)$$

The initial pulse, $prep_y$, is used to place the proton magnetization into the toggling frame appropriate for the multiple pulse sequence chosen, i.e., a $\pi/2$ pulse is chosen for MREV-8, a $\pi/4$ pulse is chosen for BR-24. The proton magnetization is then allowed to evolve due to chemical shift interactions while homonuclear interactions are removed by the multipulse sequence. In our experiments, we found it convenient to place the frequency of the transmitter off resonance and adjust n, the number of multiple pulse cycles, such that the magnetization to be selected made one complete revolution in the toggling frame, taking into account the appropriate scaling factor of the multiple pulse sequence used. The cycle was repeated k times until the desired selectivity is achieved.

Generally, transmitter offsets of ~5 to 7 KHz were required for the selection experiments compared to typical multiple pulse offsets of 2-3 KHz. The larger transmitter offsets used in the selection experiment do degrade the resolution of the spectra to a small degree. The spinning speed is also adjusted such that an integer number of rotor cycles occurs during the multiple pulse evolution period. This sequence was applied to 50/50 (w/w) blends of CA/PVP and CA/PS.

Typical ^{1}H CRAMPS spectra (without the chemical shift filter) of CA, PS, PVP, 50/50 CA/PS and 50/50 CA/PVP using a BR-24 pulse sequence are shown in Figure 1. The CA spectrum consists of a peak centered at ~2 ppm, assigned to the acetate methyl groups and at least two broader peaks centered around ~4.5 ppm assigned to cellulose hydroxyls and protons attached to the cellulose backbone. The PVP and PS spectra are characterized by 2 peaks centered around ~2 ppm, assigned to the methylene protons, and 7 ppm, assigned to aromatic protons. The phenolic hydroxyl protons of PVP should also resonate at about 7 ppm..

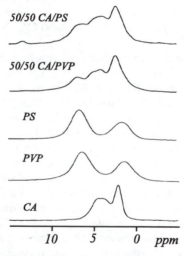

Figure 1. Typical CRAMPS spectra of cellulose acetate (CA), poly(vinyl phenol) (PVP), Poly(styrene) (PS), a 50/50 (w/w) blend of CA and PVP (50/50 CA/PVP) and a 50/50 (w/w) blend of CA and PS (50/50 CA/PS).

The extent of polymer mixing is probed using the pulse sequence diagrammed in (2) and incrementing the delay, τ_{dd}, after the aromatic magnetization from the PVP or PS has been selected. During the time period, τ_{dd}, spin diffusion will transfer magnetization to spatially nearby protons. The time required for PVP proton magnetization to be transferred to the protons of the CA is determined by the extent of mixing in the PVP/CA polymers blend. Figure 2a shows the spectrum of the 50/50 (w/w) CA/PVP blend after the application of the chemical shift filter to select the PVP aromatic protons centered around 7 ppm. The small upfield shoulder at about 5 ppm indicates that a small amount of residual magnetization from the CA remains due to inefficiencies in the

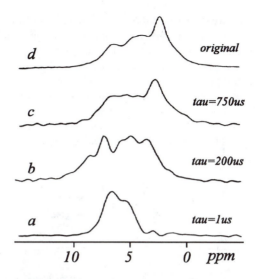

Figure 2. ^{1}H spectra of 50/50 (w/w) CA/PVP as a function of the spin diffusion mixing time, τ_{dd}. a) τ_{dd}=1 µs b) τ_{dd}=200 µs c) τ_{dd}=750 µsec and d) ^{1}H CRAMPS spectrum without application of the chemical shift filter.

chemical shift filter. Magnetization is observed to grow in rapidly between 0 and 5 ppm and after a spin diffusion mixing time of 750 µs, the spectrum (Figure 2c) looks similar to the typical ^{1}H CRAMPS spectrum of the CA/PVP blend (Figure 2d). The clear observation of spin diffusion between the PVP and CA polymers at 200 µs (Figure 2b) indicates that the polymers must be mixed at the molecular level because for spin diffusion to occur on this time scale the protons have to be within 2-5 angstroms (*19*).

In contrast are the results of the spin diffusion experiment applied to the 50/50 (w/w) CA/PS blend shown in Figure 3. Figure 3a shows the spectrum obtained immediately after the application of the chemical shift filter to select the aromatic protons of the PS. Again there is a small upfield shoulder due to residual CA magnetization. At 100 µs (Figure 3b), the residual CA magnetization has equilibrated throughout the CA spin system, but it appears that no magnetization is being transferred from the PS aromatic protons to the CA polymer. At 1000 µs (Figure 3c), the spectrum looks similar to the polystyrene ^{1}H CRAMPS spectrum (Figure 1(PS)) indicating that magnetization has been transferred throughout the polystyrene spin system. However, the lack of any observable magnetization that can be assigned to cellulose acetate in the τ=1000µs spectrum indicates that the polystyrene must not be in intimate contact with the cellulose acetate.

NMR Experiments Exploring Interactions Between Functional Groups

Changes in NMR chemical shifts can be used to probe sites of specific interactions in solids although not with the predictability that is associated with infrared spectroscopy.

Belfiore and coworkers have demonstrated that the chemical shift of the phenolic carbon of PVP is sensitive to nearest neighbor interactions that can distort the electronic environment of the phenolic carbon (22-24). Hydrogen-bonding interactions between PVP and polymers such as poly(ethylene succinate), PES, and poly(vinyl methyl ketone), PVMK, have been studied previously. The chemical shift of the phenolic carbon of PVP and the carbonyl carbons of PES and PVMK were sensitive probes of hydrogen bonds formed during mixing of the polymers. However, the magnitude of the change in the chemical shift of the phenolic carbon cannot be used to determine the strength of the hydrogen bond because other factors can cause changes in the carbon chemical shifts (24).

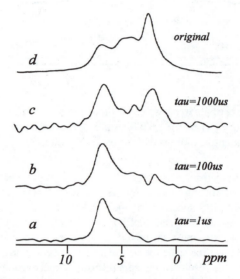

Figure 3. ^{1}H spectra of 50/50 (w/w) CA/PS as a function of the spin diffusion mixing time, τ_{dd}. a) τ_{dd}=1 µs b) τ_{dd}=100 us c) τ_{dd}=100 µsec and d) ^{1}H CRAMPS spectrum without application of the chemical shift filter.

An expansion of the ^{13}C CP/MAS spectrum showing the carbonyl and phenolic hydroxyl carbon region for a series of CA/PVP blends differing in CA/PVP composition are shown in Figure 4. The peak appearing at 153.5 ppm in the 0/100 (w/w) spectrum (pure PVP) is assigned to the carbon in the aromatic ring bearing the phenolic hydroxyl and the peak at 171.5 ppm in the 100/0 (w/w) spectrum (pure CA) is assigned to the carbonyl carbons of the acetate groups attached to the cellulose backbone.

As the amount of CA is increased in the CA/PVP blends, differences can be clearly seen in the spectra of the different blends. The first observation is that the hydroxyl carbon of PVP is seen to shift from 153.5 ppm in the 0/100 CA/PVP spectrum to 155.1 ppm in the 75/25 CA/PVP spectrum. The carbonyl peak of the CA broadens upon the addition of the PVP with the FWHH (full width at half height) increasing from

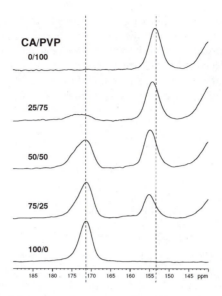

Figure 4 Expansion of the carbonyl and phenolic carbon region of the ^{13}C CP/MAS spectrum of blends of cellulose acetate (CA) with poly(vinyl phenol) (PVP). The peak heights are scaled to arbitrary intensity. The dotted lines indicate the chemical shifts of undiluted CA and PVP.

4.3 ppm in the 100/0 spectrum to 7.0 ppm in the 25/75 spectrum. The broadening of the carbonyl peak appears to be due to a shift of intensity from 171.5 ppm to about 174 ppm. The changes in chemical shift observed in the NMR spectra are consistent with changes observed in FTIR spectra obtained from the same samples (data not shown). Therefore, we conclude that the peak shifts observed in the NMR spectra reflect hydrogen-bonding interactions of different strengths similar to the conclusion made by Belfiore and coworkers (22-24).

The 90/10 CA/PVP spectrum (Figure 5) shows the appearance of a new peak at 161 ppm (this peak is also present in the 75/25 CA/PVP spectrum but cannot be clearly seen in Figure 4). The appearance of the 161 ppm peak in the 90/10 and, to a lesser extent, 75/25 CA/PVP blend spectra indicates that there may be phenolic hydroxyls that form stronger hydrogen bonds at higher CA concentrations than is observed at lower CA concentrations. At the higher CA concentrations, the phenolic hydroxyls may be able to form hydrogen bonds with the unreacted CA hydroxyl.

A second possible explanation of the 161 ppm peak may be that more than one CA carbonyl (or other hydrogen-bonding site) is interacting with a single phenolic hydroxyl and the multiple hydrogen bonds cause the larger change in the PVP phenolic carbon chemical shift. The disappearance of the 161 ppm peak at higher PVP concentrations then indicates that these multiple-hydrogen bonds (or stronger hydrogen bonds) are not present at lower CA concentrations as more phenolic hydroxyls become available for hydrogen bonding.

The FWHH of the phenolic carbon decreases from 4.0 ppm to 3.0 ppm as the concentration of CA increases from 0% to 90% . The trend in decreasing linewidth follows a trend in decreasing T_g as one lowers the PVP concentration. The measured Tg of the 25/75 CA/PVP blend is 175 °C which decreases to 155 °C in the 90/10 CA/PVP blend. A narrowing of the phenolic carbon in PVP blends with decreasing T_g has been observed previously by Qin and coworkers in their study of PVP/poly(vinyl methyl ketone) blends (24).

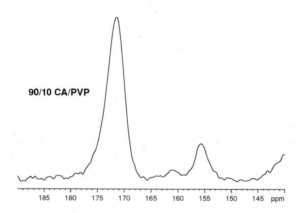

Figure 5. Expansion of the carbonyl and phenolic carbon region of the ^{13}C CP/MAS spectrum of a 90/10 blend of cellulose acetate (CA) with poly(vinyl phenol) (PVP).

Interactions between different functional groups were also explored by changing the nature of the polymers constituting the blend. One modification of the polymers studied was replacement of the acetate side groups with bulkier, more hydrophobic, butyryl sidechains that could hinder the formation of hydrogen bonds between the carbonyls on the cellulose ester and the phenolic hydroxyls on the PVP. The CAB/PVP blends form a single phase as indicated by a single T_g and FTIR spectra indicate the presence of hydrogen bonding between the phenolic hydroxyls and the carbonyls. The ^{13}C CP/MAS spectra of the CAB/PVP blends (Figure 6) show changes that are consistent with the formation of weaker hydrogen bonds between the CAB and the PVP when compared to the CA/PVP blends.

There is a smaller shift in the phenolic carbon of PVP from about 153.5 ppm to 154.2 ppm that would suggest a weaker hydrogen bonding interaction, although Belfiore and coworkers have suggested that factors other than the strength of the hydrogen bonding may also effect the phenolic carbon chemical shift. The unresolved group of peaks centered around 172.5 ppm in Figure 6 is assigned to butyryl carbonyls attached at different positions on the cellulose backbone. This assignment is consistent with liquid-state spectra of CAB that show the carbonyls bonded to different carbons on the

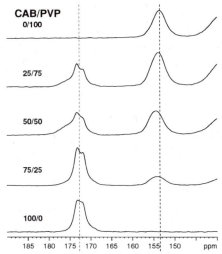

Figure 6. Expansion of the carbonyl and phenolic carbon region of the ^{13}C CP/MAS spectrum of blends of cellulose acetate butyrate (CAB) with poly(vinyl phenol) (PVP). The peak heights are scaled to arbitrary intensity. The dotted lines indicate the chemical shifts of undiluted CAB and PVP.

cellulose backbone have different chemical shifts (25,26). To our knowledge, the assignment of the unresolved peaks have not been made in the solid state. The smaller shoulder centered at ~ 169 ppm may be due to residual acetyl carbonyls present in the polymer. Formation of a broad downfield shoulder on the carbonyl peak centered around 172.5 ppm indicates that the carbonyls of the butyryl side groups are involved in hydrogen bonding with the phenolic carbons. It is apparent from the changes in the relative intensities of the different unresolved carbonyl peaks as the PVP content increases that a butyryl group attached to at least one position may be favored in the formation of hydrogen bonds although the exact substitution position can not be determined without further assignment of the carbonyl region in the solid state spectra. The change in the relative peak heights could also be caused by a change in molecular motion of the butyryl sidechains due to the hydrogen bonding. A change in molecular motion of the butyryl sidechain could alter the carbonyl linewidths and intensities.

Further evidence of a possible favored carbonyl position in the hydrogen-bonding formation can be observed in blends of the CAB with a Novolac resin. The phenolic hydroxyls in this Novolac resin are relatively hindered compared to PVP in terms of their ability to hydrogen bond because of the presence of ortho methylene bridges adjacent to the phenolic hydroxyl group. These methylene bridges also restrict the number of confirmations that the aromatic rings can assume, further hindering the formation of hydrogen bonds. Changes in the ^{13}C CP/MAS spectra of CAB/Novolac blends, shown in Figure 7, are consistent with the formation of hydrogen bonding interactions between the CAB and Novolac resin. There is a shift in the phenolic carbon from 150.6 ppm to 152.3 ppm and a broad downfield shoulder appears in the carbonyl region centered around 172.5 ppm suggesting that the phenolic hydroxyls and butyryl carbonyls are forming new hydrogen bonds. The unresolved butyryl carbonyl peaks clearly show changes in the relative intensities as the Novolac content is increased. These changes in intensity again indicate that one or more of the butyryl positions may be favored. It is interesting that the changes in the relative intensities of the unresolved carbonyl peaks appear to be greater in the CAB/Novolac blends than the CAB/PVP blends.

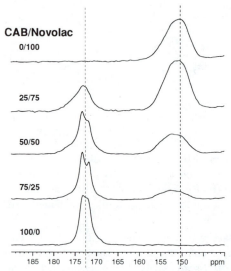

Figure 7. Expansion of the carbonyl and phenolic carbon region of the ^{13}C CP/MAS spectrum of blends of cellulose acetate butyrate (CAB) with an uncured Novolac resin. The peak heights are scaled to arbitrary intensity. The dotted lines indicate the chemical shifts of undiluted CAB and Novolac resin.

Conclusions

Based on relaxation and spin diffusion measurements the results of this study clearly show that cellulose ester and phenolic polymers are intimately mixed at the molecular level with interproton separations of the two polymers of about 5 angstroms. The same experiments performed on blends of cellulose acetate and poly(styrene) showed no evidence of molecular-level mixing between the polymers and that the domains must be larger than 70 nanometers. Changes in the peak positions of the carbonyl peak of the cellulose esters and the phenolic hydroxyl carbon of the phenolic polymers were consistent with the formation of hydrogen bonds between the two different polymers. ^{13}C CP/MAS NMR spectroscopy of blends of CAB/PVP and CAB/Novolac suggested one or more butyryl side groups attached to a specific position along the cellulose backbone may have been more efficient at forming hydrogen bonds.

Literature Cited

(1) Krause, S. In *Polymer Blends*; Paul, D. R.; Newman, S., Eds.; Adademic Press, Inc., New York, New York, 1978, Vol. 1; 16-106.

(2) Masson, J.; Manley, R St. J.. *Macromolecules* **1992**, 25, 589.

(3) White, A. W.; Buchanan, C. M.; Pearcy, B.G.; Wood, M. D. *J. Appl. Polym. Sci.* **1994**, 52, 525.

(4) Sun, J.; Cabasso, I. *Macromolecules*, **1991**, 24, 3603.

(5) Aptel, P.; Cabasso, I. *J. Appl. Polym Sci.* **1980**, 25, 1969.

(6) Kondo, T.; Sawatari, C.; Manley, R St. J. .; Gray, D. G. *Macromolecules* **1994**, 27, 210.

(7) Goh., S.H.; Slow, K.S. *Polym. Bull.* **1987**, 17, 453.

(8) Serman, C. J.; Xu, Y.; Painter, P.C.; Coleman, M. M. *Macromolecules* **1989**, 22, 2019.

(9) Jong, L.; Pearce, E. M.; Kwai, T. K. *Polymer*, **1993**, 34, 48.

(10) Fahrenholtz, S. R.; Kwei, T. K. *Macromolecules* **1981**, 14, 1076.

(11) Pennacchia, J. R.; Pearce, E. M.; Kwei, T. K.; Bulkin, B. J.; Chen, J-P. *Macromolecules* **1986**, 19, 973.

(12) Serman, C. J.; Xu Y.; Painter, P. C.; Coleman, M. M. *Polymer* **1991**, 32, 516.

(13) Moskala, E. J.; Varneli, D. F.; Coleman, M. M. *Polymer* **1985**, 26, 228.

(14) Caravatti, P.; Neuenschwander, P.; Ernst, R.R. *Macromolecules* **1985**, 18, 119.

(15) Caravatti, P.; Neuenschwander, P.; Ernst, R.R. *Macromolecules* **1986**, 19,1889.

(16) Campbell, G. C.; VanderHart, D. L. *J. Magn. Reson.* **1992**, 96, 69.

(17) Clauss, J.; Schmidt-Rohr K.; Spiess H. W. *Acta Polymer* **1993**, 44, 1.

(18) Yang, H.; Kwei, T. K.; Dai, Y. *Macromolecules* **1993**, 26, 842.

(19) VanderHart, D. L. *Macromolecules* **1994**, 27, 2837.

(20) VanderHart, D. L. *Macromolecules* **1994**, 27, 2826.

(21) Tekely, P.; Brondeau, J.; Elbayed, K.; Retournard, A.; Canet, D. *J. Magn. Reson.* **1988**, 80, 509.

(22) Belfiore, L. A.; Qin. Q; Pires, A.; Ueda, E. *Polym. Prep. (Am. Chem. Soc., Div. Polym. Chem.)* **1990**, 31, 170.

(23) Belfiore, L. A.; Lutz, T. J.; Cheng, C.; Bronnimann, C. E. *J. Polym. Sci., Polym. Phys. Ed.* **1990**, 28, 1261.

(24) Qin, C.; Pires, T. N.; Belfiore, L. A. *Macromolecules* **1991**, 24, 666.

(25) Buchanan, C. M.; Hyatt, J. A.; Lowman, D. W. *Macromolecules* **1987**, 20, 2750.

(26) Tezuka, Y. *Carbohydrate Res.* **1993**, 241, 285.

Chapter 21

Interchain Hydrogen Bonds in Cellulose–Poly(vinyl alcohol) Characterized by Differential Scanning Calorimetry and Solid-State NMR Analyses Using Cellulose Model Compounds

Tetsuo Kondo[1] and Chie Sawatari[2]

[1]Forestry and Forest Products Research Institute (FFPRI), P.O. Box 16, Tsukuba Norin Kenkyu, Ibaraki 305, Japan
[2]Faculty of Education, Shizuoka University, Shizuoka 422, Japan

To characterize the nature of the interchain hydrogen bonds involved in cellulose/PVA blends, DSC and solid-state NMR analyses were carried out on mixtures of two cellulose model compounds (2,3-di-*O*-methycellulose; 23MC and 6-*O*-methyl-cellulose; 6MC) and PVA. The conclusions reached, based on our experimental data, are as follows: first, the thermodynamic data confirms that the interchain hydrogen bonds between the cellulose ring oxygen and the OH group of PVA are favored over bonds formed between the hydroxyl groups on each homopolymer. Second, the proton spin-lattice relaxation time in rotating frames, $T_{1\rho}H$, showed that both MC/PVA blends examined were homogeneous down to a scale of 17.5 nm or less at most concentrations, suggesting that blends of pure cellulose and PVA may also be miscible on a molecular scale. Finally, we comment on the correlation between the interaction parameter, χ, assessed from thermodynamic data obtained using DSC and the domain size for the hydrogen bonding pairs in the blends.

Since the late 1970's cellulose/synthetic polymer blends have been extensively studied while more recently attention has been focussed on biodegradable polymer blends(*1-15*). The polymer-polymer interactions believed to be reponsible for the miscibility are mainly attributed to hydrogen bonds formed between the three hydroxyl groups on the anhydroglucose units of cellulose and the functional groups present on the synthetic polymers. More recently, the domain size or good-mixing scale was estimated by measuring the T_1 and $T_{1\rho}$ proton with solid state CP-MAS ^{13}C-NMR (*16-18*). However, in cellulose/PVA blends it has been difficult to accurately characterize the hydrogen bonds involved in the process because there are three hydroxyl groups in each repeating unit of the cellulose. In a previous paper (*19*), the authors reported on the FT-IR differences of the hydrogen bonds formed depending on the regiochemistry with regard to the miscibility in certain cellulosic blends. The work was carried out using cellulose model compounds with regioselectively methylated hydroxyl groups, namely 2,3-di-*O*-methylcellulose (23MC)(*20*) and 6-*O*-methylcellulose (6MC)(*21*) (Figure 1). From the FT-IR study of the MC/PVA blend samples, it was revealed that two types of hydrogen bonds can be formed in cellulose /PVA blends: i) hydrogen bonds between the OH groups at either the C-2 or the C-3 position of cellulose and the side chain OH groups of PVA (these interactions are postulated to form mainly at the C-2 position of cellulose), and ii) hydrogen bonds between the ring oxygen (O-5) of cellulose and the OH groups in PVA (Figure 1). In the present study we want to clarify further the regiochemical effects on the pure cellulose/PVA blend previously investigated (19). We also estimated quantitatively the thermodynamic interaction of the hydrogen bond formation between the two regioselectively methylated cellulosics and PVA using DSC analysis and we extrapolated these results to the case of pure cellulose/PVA blends. Further, the domain size was also evaluated by spin diffusion from solid-state NMR measurements.

Experimental
Materials. The two cellulose model samples, 2,3-di-O-methylcellulose (23MC)(20) and 6-O-methylcellulose (6MC)(21), were prepared according to ref19. Each polymer had a uniform structure having every structural unit substituted regioselectively. Poly(vinyl alcohol) (PVA) was purchased from Polyscience Inc. and it had a nominal molecular weight of 2.5 x 10^4. HPLC-grade N,N-dimethylacetamide (DMAc) (Aldrich Chemical Co., Inc.) was used without further purification.
Preparation of film specimens. MC/PVA blends were prepared from mixed polymer solutions in DMAc according to the film casting method described in an earlier paper[19]. The two film blend systems subjected to DSC measurements with a Perkin-Elmer DSC-7 apparatus had relative compositions of the two polymers (MC / PVA) of 80/20, 65/35, 50/50, and 30/70 by weight. For the solid-state NMR measurements, the compositions were 75/25, 50/50 and 25/75 by weight, respectively for both the 23MC/PVA and 6MC/PVA mixtures. Wide-angle X-ray diffraction patterns (19) verified that the blended films were predominantly amorphous. The densities for the two highly amorphous cellulosic homopolymers (23MC and 6MC) and for the PVA films were measured by pycnometry in a mixed medium of p-xylene and carbon tetrachloride. The characterization data for the three films studied are listed in Table I.
DSC Measurements. Film specimens weighing from 5 to 17 mg were placed in aluminum sample pans which were heated to 242° C at the heating rate of 10°C/min and maintained at this temperature for 5 minutes in an atmosphere of nitrogen to eliminate PVA crystalline residues. The samples were quenched at the selected isothermal crystallization temperature T_{ic}, were then held at T_{ic} for 7 hours to allow complete crystallization and then were cooled to 20° C. After each sample was isothermally crystallized, the melting point T_m was then measured using a heating rate of 10°C/min. Subsequently by using Hoffman-Weeks plots (22) of the Tm's thus obtained, the equilibrium melting point (Tm^{eq}) was determined for each film.
Solid-state NMR Measurements. Before NMR measurements the samples were completely dried for more than a week under vacuum at 80° C. NMR spectra were obtained on a JEOL JNM-GSX 400 instrument equipped with a dedicated solids accessory. The CP/MAS measurements were obtained at 9.4T, which corresponds to frequencies of 100.4 MHz for ^{13}C and 400 MHz for protons at room temperature. Spinning rates were 5.0 to 6.0 kHz, and the Hartmann-Hahn match was adjusted prior to each accumulation. Cross-polarization times were typically 1 to 2 ms. The recycle time between pulses was 10s. The spectra were accumulated ca. 500 times. The CH signal of adamantane was used as an external reference to determine chemical shifts. The pulse sequence for $T_{1\rho}H$ measurements was done as described in a previous paper (18). Measurements of proton spin-lattice relaxation times in the

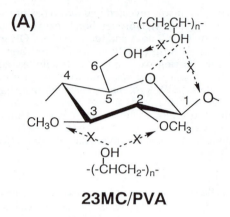

23MC/PVA

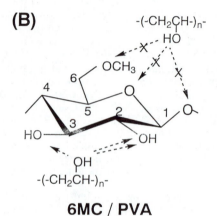

6MC / PVA

Figure 1. Proposed hydrogen bonding schemes for the two cellulosics blended with PVA, (A) 23MC/PVA and (B) 6MC/PVA.

Table I Characterization data for the homopolymer components.

Sample	Source	Molecular weight	Density (gcm^{-3})
23MC	Synthesized	4.0×10^4	1.30
6MC	Synthesized	3.5×10^4	1.30
PVA	Polyscience	2.5×10^4	1.27

rotating frame ($T_{1\rho}H$) were obtained by using a computer-generated best fit for the intensity of the ^{13}C-NMR spectra to the single-exponential equation $M(t)=M(0)exp(-\tau/T_{1\rho}H)$. The proton spin lock time (delay times), τ, ranged from 1 to 35 ms depending on the blend under investigation and eight τ values were employed to determine each relaxation time. Each plot used to obtain $T_{1\rho}H$ was a single exponential.

Results and Discussion
DSC Characterization.
Since both amorphous 6MC and 23MC homopolymers showed diluent effects in our blend systems, we have assumed the change in the chemical potential with increasing amount of the MC components (*4,6,9, 23-29*) reflected itself as a depression in the melting point of the PVA crystallized isothermally in the MC/PVA blends. The thermodynamic mixing of the two polymers has been dealt with by Scott (*30*) using the Flory-Huggins approximation (*31*). The conventional equation for the thermodynamic depression of the melting point caused by a diluent is as follows:

$$1/T_m - 1/T_m^0 = -R(V_{2u}/\Delta H_{2u})$$

$$X \ \{ \ lnv_2 \ / V_2 + (1/ V_2 - 1/ V_1)v_1 + Bv_1^2/RTm \} \tag{1}$$

where T_m^0 is the melting point of PVA and T_m is the observed melting point of the blended PVA. In this equation, 1 and 2 refer to MC and PVA, respectively. v_1 and v_2 are the polymer volume fractions while V_1 and V_2 are molar volumes. V_{2u} is the molar volume of the repeating units of 2, ΔH_{2u} is the enthalpy of fusion per mol of the repeating units of 2, and v_1^2 is the square of the volume fraction of non-crystallizable component 1 (MC).

Since V_1 and V_2 are on the order of 10^4 for 23MC, 6MC and PVA the entropy term in eq 1 can be entirely neglected[24]. Eq 1 can then be rearranged into the following form (eq 2) where the enthalpic contribution to the melting point depression can be evaluated:

$$\Delta T_m = T_m^0 - T_m = -T_m^0 (V_{2u}/\Delta H_{2u})Bv_1^2 \tag{2}$$

where ΔT_m is the melting point depression of the PVA component and R is the universal gas constant, 1.986 cal·deg^{-1}·mol^{-1}. Eq 3 relates the B parameter, or interaction energy density characteristic of the two polymers, to the Flory-Huggins interaction parameter, χ_{12}, which describes the enthalpy of mixing:

$$B=RT(\chi_{12}/V_{1u}) \tag{3}$$

However, there are some morphological effects which must be considered; specifically the effects are known to be mainly due to the degree of perfection and the finite size of the crystals (*22*). To cancel these effects, the equilibrium melting points, Tm^{eq0} and Tm^{eq}, obtained from the Hoffman-Weeks plots (*22*) were used instead of Tm^0 and Tm respectively in eq.2. In order to use the Hoffman-Weeks plots, it is necessary that the PVA component be completely crystallized isothermally. Thus, the Tm^0 and Tm^{eq} were determined from the Hoffman-Weeks plots using the melting temperatures, Tm's. In Figure 2, the experimental data for the melting point depression, ΔT_m, for each system were plotted versus the square of the volume fraction of the cellulosic component, v_1^2, so we could use eq. 2. The solid line was drawn by using a least-square fitting method assuming there is a linear relationship between ΔT_m and $v_1{}^2$. The volume fraction was calculated using density data for the cellulosics and PVA as noted in Table I. Both straight lines yielded positive intercepts. Residual entropic effects might be responsible for the non-zero intercept (23-29). However, in the present cases the low magnitude of the intercepts makes their contribution negligible. From the slopes of the two ΔT_m versus $v_1{}^2$ plots, we can assess values for the B parameter and χ_{12} by using eq. 2 and 3 in combination with other known necessary quantities as follows; the heat of fusion per unit volume $(\Delta H_{2u}/V_{2u})$ of PVA[6] is 45.4 cal·cm^{-3} and V_{1u}=146.15 and 135.38 (cm^3·mol^{-1}) for 23MC and 6MC, respectively. V_{1u} values are calculated from the molar masses (190 and 176 for 23MC and 6MC) and densities (Table I) for each cellulosic. The slopes of the two straight lines were 155.5 deg and 110.0 deg for the 23MC/PVA and 6MC/PVA, respectively.

From these calculations, the thermodynamic interaction parameters were evaluated (Table II). The fairly large negative values for the B and χ_{12} parameters for both of the MC/PVA blends can be explained by the two components interacting thermodynamically in the mixture. The values were also greater than that for a pure cellulose/PVA blend, indicating that the two polymer pairs present were more favorably miscible than a pair consisting of pure cellulose and PVA. In other words, the two model systems showed enhanced miscibility. The large negative values of B for the model systems were estimated to be 2 to 6 times larger than those reported for other polymer pairs (23-29) while the B value of a cellulose/PVA system was only 1.5 to 4 times (6) the B values for other polymers. This result strongly suggests the presence of favorable interactions presumably due to the formation of interchain hydrogen bonding. This conclusion is also supported by our previous FT-IR results (19).

The difference in B values between 23MC/PVA and 6MC/PVA is assumed to reflect mainly the difference in strength of the interchain interactions involved in hydrogen bonding in the blends. In a previous paper using FTIR (19), we reported that in 23MC/PVA hydrogen bonding takes place between an ether oxygen in the anhydroglucose ring and a hydroxyl group of PVA, while the hydrogen bonds in 6MC/PVA were formed mainly between the hydroxyl groups of each component. Thus, in cellulosic blends the formation of hydrogen bonds between ring oxygens and OH groups is considered to be more favorable than those between component hydroxyl groups. In addition, since the former type of hydrogen bond (O-HO) is considered to be very similar to one forming intramolecular hydrogen bonds between ring oxygens and hydroxyls at the C-3 position in cellulose molecules, this result also indicates that the intramolecular hydrogen bonds formed should be strong as is well-known in cellulose chemistry.

Proton T_1 and $T_{1\rho}$ Measurements.   The relaxation of proton spins in the spin-lattice (T_1H) and the rotating frame ($T_{1\rho}H$) is observed by its effect on the NMR of the carbon nucleus to which the proton is bound. If the different protons are not in contact, each has its own characteristic relaxation time. At the other extreme, if the protons are closely coupled by rapid spin-spin exchange dipole-dipole interactions, i.e. spin diffusion, then these protons share the same $T_{1\rho}H$. In miscible polymer blends on the scale characterized by the relaxation time, the measured proton relaxation rate is an average of the proton relaxation rates for the constituent polymers (16-18,38). The relaxation measurement method is complementary to CP-MAS since the relaxation times are sensitive to homogeneity scales which differ from the scale to which CP-MAS is sensitive. It has been previously reported that the

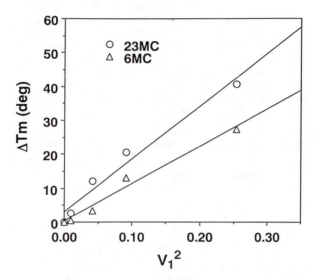

Figure 2. A plot of the melting point depression, ΔT_m, *vs.* the square of the volume fraction of the cellulosics, $v_1{}^2$, for 23MC/PVA (◯) and 6MC/PVA(△).

Table II Values of the interaction energy density, B, and of the interaction parameter, χ_{12}, for binary blends with components compatible in the melt.

	Cellulose/PVA[6]	23MC/PVA	6MC/PVA
B (Cal. cm-3)	-9.38	-13.65	-9.65
χ_{12} (at 517 K)	-0.985 (513K)	-1.94	-1.32

$T_{1\rho}H$ values for cellulose/PVA blends could not be used to calculate a domain size, because the $T_{1\rho}H$ value for both blend components were virtually identical within the experimental error. In contrast, in the present study such a problem did not exist since both 23MC and 6MC had significantly different $T_{1\rho}H$ values from PVA.

The measured $T_{1\rho}H$ values for the protons attached to the characteristic carbon nuclei at 107, 104 and 45 ppm for 23MC, 6MC and PVA, respectively, are listed in Table III for each blend studied. As denoted by the underlined values in Table III, the compositions which gave identical component relaxation times for 23MC/PVA and 6MC/PVA were 75/25 and 25/75, respectively. Thus, at these compositions both blends were thoroughly-mixed at this measurement level. Since the 75/25 composition of 23MC/PVA is the closest to a molar ratio of 1/1 among the studied compositions, the miscible 23MC/PVA blend was considered to be equivalently mixed in the two components. On the other hand, 6MC and PVA were well-mixed only when an excess of PVA was present. This behavioral difference might be due to the distinctive type of hydrogen bond interactions, previously described, in the two blends. In the above cases when the protons for the two components in the blend have the same $T_{1\rho}H$ values, then eq. 4 (39) can be used if the molecule are in contact at an intimate molecular distance, L, over which the protons can effectively diffuse in a given time, t.

$$<L^2> \approx (t / T_2) \cdot < l_0{}^2>$$ (4)

where l_0 is the distance between protons, typically 0.25nm and T_2 is the proton spin-spin relaxation time, ca.10 μs below Tg (39). The available time for spin diffusion, t, is usually equal to $T_{1\rho}H$. Using the average $T_{1\rho}H$ values of t=30 ms for 23MC/PVA and 49ms for 6MC/PVA, we calculated the observation diameter for spin diffusion to be about 13.7 nm (23MC/PVA) and 17.5 nm (6MC/PVA). Hence, each pair in 23MC/PVA and 6MC/PVA is homogeneous on a scale of 13.7 and 17.5 nm or less, respectively. These values are significantly different considering the error range of ± 5%.

Despite being highly amorphous, as described in the experimental section, the PVA showed a much longer $T_{1\rho}H$ time of 15.9 ms, than reported previously by Masson and Manley (18) (4.3 ms), 9.7 ms by Zhang et al. (34) (9.7 ms),, and by Horii et al. (36) (8.2 ms). In observing PVA by CP-MAS spectroscopy, the characteristic methine signal I due to the mm triad exhibited a relatively higher and sharper peak when compared to the same signals in the above previous reports (18,34,36). This signal has already been attributed (35) to the formation of intramolecular hydrogen bonds by the mm triad. Thus the longer $T_{1\rho}H$ can be attributed to the formation of a larger amount of intramolecular hydrogen bonds. Furthermore, for the 25/75 composition of 23MC/PVA, the $T_{1\rho}H$ value for the proton bound to the C-1 carbon of 23MC is shorter than that of the PVA which has the shortest $T_{1\rho}H$ of the homopolymers, suggesting that a conformational change in the skeleton of the 23MC may occur (38). As our proposed interaction for the 23MC/PVA blend was a hydrogen bond between an ether ring oxygen of 23MC and an OH of PVA (Figure 1A), the conformation of the 23MC glucose ring can reasonably be changed by the interaction.

Table III   Proton T_1 and $T_{1\rho}$ relaxation time for 23MC, 6MC, PVA and their blends.

	$T_{1\rho}{}^H$, * ms			
blend composition	23MC (107 ppm)	PVA (45 ppm)	6MC (104 ppm)	PVA (45 ppm)
0/100		15.9		15.9
25/75	10.0	31.2	50.5	47.1
50/50	40.2	25.2	16.7	25.2
75/25	30.2	29.6	27.4	67.9
100/0	76.4		125.9	

* Accuracy is ±5%.

Conclusions

We have attempted to characterize the interchain interactions engaged in hydrogen bonding in a cellulose/PVA blend systems using two cellulose model compounds, 23MC and 6MC, to represent the cellulosic component.

Thermodynamic DSC data allowed us to quantify the interchain interaction in the blends and mixtures by using interaction parameters, χ. The χ values were all negative and very similar for the MC/PVA blends and for the cellulose/PVA systems, indicating that the cellulosic components in both systems interact favorably with PVA. In light of our previous FTIR results (19), these interactions can be assumed to be unique for each blend. Furthermore, an evaluation of the interaction parameters, χ, for each of the two blends suggests that hydrogen bond interactions between the ring oxygen and the OH groups of the PVA are more favorable than those between the hydroxyl groups on each component homopolymer. Thus the large negative value of χ values in the cellulose/PVA system implies that the system is themodynamically miscible, which means down to the molecular level.

To evaluate in more detail the domain sizes for the two MC/PVA blends, we observed changes in the $T_{1\rho}H$ relaxation times with blend composition. These results gave domain sizes of 13.7 nm for 23MC/PVA and 17.5 nm for 6MC/PVA. Masson *et al.* (18) reported that the domain of the cellulose and PVA pair is on a scale of 36 nm or less. Since their cellulose and PVA showed equal $T_{1\rho}H$ values within the experimental error, they could not use these $T_{1\rho}H$ values in their calculations for the blends. However, fortunately for our model compounds the $T_{1\rho}H$ measurements permitted us to analyze the blend domain size on a molecular level scale. Therefore the above calculated domain sizes for the MC/PVA blend prove that the cellulose/PVA blend is miscible on a molecular level as already suggested by FTIR.

Finally in addition to the investigation on a pure cellulose/PVA blend, some interesting aspects of hydrogen bond interactions in the two model systems themselves were revealed. From the DSC analysis, we quantitatively propose that the more favorable formation of interchain hydrogen bonds occurs between the ether ring oxygen and the OH groups rather than between the two hydroxyls in the blends. In interpretating the $T_{1\rho}H$ values in light of the above proposal, the stronger interaction in 23MC/PVA seems to make the miscible domain of the blend smaller, while the weaker interaction in 6MC/PVA can be attributed to the larger domain size. Namely, a strong interaction makes the molecular assembly more compact and hence the density may be higher, which appears reasonable. In addition, when considering that the interaction parameter, $\chi_{23MC/PVA}$ was almost 50 % higher than $\chi_{6MC/PVA}$, whereas the domain size for 23MC/PVA was almost 20 % smaller than that for 6MC/PVA, it is concluded that the interaction strength in the blend is the dominant factor directly contributing to the domain size.

Acknowledgment. The authors wish to thank Dr. Rita S. Werbowyj of the Pulp and Paper Research Institute of Canada for editing the text.

Literature Cited

1. Jolan, A. H.; Prud'homme, R. E. *J. Appl. Polym. Sci.* **1978**, *22*, 2533.
2. Field, N. D.; Song, S. S. *J. Polym. Sci. Polym. Phys. Ed.* **1984**, *22*, 101.
3. Nishio, Y.; Roy, S. K.; Manley, R. St. J. *Polymer* **1987**, *28*, 1385.
4. Nishio, Y.; Manley, R. St. J. *Macromolecules* **1988**, *21*, 1270.
5. Morgensten, B.; Kammer, H. M. *Polymer Bull.* **1989**, *22*, 265.
6. Nishio, Y.; Haratani, H.; Takahashi, T.; Manley, R. St. J. *Macromolecules* **1989**, *22*, 2547.

7. Nishio, Y.; Roy, S. K.; Manley, R. St. J. *Polym. Eng. and Sci.* **1990**, *30*, 71.
8. Jutier, J.-J.; Lemieux, E.; Prud'homme, R. E. *J. Polym. Sci. Polym. Phys. Ed.* **1988**, *26*, 1313.
9. Nishio, Y.; Hirose, N.; Takahashi, T. *Polym. J.* **1989**, *21*, 347.
10. Nishio, Y.; Hirose, N.; Takahashi, T. *Sen-i Gakkaishi* **1990**, *46*, 441.
11. Sakellariou, P.; Hassen, A.; Rowe, R. C. *Polymer* **1993**, *34*, 1240.
12. Scandola, M.; Ceccorulli, G.; Pizzoli, M. *Macromolecules* **1992**, *25*, 6441.
13. Ceccorulli, G.; Pizzoli, M.; Scandola, M. *Macromolecules* **1993**, *26*, 6722.
14. Buchanan, C. M.; Gedon, S. C.; White, A. W.; Wood, M. D. *Macromolecules* **1992**, *25*, 7373.
15. Buchanan, C. M.; Gedon, S. C.; Pearcy, B. G.; White, A. W. *Macromolecules* **1993**, *26*, 5704.
16. Masson, J.-F.; Manley, R. St. J. *Macromolecules* **1991**, *24*, 5914.
17. Masson, J.-F. ; Manley, R. St. J. *Macromolecules* **1991**, *24*, 6670.
18. Masson, J.-F. ; Manley, R. St. J. *Macromolecules* **1992**, *25*, 589.
19. Kondo, T.; Sawatari, C.; Manley, R. St. J.; Gray, D. G. *Macromolecules* **1994**, *27*, 210.
20. Kondo, T.; Gray, D. G. *Carbohydr. Res.* **1991**, *220*, 173.
21. Kondo, T. *Carbohydr. Res.* **1993**, *238*, 231.
22. Hoffman, J. D.; Weeks, J. J. *J. Res. Natl. Bur. Stand., Sect. A* **1962**, *66*, 13.
23. Nishi, T.; Wang, T. T. *Macromolecules* **1975**, *8*, 909.
24. Imken, R. L.; Paul, D. R.; Barlow, J. W. *Polym. Eng. Sci.* **1976**, *16*, 593.
25. Kwei, T. K.; Patterson, G. D.; Wang, T. T. *Macromolecules* **1976**, *9*, 780.
26. Paul D. R.; Barlow, J. W.; Bernstein, R. E.; Wahrmund, D. C. *Polym. Eng. Sci.* **1978**, *18*, 1225.
27. Ziska, J. J.; Barlow, J. W.; Paul, D. R. *Polymer* **1981**, *22*, 918.
28. Martuscelli, E.; Pracella, M.; Yue, W. P. *Polymer* **1984**, *25*, 1097.
29. Martuscelli, E. *Polym. Eng. Sci.* **1984**, *24*, 563.
30. Scott, R. L. *J. Chem. Phys.* **1949**, *17*, 279.
31. Flory, P. J. "Principles of Polymer Chemistry" **1953**, Cornell University Press, Ithaca.
32. Kondo, T.; Sawatari, C. *Polymer* **1994**, *35*, 4423.
33. Grobelny, J.; Rice, D. M.; Karasz, F. E.; MacKnight, W. J. (a) *Macromolecules* **1990**, *23*, 2139; (b) *Polym. Commun.* **1990**, *31*, 86.
34. Zhang, X.; Takegoshi, K.; Hikichi, K. *Polym.J.* **1991**, *23*, 87
35. Terao, T.; Maeda, S.; Saika, A. *Macromolecules* **1983**, *16*, 1535.
36. Horii, F.; Hu, S.; Ito, T.; Odani, H.; Kitamaru, R.; Matsuzawa, S.; Yamaura, K. *Polymer* **1992**, *33*, 2299.
37. Kondo, T. *J. Polym. Sci., B: Polym. Phys.* **1994**, *32*, 1229.
38. Dickinson,L. C.; Yang, H.; Chu, C.-W.; Stein, R. S.; Chien, J. C. W. *Macromolecules* **1987**, *20*, 1757.
39. McBrierty, V. J.; Douglass, D. C. *J. Polym. Sci., Macromol. Rev.* **1981**, *16*, 295.

Chapter 22

Supramolecular Architectures of Cellulose Derivatives

M. Schulze[1], M. Seufert, C. Fakirov, H. Tebbe, V. Buchholz, and G. Wegner

Max Planck Institute for Polymer Research, P.O. Box 3148, D-55021 Mainz, Germany

Abstract: Several cellulose derivatives belong to a special class of polymers called *hairy-rod macromolecules* that are used to generate well-defined supramolecular architectures by the Langmuir-Blodgett (LB) technique.   In particular, trimethylsilyl-cellulose (TMSC) forms monomolecular films on the Langmuir-trough and is transferred onto hydrophobic substrates with a constant transfer ratio, as it does not undergo chemical changes in the film-building process.   Silylated celluloses were regenerated, a convenient method for the generation of homogenous ultrathin films with hydrophilic surfaces.   The adsorption of polymers and dyes as well as biomolecules onto regenerated and modified cellulose LB films have been studied.   In addition, chemical reactions - such as cycloaddition, desilylation, and crosslinking reactions within single monolayers - have been performed.

Hydrophobic ethers of cellulose can be spread on a water surface to form monomolecular layers *(1, 2)*. These layers can be repeatedly transferred onto solid hydrophobic substrates by the Langmuir-Blodgett (LB) technique. Within the so formed multilayer systems, the *hairy rod* macromolecules are oriented in one direction and distributed homogeneously over a large area in the layer plane *(3, 4)*. The layer distance, which is in the range of 10-20 Å, is also well-defined by the side-chain interactions. Because the alkyl side-chains of the cellulose derivatives are in a fluid state, diffusion of small molecules is still possible. Small molecules are able to enter the layered systems from a gas or liquid phase, swell it, and react with functions at the side-chains. The present paper will briefly review (i) chemical reactions performed between adjacent monolayers such as [2+2] cycloaddition, Diels-Alder, crosslinking, and desilylation reactions; (ii) investigations of crosslinked multilayers as liquid separation and transport membranes; and (iii) adsorption studies on ultrathin films of regenerated cellulose derivatives.

[1]Current address: Fraunhofer Institute of Applied Materials Research, Kantstrasse 55, D-14513 Feltow, Germany

Layered cellulose assemblies as media of chemical reactions *(5)*.

Reactions in layered assemblies of organic molecules as obtained by the LB-technique are frequently of a topochemical nature *(6, 7)*, that is, the course of the reaction and the direction in which the product is formed are completely controlled by the packing of the starting material in the lattice *(8)*. The monolayers of which LB assemblies are composed extend in the *x,y* plane; the assembly direction defines the *z* axis (Figure 1a). All reactions occurring in such systems can be characterized by a *dimensionality*, where *dimensionality of the reaction* should be differentiated from *product dimensionality (5)*. The latter refers to the spatial structure of the product, that is, whether a linear macromolecule or a two- or three-dimensional network is the reaction product. In contrast, the *dimensionality of the reaction* describes the tensorial character of the reaction. The polymerization of amphiphilic monomers within LB-layered assemblies is a reaction that proceeds in one direction and gives polymeric chains as one-dimensional products. Reactions with zero-dimensionality, the type of reaction that takes place at the site of individual molecules only, are exemplified, for example, by *cis-trans* isomerizations.

Layered assemblies of cellulose derivatives, which may be regarded as copolymers composed of anhydroglucose units randomly substituted by ether and ester groups are investigated. In particular, isopentylcellulose **1** with a degree of substitution (DS) of 2.9, [5-(9-anthrylmethoxy)pentyl]isopentyl cellulose **2**, (DS = 2.8 with respect to the isopentyl and 0.1 with respect to the anthryl residues, and a partially modified polymer containing fumarate **3** (DS = 2.5 with respect to fumarate groups) have been used (Figure 2). The average degree of polymerization P_n of the cellulose derivatives **1**, **2** and **3** was 87. The anthryl residues undergo 4,+4, cycloaddition when irradiated, a process that leads to a network *(9)*. This reaction can be monitored by the decrease of the anthracene absorption at 250 nm. The reaction in the z direction is prevented by embedding single layers of **2** between layers of **1**. This lack of reaction partners in *z* direction restricts the reactivity of the anthryl groups to the *x,y* plane and results in the formation of a two-dimensional network. Since the self-diffusion coefficient of the macromolecules is immeasurably small in the layered assemblies *(10)*, a photoreaction involving lateral diffusion of the macromolecules can be excluded.

A further strategy to create two-dimensional products involves selective cross-linking of adjacent layers. This reaction, which ought to proceed one-dimensionally in the *z* direction, nevertheless results in a two-dimensional network, as can be seen in Figure 1c. Reactions occurring in the *x,y* plane can be prevented by functionalization of adjacent layers. Single layers of **2** and **3** were each separated by two layers of **1**. If this periodic $(112113)_{15}$ multilayer was annealed at 120-150 °C for more than five hours, the reaction stopped occuring. If, however, an analogous layered assembly was prepared in which adjacent layers of **2** and **3** were each embedded between four layers of **1**, a substantial decrease of the intensity of anthracene absorption could be observed. The layered assembly exhibits a thickness of D = 817 Å immediately after preparation, as determined by X-ray reflection; the thickness shrinks by about 3% to 790 Å in the course of the reaction. The stability of the two-dimensional network obtained by the Diels-Alder reaction was checked by washing the whole sample with solvent. It is important to note that a periodic layer structure remains intact even after removal of the cellulose derivative **1**. Thus, a formation of *extended 2-d-networks* has been achieved.

Ultrathin membranes of cellulose assemblies on porous substrates *(11)*.

Hairy rod macromolecules based on cellulose alkylethers can be used to construct membranes if transferred by the LB-technique to a porous Celgard 2400® film as the substrate. Celgard 2400 polypropylene film is a well-characterized commercial

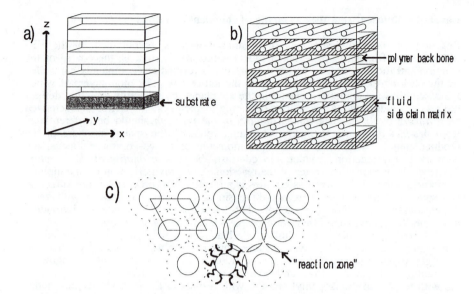

Figure 1. a) General architecture of a LB layered assembly; b) layered assemblies composed of *hairy rod* macromolecules; c) definition of the reaction zones within a hexagonally dense packing relevant for intermolecular reactions.

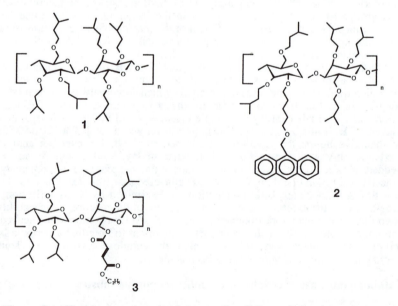

Figure 2. Chemical structure of the used cellulose derivatives: isopentyl cellulose **1**, [5-(9-anthrylmethoxy)pentyl]isopentyl cellulose **2**, and cellulose fumarate **3**.

product with nearly uniform pore size distribution that can be used as a substrate without further treatment *(12)*. Because of the small thickness (less than 10 Å per layer) and the possibility of introducing functional groups into the cellulose molecules, cellulose ethers seem to be promising molecules for the design of membranes via LB technique (Figure 3).

Sites for photo-crosslinking are introduced by attaching a few cinnamyl residues to the cellulose backbone. A mixed hydrophobic (isopentyl/cinnamyl) cellulose ether is prepared and spread from a dilute chloroform solution. LB layers are formed with transfer ratios of 100%. Membrane assemblies consisting of different numbers of monolayers on porous Celgard 2400 polypropylene film were prepared and the multilayer assembly was crosslinked by photoirradiation according to the reaction shown in Figure 4 *(13)*.

The morphology and quality of the films on the porous substrate were investigated by TEM (Zeiss 902, 80 kV) using a simple one-stage replication technique. A defect-free membrane surface was obtained after deposition of 40 monolayers. The pores in the substrate film were completely coated by the LB-film. Photocrosslinked and thus insoluble LB assemblies of cellulose derivatives can be reversibly swelled by organic solvents, which are good solvents for the uncrosslinked materials *(13)*. The osmotic pressure exerted on the membranes exposed to swelling solvents does not change the morphology or damage the membranes. The transport properties of the composite membranes were tested in a conventional membrane osmometer based on the time dependence of the osmotic pressure of toluene solutions of polystyrene molecular weight standards. Samples of degree of polymerization $8 < N < 1632$ were used. The decay of osmotic pressure with time could be represented by a monoexponential function in a first approach to this effect *(14)*. The half-life time $t_{1/2}$ of the initial pressure was determined for each polymer sample, where the concentration of the initial polymer solution was fixed to 10 g/l. The results are plotted in Figure 5 in a double logarithmic scale, together with data describing the case of uncoated Celgard as the separation of the osmotic cell (lower solid line).

Coating of the porous substrate by 60 or 70 cellulose monolayers strongly changes the permeation behavior. The porous substrate alone shows an approximate dependence of $t_{1/2} \approx N^{0.6}$ for good solvents. For a freely diffusing polymer in solution, the behavior expected is between $t_{1/2} \approx N^{0.5}$ for θ-conditions and $t_{1/2} \approx N^{0.6}$ for good solvents.

The membrane covered by 60 monolayers shows an approximate relationship of $t_{1/2} \approx N^{1.03}$, while the one covered by 70 monolayers shows an unexpected dependence with strongly nonlinear behavior over the range of polystyrene samples tested for permeation. A dependence of $t_{1/2} \approx N^{1.57}$ is found for the higher molecular weight samples.

Whether the data found for the migration of polymer of P_n^2 80 through this particular membrane is caused by residual micro- or other defects is unknown at present. In any case, it is significant that a membrane built by deposition of 50 layers poly(glutamate) on top of Celgard exhibits $t_{1/2} \approx N^{1.78}$ behavior. It should be mentioned that the thickness of an individual monolayer of the cellulose derivatives is 0.96 nm, as compared to 1.74 nm for the poly(glutamate). Thus, the absolute value of $t_{1/2}$ is reduced by a factor of 10 comparing the copoly(glutamate) with the cellulose system of 70 monolayers.

The last point in each curve corresponds to the molecular-weight-cut-off (MWCO) of each particular membrane. In other words, the membrane was impermeable for polymers of a degree of polymerization $N > N_{MWCO}$. The transport of macromolecules through ultrathin layers of the type discussed here (Figure 1) is also interesting from a theoretical point of view. The solvent-swollen multilayer part of the membranes, which obviously controls the transport kinetics, has a thickness d that is ca. 10-100 times larger than the radius of gyration of the macromolecules

310

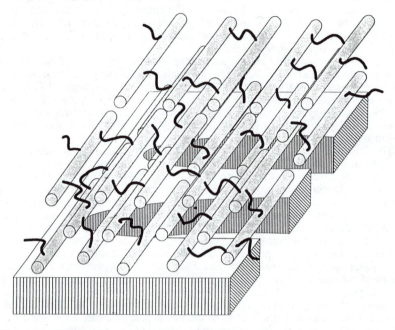

Figure 3. Schematic presentation of the crosslinked layer system of *hairy rods* on top of the Celgard 2400® polypropylene membrane surface.

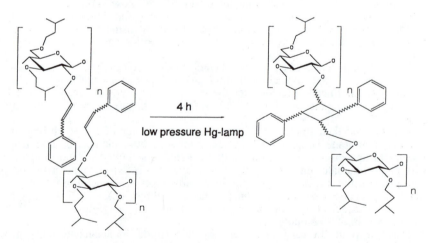

Figure 4. Crosslinking reaction of the isopentyl cinnamyl cellulose ether **1**; only one of several possible types of crosslinks is shown.

migration. However, adjacent backbones within the layers form a grid with a distance between the backbones rather rigidly fixed by the side chains and crosslinks to a value of ca. 05.-1 nm in the swollen state. Thus, transport of the polymer chains should only be possible by reptation *(15)*. In what way the structure of the layered assemblies enforces distortions of the unperturbed dimensions of the macromolecules while they are migrating is not known at present. It will be particularly interesting to study cases in which the total layer thickness d becomes smaller or comparable to the radius of gyration, as one could then probably observe a crossover in the transport behavior.

Adsorption on regenerated cellulose LB films *(16-18).*

Silylated cellulose derivatives such as trimethylsilyl cellulose (TMSC) can be regenerated to cellulose by exposure to aqueous acids and thus allow *in situ* conversion of TMSC hydrophobic films to hydrophilic films of regenerated cellulose (Figure 6a) *(16, 17)*. The regeneration is quantitative after 30 s and leads to a well-defined ultrathin cellulose film. Earlier studies could show that (i) in the TMSC LB-film, as well as in the regenerated cellulose film, the polymer backbones are preferentially oriented parallel to the dipping direction, and (ii) both films contain regular and homogeneous structures [10].

Figure 7 shows the regeneration process for a film of 100 layers TMSC. In contrast to the hydrophobic LB films of silylated cellulose, the regenerated cellulose films have hydrophilic surface properties, as evident from the static contact angle with water (78° for hydrophobic TMSC and 23° for hydrophilic cellulose). In addition, the cellulose multilayer systems are insoluble in most common organic solvents and water, and are stable against oxidation and thermal degradation. The obtained regenerated cellulose films have been used as substrates for adsorption of monomers (dyes) and polymers *(18)*.

In addition, modified cellulose materials have been prepared in order to introduce carboxyl functionalities. Cellulose with a higher content of carboxyl groups should show a different adsorption behavior from native analogues. Thus, cellulose succinates have been prepared using ring-opening esterification with succinic anhydride, as shown in Figure 6b, *(18)*. The adsorption properties of unmodified and modified films were investigated using high molecular weight cationic polyacrylamide, (poly(acrylamido propyl)trimethylammonium chloride APTAC-C, see Figure 6c), with a content of 6% randomly distributed cationic side-groups.

The polyelectrolyte APTAC-C is a typical polymeric additive used in papermaking processes; it works as a flocculating agent helping to retain filler particles and fiber fragments in the sheet being formed. Analysis of the surface plasmon resonance (SPR) spectra of APTAC-C adsorbed on cellulose and modified cellulose films reveals that the latter films adsorb much more APTAC-C than does the unmodified cellulose, and the adsorbed amount increases as the reaction time with SA is prolonged. The amount of adsorbed polyelectrolyte was doubled when a film of 20 layers was treated with SA for 3 hours. Films modified with SA for 13.5 h adsorb more than four times the amount of APTAC-C compared to the unmodified films (Figure 7). It is noteworthy that the time for reaching the plateau region increases with the length of the reaction time. When the number of cellulose monolayers is increased to 100 and the film is treated with SA for 13.5 h, the system absorbs more than 18.7 mg/m^2 and reaches the plateau level after two hours. In the case of thicker films, one should consider that the modification could also occur in deeper layers and that the polyelectrolyte could not only adsorb on the surface but also penetrate into the modified multilayer system. Preliminary investigations on the influence of salts and an additional polyelectrolyte have shown that it is possible to detect double-adsorption processes. Following this concept, a simulation of a two-

312

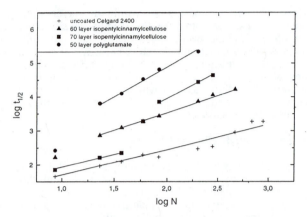

Figure 5. Double logarithmic plot of the half-life time $t_{1/2}$ of osmotic pressure versus degree of polymerization N of the polystyrene molecular weight standards.

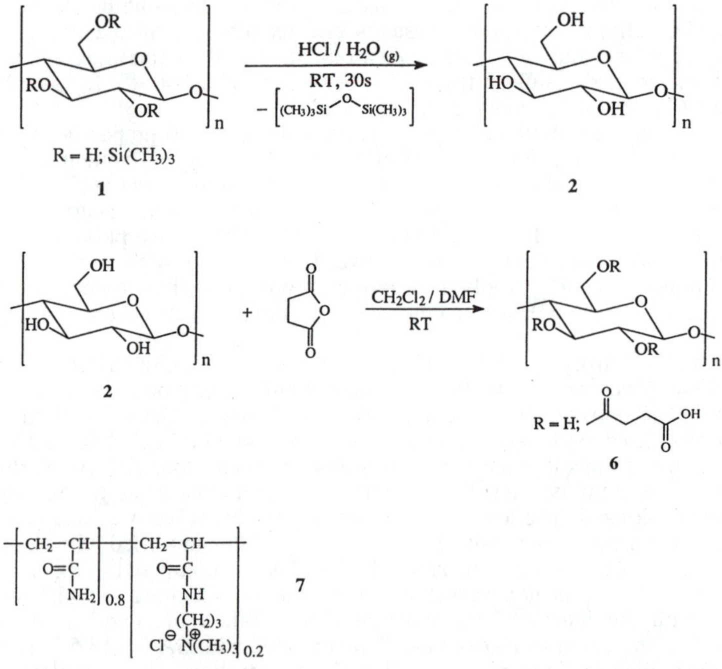

Figure 6. a) Desilylation reaction to obtain regenerated cellulose; b) treatment of regenerated cellulose with succinic anhydride (SA) giving cellulose succinate derivatives; c) chemical structure of the polyelectrolyte APTAC-C.

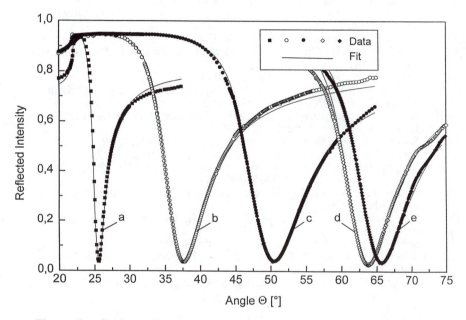

Figure 7. Surface plasmon resonance spectra of (a) the substrate; (b) regenerated cellulose; (c) TMSC vs. air; (d) cellulose modified with SA for 13.5 h against buffer (1 mM $Na_2(CO_3)$) and (e) the corresponding measurement after adsorption of APTAC-C.

component retention system consisting of a combination of cationic and anionic polymers should be possible.

Literature Cited

(1)   Wegner, G. *Mol. Cryst. Liq. Cryst.* **1993**, *235*, 1.
(2)   Wegner, G.; Schulze, M.; Seufert, M.; Schaub, M.; Vahlenkamp, T. *Polym. Prepr.* **1995**, 598.
(3)   Schwiegk, S.; Vahlenkamp, T.; Xu, Y.; Wegner, G. *Macromolecules* **1992**, *25*, 2513.
(4)   Ferencz, A.; Armstrong, N. R.; Wegner, G. *Macromolecules* **1994**, *27*, 1517.
(5)   Seufert, M.; Schaub, M.; Wenz, G.; Wegner, G. *Angew. Chem. Int. Ed. Engl.* **1995**, *34*, 340.
(6)   Whitten, D. G. *Angew. Chem. Int. Ed. Engl.* **1979**, *18*, 440.
(7)   Eggl, P.; Pink, D.; Quinn, B.; Ringsdorf, H.; Sackmann, E. *Macromolecules* **1990**, *23*, 3472.
(8)   Tieke, B. *Adv. Polym. Sci.* **1985**, *71*, 79.
(9)   Müllen, K.; Böhm, A.; Fiesser, G.; Garay, R. O.; Mauermann, H.; Stein. S. *Polym. Prepr.* **1993**, *34*, 195.
(10)   Schaub, M.; Mathauer, K.; Schwiegk, S.; Albouy, P.-A.; Wenz, G.; Wegner, G. *Thin Solid Films* **1992**, *210/211*, 397.
(11)   Seufert, M.; Fakirov, C.; Wegner, G. *Adv. Mater.* **1995**, *7*, 52.
(12)   Sarada, T.; Sawyer, L.; Ostler, M. *J. Membr. Sci.* **1983**, *15*, 97.
(13)   Iida, S.; Schaub, M.; Schulze, M.; Wegner, G. *Adv. Mater.* **1993**, *5*, 564.
(14)   Coll, H. *Makromol. Chem.* **1967**, *109*, 38.
(15)   de Gennes, P. G. *Scaling Concepts in Polymer Physics*, Cornell University Press, Ithaca **1979**.
(16)   Schaub, M.; Wenz, G.; Wegner, G.; Stein, A.; Klemm, D. *Adv. Mater.* **1993**, *5*, 919.
(17)   Schaub, M.; Fakirov, C.; Lieser, G.; Wenz, G.; Wegner, G.; Albouy, P.-A.; Schmidt, A.; Majrkzak, C.; Satija, S.; Wu, H.; Foster, M. D. *Macromolecules* **1995**, *28*, 1221.
(18)   Buchholz, V.; Wegner, G.; Stemme, S.; Ödberg, L. *Adv. Mater.* **1996**, *8*, 399.

Chapter 23

Surface Segregation Phenomena in Blends of Cellulose Esters

Ulrike Becker, Jason Todd, and Wolfgang G. Glasser

Biobased Materials/Recycling Center and Department of Wood Science and Forest Products, Virginia Polytechnic Institute and State University, Blacksburg, VA 24061

The principle of surface segregation and subsequent self-assembly of mixtures of molecules has been utilized to create cellulosic films that have properties characteristic of smart materials. This has been achieved by solvent-casting a blend of cellulose propionate and an F-containing cellulose ester derivative. Evidence suggests that surface segregation of the F-containing species takes place spontaneously in cellulosic systems. Cellulose propionate films containing a small amount of F-containing cellulose derivative exhibit a much lower surface free energy (i.e. are more hydrophobic) than predicted by the rule of mixing, and this indicates surface segregation. The current research examined the influence of the chemistry of the fluorine-containing component (type of derivative and chemistry of the F-containing group) on the segregation process.

A smart material is defined as a structure "…having the following attributes: it senses the environment, it responds in a purposeful way through design or by means of a logic controller element" (1). The surface segregation and subsequent self-assembly observed in some multicomponent films can be classified as a smart process, and thus the resulting film is a smart material.

The principle of surface segregation (see below) allows constructing films with surface energetics distinctly different from the bulk phase. Natural polymers with hydrophobic surfaces can be created in this manner, although the bulk retains the usual hydrophilic nature which is due to a high oxygen content. This creates new opportunities for applications of natural polymers. For example, cellulose and its derivatives have achieved prominence in the biomedical field as kidney dialysis membranes. However, advancements in this field have been limited because of the limited surface energetics of cellulosic materials. As polymers of natural origin, cellulose and its derivatives are prime candidates as raw materials for biomedical implants because of their general biological compatibility and biodegradability.

However, for an implant material to function at the interface with living cells and body fluids, characteristic surface requirements must be met which prevent the spontaneous formation of blood clots. The general ability of foreign materials to cause thrombosis constitutes one of the principal limitations of biomedical implant materials. While it is recognized that the process of blood clotting is initiated by the adsorption of proteins on the surface of the implants, consensus has not yet been reached over whether this process is aggravated by excessively hydrophilic or hydrophobic surfaces (2,3). General agreement has, however, been reached over the issue of the importance of surface engineering in terms of polarity. Thus, the ability to engineer the critical surface free energy (γ_C) of implant materials is pivotal to the development of biomedical implants (4). Most studies indicate that thrombogeneity increases with increasing γ_C (5), and that simultaneously, clotting time decreases (6). This suggests that a more hydrophobic surface is less prone to thrombosis. It is therefore of interest to examine methods of engineering the surface characteristics of cellulose-based materials in terms of their surface free energy.

While fluorine plasma treatment represents the easiest and most widely practiced method of surface modification, negative toxicological effects have been noted from this method which result from the partial molecular breakdown and leaching of toxic degradation products into the surrounding medium (7). Utilizing the smart process of self-assembly, it is possible to create cellulose-based materials with hydrophobic surfaces for potential use as biomedical implant materials.

The Principle of Spontaneous Self-Assembly by Surface Migration

The principle of spontaneous segregation of blend components by migration of one component to the surface on the nano-scale has been employed for engineering bi-layered materials in the past. By mixing two incompatible polymers with each other in a blend, molecular composites with a distinct surface layer, which is different from the bulk phase, can be obtained.

Surface migration is a thermodynamically-driven process. Fundamental thermodynamics dictates that a system adopt the state of lowest possible total free energy under equilibrium conditions. In the case of a film consisting of at least two polymers, three quantities have to be taken into account: 1. entropy of mixing; 2. interaction energy between polymers; and 3. surface free energy (8). Quantities 1 and 2 are given by the composition of the system. At a given film composition, however, the surface free energy, which is the interfacial energy between the film and air (which is considered to be hydrophobic) can be subjected to changes. During film formation the system rearranges itself so that the lowest free energy component resides preferentially at the surface, and therefore the lowest free energy of the entire system is gained. Therefore, surface migration can be used to create a system with a low surface free energy, i.e. a hydrophobic surface.

Since fluoro-carbon molecules are known to have low surface free energies, they are likely to migrate to the surface in a mixture of F-free and F-containing molecules. The goals of this study were to examine the surface migration effects of two types of novel F-containing cellulose derivatives in a blend with cellulose propionate. The first type of F-containing component was a mixed ester of cellulose propionate (CP) with fluoroalkoxy acetate groups whereby the F-containing substituents were randomly distributed, and the second type of component was a fluorine-terminated cellulose propionate oligomer with exactly one F-containing endgroup (ie, with a blocky type of architecture). Both F-containing species were expected to surface-segregate, but differences in migration behavior were expected depending on the difference in the chemistry of the derivative and the type and the distribution of F-containing group. Two different F-containing groups were used, trifluoroethyl (with CF_3 terminus) and octafluoropentyl (with CF_2H terminus) substituents.

Energetic effects that stem from the presence of atoms other than hydrogen (X) in hydrocarbon containing molecules are the cause for the surface migration of F-containing molecules. These effects are described with nearest neighbor interactions. χ_s is defined as the difference between the adsorption energies (U_α) of the segments of the components ($\alpha = 1$ or 2 in a two component system),

$$\chi_s = U_1 - U_2 \qquad \text{(equation 1)},$$

where

$$U_\alpha \equiv u_\alpha / k_B T \qquad \text{(equation 2)},$$

and

$$u_\alpha \equiv \omega_{\alpha s} - 0.5\omega_{\alpha\alpha} \qquad \text{(equation 3)}$$

with k_B as the Boltzmann constant, T the temperature, $\omega_{\alpha\alpha}$ as the intracomponent interaction energy, and $\omega_{\alpha s}$ the surface interaction energy. u_α represents the difference in adsorption energies between the surface and the bulk (9). Thus χ_s is a measure for the adsorption energy difference between the segments of the two constituent chains, and this is a measure for the preference of one component to be at the surface over the other (10). In the case of a mixture at an energetically neutral surface (like air) the term simplifies to

$$\chi_s = - (\omega_{11} - \omega_{22})/k_B T \qquad \text{(equation 4)}$$

with ω_{11} and ω_{22} as the interaction energy between the monomers in the bulk where ω_{11} is larger than ω_{22}. A positive χ_s is an expression for the fact that the system partitions the component with the relatively less favorable monomer interaction in the bulk, i.e. less attraction forces between the monomers, to the surface (11). This species experiences less energy loss due to the smaller number of neighbors at the surface than a species with a large interaction energy. In systems like C-H and C-F, possible interactions arise from London forces. The C-F bond has a high polarity, but at the same time, due to the high electronegativity of the fluorine atom, a low polarizability. London forces are proportional to polarizability. The C-F containing species has therefore weaker London forces and thus less interaction energy when compared to a C-H containing species. This results in a positive χ_s-parameter and it drives the partitioning of the fluorinated component to the surface. Altogether, the result is a lower surface free energy. The more F-atoms present, the weaker are the London forces. As a result, the χ_s parameter becomes more positive. This means that the driving force for surface migration increases.

The F-containing, randomly-substituted cellulose esters have more F-containing substituents per molecule than an F-terminated oligomer, which has exactly one F-containing substituent (Figure 1). Thus the driving force for the random copolymer is larger on a molecular basis. But at the same time, the F-containing groups of the terminated oligomers are independent from each other, whereas in the random copolymer many F-substituents are connected through the cellulose backbone. In addition the oligomers are more mobile and encounter fewer entanglements due to their smaller size. Thus it is expected that the surface migration of the oligomers is more efficient when compared to the random copolymers on the basis of number of F-containing groups in the system. Consequently, blends containing F-terminated oligomers are expected to yield materials with lower surface free energy than blends with randomly substituted F-containing copolymers.

Comparing the F-containing substituents alone shows two effects depending on the chemistry of the substituent. First, the χ_s parameter increases with increasing number of F-atoms in the group. In the case of an energetically neutral surface, the value of χ_s can be deduced from the cohesive energy density (CED) since the CED is proportional to $\omega_{\alpha\alpha}$ (12). For polymers, the CED can be calculated through group contribution as the sum of the molar attraction constants, F, for the group (13). For every proton that is replaced by a fluorine atom, the interaction energy decreases. Therefore, the more H-atoms are replaced by F-atoms, the lower is the interaction energy of the species. Consequently, χ_s increases. As χ_s represents the driving force for the surface segregation, a larger value of χ_s indicates a stronger "pulling force" for the surface segregation and the equilibrium state is expected to be reached in a shorter time. Accordingly, derivatives containing the octafluoropentyl group are expected to show a larger degree of surface segregation and therefore a more hydrophobic surface when compared to the trifluoroethyl-containing derivative of the same type.

The overall objectives of this study were to synthesize novel types of F-containing cellulosic derivatives and to examine their surface segregation properties in relation to fluorine content and polymer architecture. The F-containing components were either randomly-substituted cellulose esters with variable F-content and with either CF_2H- or CF_3-terminal groups, or they were F-terminated cellulose propionate oligomers.

Experimental

I. Materials. Micro-crystalline CF-11 cellulose was purchased from Whatman Chemicals. Cellulose propionate was obtained from Eastman Chemical Co. All solvents were obtained from Fisher Scientific and used as received, except tetrahydrofuran for the synthesis of the F-terminated oligomers, which was used freshly distilled over sodium. P-toluene-sulfonyl chloride (TsCl), toluene diisocyanate (TDI), and other reagents were obtained from Aldrich Chemical Co. and used as received.

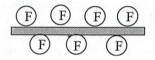

F
F
F
F
F
F
F
F
Random Copolymer
F-terminated oligomer
("blocky" structure)
high molecular weight
high molar F-density
low molecular weight
low molar F-density

Figure 1. Schematic illustration of F-containing cellulose ester options.

II. Methods

1. Copolymer synthesis. Cellulose mixed esters with fluorine-containing substituents randomly-distributed along the cellulose backbone were synthesized via homogeneous phase reaction of cellulose solution in dimethylacetamide / lithium chloride (DMAc/LiCl). This reaction is described in detail elsewhere (14). Briefly, a fluorinated acid was reacted with cellulose in solution via a mixed anhydride intermediate formed with p-toluene sulfonyl chloride (TsCl). The fluorinated acids used were trifluoroethoxy acetic acid and octafluoropentoxy acetic acid. The acids were synthesized by reaction of either 2,2,2-trifluoroethanol or 2,2,3,3,4,4,5,5-octafluoropentanol, respectively, with chloroacetic acid in water under reflux with strong alkali. Pyridine, followed by fluorinated acid, and finally TsCl, were added to cellulose in DMAc/LiCl solution. The reaction mixture was stirred overnight at 40-60°C. The resulting randomly-substituted fluorinated cellulose esters had a degree of substitution of fluorinated substituent (DSF) of approximately 1.5 (by NMR and/or elemental analysis). The remaining cellulose hydroxyl grops were subsequently acylated via reaction with acetic or propionic anhydrides, resulting in cellulose mixed esters with a random distribution of fluorinated and unfluorinated ester substituents. Molecular weight distributions of the random mixed esters were determined by GPC.

Fluorine-terminated cellulose propionate oligomers were prepared using monofunctional (OH-terminated) cellulose esters of variable degree of polymerization and coupled with fluorine-containing alcohols. In brief, the procedure involves first the hydrolytic degradation of cellulose propionate into CP oligomers of various degrees of polymerization (DP) as described elsewhere (15). The resulting oligomers were endcapped with toluene diisocyanate according to deOliveira and Glasser (16). The isocyanate terminated oligomers were immediately coupled with a fluorinated alcohol. Typically, 3 grams of the predried endcapped oligomer were placed in a 250 ml three neck round bottom flask with 0.01 ml of stannous octanoate as a catalyst. Two necks of the flask were closed with rubber septa, then the flask was placed in a polycarbonate desiccator with rubberseal and connected to a vacuum pump overnight. The flask was then connected to a condenser and flushed with prepurified nitrogen gas. Subsequently, 60 ml of freshly distilled THF were introduced into the flask using syringe and needle. The flask was put into an oilbath and heated to 50°C. At that time, a 5 molar excess of the F-containing alcohol was added via syringe. The flask was stirred at 50°C for 24 hours under nitrogen. After that, it was allowed to cool to room temperature. The product was precipitated in petroleum ether under vigorous stirring. The precipitate was filtered and washed several times with petroleum ether. The product was dried and kept until further use in a small polypropylene container.

2. Copolymer characterization. Copolymer characterization involved ^{1}H and ^{19}F-NMR spectroscopy, gel permeation chromatography (GPC), differential scanning calorimetry (DSC), and solubility studies. The molecular weight and the distribution of hydroxyl-terminated CP oligomers were determined by GPC with a differential viscosity detector (Viscotek Model No. 100) and a differential refractive index (concentration) detector (Waters 410) in sequence. The system was controlled by

Viscotek software (Unical GPC software, Version 3.02). The CP oligomers were dissolved in THF and analyzed using a high pressure liquid chromatography system. The calculations were based on an universal calibration curve using polystyrene standards.

^{19}F-NMR spectra were recorded on a Varian 400 MHz spectrometer. The samples were dissolved in protonated THF and run without lock. In order to accurately determine peak shifts, all samples were run with the addition of 3-(trifluoro)methyl benzophenone as an internal standard.

^{1}H-NMR was used to determine the degree of polymerization through endgroup determination. The analysis was conducted on a Varian 400 MHz instrument with deuterated chloroform as the solvent. In order to increase the detection sensitivity of the hydroxyl proton, the oligomers were silylated at the lone terminal hydroxyl group as described elsewhere (17). The DP was calculated as a ratio of the peak areas for the silyl protons to the cellulose backbone protons.

DSC measurements were conducted on a Perkin Elmer model DSC4 with a Perkin Elmer Thermal Analysis Data Station. The temperature was scanned between -30°C and +270°C at a heating rate of 10°C/min. The samples were subjected to three heating and three cooling cycles. The glass transition temperatures (T_g) were taken as the mid-point of the step-function change in slope of the baseline and the melting transition was taken as the temperature corresponding to the maximum point of the endothermic peak.

The solubility of the F-containing samples was tested in various solvents. Typically a small amount of the sample was transferred into a 4 ml glass vial and solvent was added. If completely dissolved, the resulting solution had a concentration of about 0.5-1%. The vial was equipped with a magnetic stir bar and stirred overnight. At that point solubility was determined visually.

3. Blend Preparation. Cellulose propionate/F-containing cellulose propionate (CP/CP-F) blends were prepared volumetrically. This involved initially the preparation of stock solutions of CP and CP-F in THF. Blends were prepared by mixing these solutions to the desired content of F-containing species. The total solids content of all solutions was 5% w/v.

4. Contact Angle Measurements. Cellulose ester and cellulose ester blend films were characterized by contact angle (CA) measurements. This was based on a modified Wilhelmy Plate method. Microscope cover slides were cleaned thoroughly in hexanes, dried and subsequently dip-coated with the respective solutions. The solvent was evaporated at 4°C over night and all films were stored for three days before the contact angle measurements. Great care was taken to ensure that all films were prepared in the same fashion. The contact angle measurements were conducted with deionized water as the wetting medium using a CAHN Analyzer, controlled both manually and by CAHN control software DCA2d Version 2.0. The instrument recorded the force depending on the immersion depth, and the water contact angle was calculated using the force and the surface free energy of water. Six measurements were performed per sample.

Results and Discussion

1. Synthesis. The synthesis of F-containing random copolymers and block copolymers is schematically illustrated in Figure 2. Monofunctional, OH-terminated cellulose propionate oligomers with variable degree of polymerization were prepared by partial hydrolytic degradation with HBr under esterification conditions.

The molecular weights of the cellulose propionate segments were determined by GPC and NMR spectroscopy (Table I). The two methods are mutually supportive. For NMR, the signal of the one OH end group was enhanced by reaction with chloro-triethylsilane. The 15 ethylsilane protons show a characteristic and easily distinguished proton peak at 0.55 ppm. Integration of this peak and the peaks associated with the cellulose backbone protons is used to calculate the degree of polymerization. At high block sizes (large DP), the enhanced ethylsilane signal becomes less distinct and block size determination by NMR becomes less accurate. A comparison of DP determinations by NMR and GPC (Figure 3) reveals that NMR and GPC produce almost identical results at low DP-values whereas at high block sizes (large DP), the determination by NMR yields a higher DP value. At higher molecular weights, a complete reaction of the chloro-triethylsilane becomes more difficult. This accounts for the fact that the DP values obtained by NMR are higher than those found by GPC.

The OH-terminated oligomers were subsequently end-capped with toluene diisocyanate resulting in a monofunctional terminated oligomer. The resulting NCO-terminated cellulose propionate oligomer was immediately (to avoid degradation of the remaining isocyanate groups) reacted with F-containing alcohols in which both F-content and terminal functionality (CF_2H versus CF_3) vary (Figure1). The alcohols were trifluoroethanol (CF_3-terminus) and octafluoropentanol (CF_2H-terminus). The reaction of the F-containing alcohols and the cellulose propionate oligomer was confirmed by ^{19}F-NMR spectroscopy. As a result of the formation of a terminated

Table I. Molecular weight characteristics of cellulose propionate oligomers

Oligomer Designation	Molecular Weight (Mn)		Molecular Weight Distribution (MWD)
	by GPC[2]	by H-NMR[3]	
B-7[1]	2375	2150	1.4
B-14	4480	5000	1.6
B-28	9380	11250	1.5
B-70	24100	26300	1.4

[1] The number behind B- denotes the DP of the segment.;
[2] Using THF as solvent and a universal calibration curve.
[3] Following OH-modification with chloro-triethyl silane (see Experimental Section).

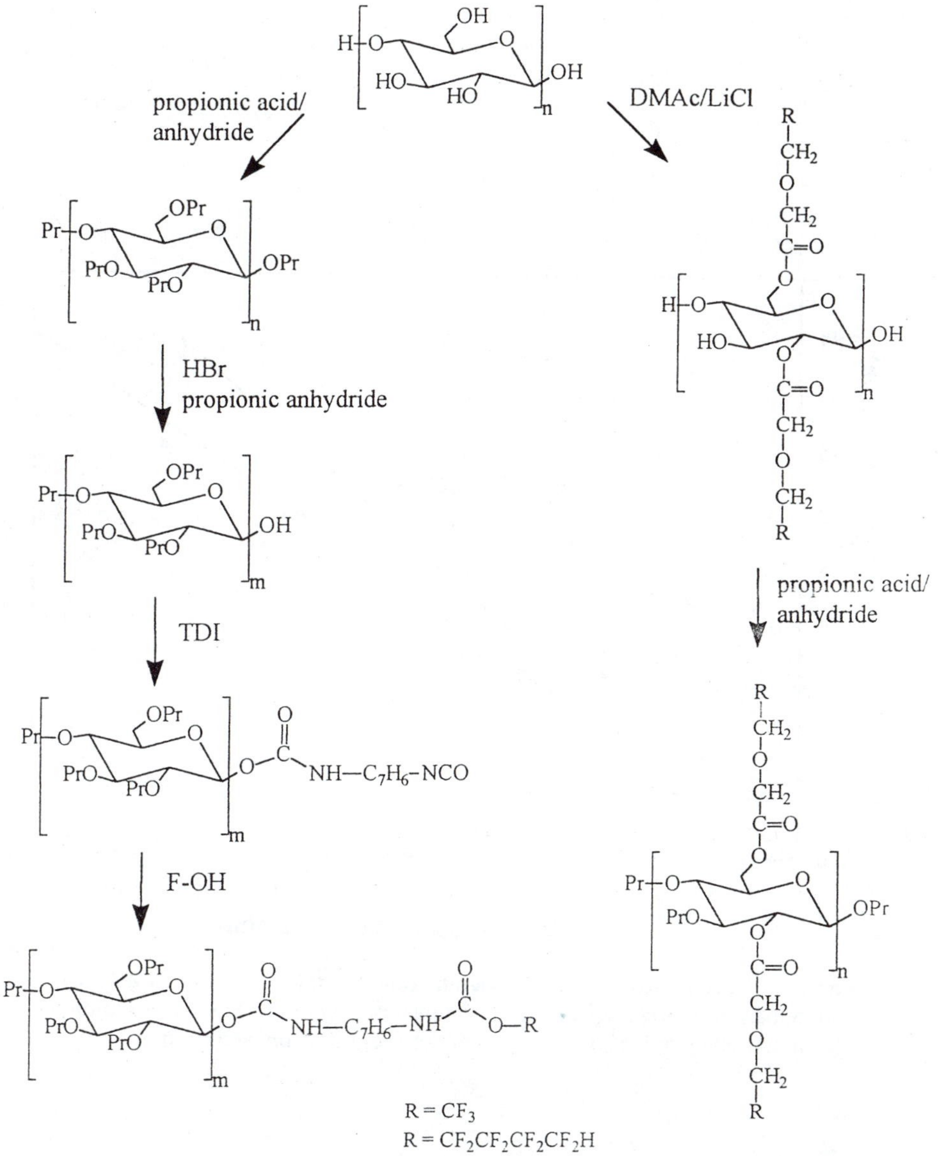

Figure 2. Reaction schemes for blocky oligomer (left side) and random copolymer (right side); n indicated the high degree of polymerization of cellulose propionate, which is maintained in the random copolymer, whereas m indicated the reduced degree of polymerization in the blocky oligomer which is achieved by hydrolytic degradation.

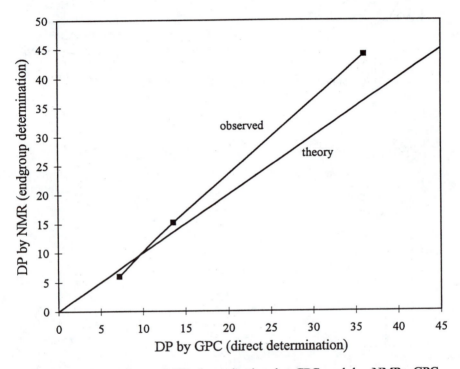

Figure 3. Comparison of DP determination by GPC and by NMR; GPC determination represents direct determination, while NMR is an indirect determination by means of determining the endgroup concentration.

oligomer, the CF_3-peak in the ^{19}F-NMR spectrum shifted downfield from -14.4 ppm to -11.5 ppm (Figure 4). All ^{19}F-NMR samples were run with an internal standard (3-(trifluoro)methyl benzophenone). This was necessary because ^{19}F-resonances often shift, consequently, the shifts can only be accurately measured when the CF_3-peak position is observed in relation to a standard peak. The F-containing cellulose derivatives used in this study are summarized in Table II.

2. Characterization of F-containing cellulose derivatives. The characteristics of random and block copolymers containing fluorinated substituents involve solubility testing and thermal analysis. The solution characteristics of copolymers in various solvents are shown in Table III. The results reveal significant differences between the random copolymers and the blocky oligomers. Among the random copolymers, differences in solution behavior depended on the nature of the F-containing substituent. Copolymers containing a substituent with a CF_3-terminus were soluble in non-polar organic solvents like chloroform and dichloromethane, as well as in acetone. In contrast, copolymers with an F-containing group with a CF_2H-terminus are not only soluble in these solvents but also in such alcohols as methanol and ethanol.

This behavior is attributed to the lone proton of the CF_2H-terminus. This proton is covalently linked to a carbon to which two F-atoms are attached. Due to the high electronegativity of the F-atoms, the carbon has a strong positive partial charge. This charge increases the negative inductive effect of the carbon onto the hydrogen atom. Altogether, the proton becomes very electron deficient and as such is able to engage in secondary interaction, like for example hydrogen bonding, with the hydroxyl-groups of the alcohols. These interactions lead to the dissolution of the copolymer in alcoholic solvents.

The end-terminated oligomers, on the other hand, do not exhibit a solubility behavior that depends on the chemistry of the F-containing endgroup. All oligomeric derivatives were formed turbid solutions in all tested solvents and at all concentrations, and did not settle out in several weeks. This behavior also did not depend on the molecular weight of the cellulose propionate segments or the type of F-containing endgroup.

Table II. Identification of F-containing derivatives used in this study

Designation	Description
B-14-CF$_3$	oligomer with DP = 14, terminated with trifluoroethyl group
B-14-CF$_2$H	oligomer with DP = 14, terminated with octafluoropentyl group
R-100-CF$_3$	random copolymer (DP = 100) with trifluoroethoxy acetate substituent (DSF = 1.5), peracetylated
R-100-CF$_2$H	random copolymer (DP = 100) with octafluoropentoxy acetate substituent (DSF = 1.5), perpropionated

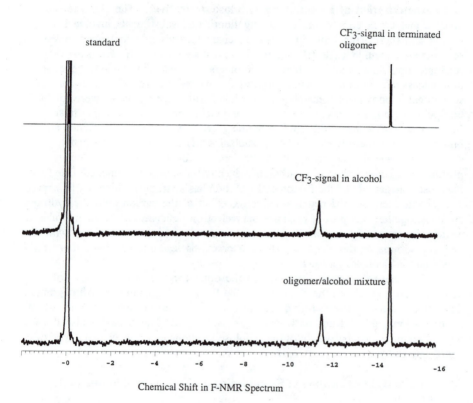

Figure 4. 19F-NMR spectra of B-14-CF₃ (top); trifluoro ethanol (middle); and B-14-CF₃ spiked with trifluoro ethanol. The standard (3-(trifluoro)methyl benzophenone) is arbitrarily set to 0 ppm.

Table III. Solution behavior of fluorinated cellulose derivatives; sample designation is as outlined in Table II.

sample	solvent				
	acetone	dichloromethane	chloroform	ethanol	methanol
B-14-CF3	T	T	T	T	T
B-14-CF2H	T	T	T	T	T
R-100-CF2H	++	+	+	++	++
R-100-CF3	++	+	+	-	-

T = turbid solution
+ = soluble with some difficulty
++ = rapidly soluble
- = insoluble

Consistent with the solution behavior of other amphiphilic substances, it is at this point hypothesized that this behavior reflects the formation of micelles where the F-containing endgroups form the core, and the corona consists of cellulose propionate residues (Figure 5).

The examination of thermal properties of copolymers by DSC produces results consistent with the generic architecture of the different derivatives (Figure 6). Whereas the random copolymers revealed a gradual, transitionless decline in T_g and T_m with increasing content of fluorinated substituents, the block copolymers revealed thermal transitions that were identical to those of the parent cellulose propionate. The fluorinated end groups in the block copolymers, constituting only a minor part of the molecule, proved to have an insignificant influence on the thermal copolymer properties.

3. Contact angle measurements of blended films. Films prepared from blends of cellulose propionate and fluorinated cellulose derivatives were characterized by contact angle measurements. Preliminary results revealed that water contact angles varied between approximately 75 and 110°, and the degree of variation from the CP control depended on fluorine content, substituent type, and polymer architecture in a significant way (Figure 7). The pure random copolymer film with a CF$_2$H-terminus achieves a maximum advancing water contact angle of 110°. This angle is lower than values found in the literature for fluorinated surfaces (18). However, the literature values describe surfaces with CF$_3$-groups, and the present case examines a surface containing CF$_2$H-groups. The lower contact angle is again explained by the presence of the lone proton at the CF$_2$H-terminus which can engage in hydrogen bonding with the water molecules in the same fashion as outlined above.

The copolymer shows an exponential increase in contact angle at low blend contents of F-containing component. This behavior represents a deviation from the rule of mixing which validates the segregation phenomenon of the F-containing species at the surface.

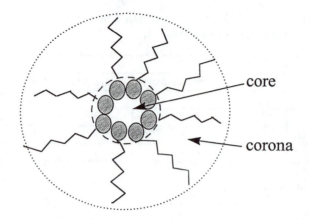

Figure 5. Schematic representation of micelles believed to be formed by F-terminated cellulose propionate oligomers. The circles represent the endgroups which from the core of the micelle and the lines represent the cellulose propionate chains in the corona.

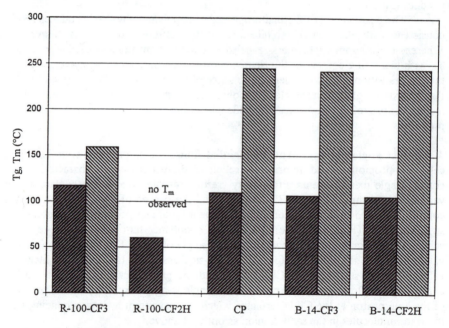

Figure 6. Melting transitions of cellulose propionate (CP) compared to blocky and randomly fluorinated derivatives: thermal behavior of the blocky oligomers is similar to CP and independent of the chemistry of the F-containing endgroup but random copolymers show a clear dependence of thermal transitions on the chemistry of the F-containing substituent. No melting transition for R-100-CF$_2$H was observed. Sample abbreviation is as outlined in Table II.

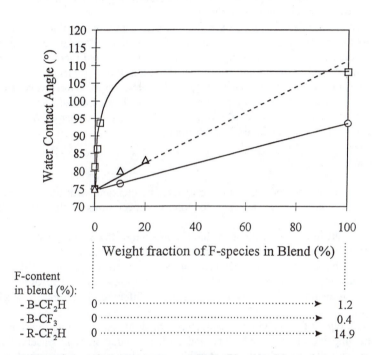

Figure 7. Experimental water contact angle for blends with random copolymer R-CF₂H (with octafluoropentyl substituent) (□); and for two blends with blocky oligomers: B-CF₃ (trifluoroethyl-terminated) (Δ) and B-CF₂H (octafluoropentyl-terminated) (o). The dashed line for the B-CF₃ indicates extrapolated values. Pure B-CF₃ could not be cast into a stable film. Note that the F-content (in %) is higher in the random copolymer than in the blocky oligomers

The terminated oligomers, however, show a linear increase of contact angle with increasing amount of CP-F in the blend. In this case the rule of mixing is followed. The apparent lack of surface enrichment is explained by micelle formation. The F-containing endgroups are "locked" into the core of the micelles and are therefore prevented from orienting themselves to the surface. This also explains the fact that the oligomers are less effective in increasing the water contact angle when compared to the random copolymer. The pure blocky oligomer with a CF_2H-terminus achieves a water contact angle of only 93°.

The effect of the nature of the F-containing group is examined in two blocky oligomers having an octafluoropentyl and a trifluoroethyl endgroup, respectively. The octafluoropendyl group is comparatively F-rich and is expected therefore to result in a higher contact angle; but preliminary results show that the oligomer with the trifluoro ethyl endgroup has a higher contact angle. This behavior is again explained by the secondary interaction of the lone proton with water. The expected increase in water contact angle in the F-rich octafluoropentyl terminated oligomer is off-set by the interactions of the proton of the CF_2H-terminus.

Conclusion

Surface engineering of cellulose ester derivatives by blending with fluorine containing cellulose ester copolymers appears to follow expectations within the following framework.

The nature of the terminus on the F-containing substituent, CF_2H versus CF_3, substantially determines surface characteristics. The observed differences are explained with the presence of an electron deficient proton in the CF_2H-terminus which readily engages in secondary interactions, like for example hydrogen bonding. This overshadows any effects that stem from the difference in atomic F-density per substituent.

The type of copolymer architecture, random versus block design, has a significant impact on both, solubility and surface character. The observations are consistent with the view that fluorinated block copolymers can form micelles of the type shown in Figure 5. The micelles are thought to consist of a fluorine-rich core with a cellulose propionate corona. They are virtually insoluble in all common solvents, and they form stable colloids in solution and in solid state. This accounts for the observed modest influence of the blocky copolymers on contact angle. Random copolymers cannot form micelles because of the delocalized distribution of the F-containing groups. Therefore they are both soluble in appropriate solvents and more effective in increasing the contact angle.

Acknowledgment

This study was financially supported by a grant from the U. S. Department of Agriculture, CSREES Contract #96-35103-3835. This support is acknowledged with gratitude. Thanks is also given to Ms. Jody Jervis, Dep. WSFP, for the help with the molecular weight determination.

This paper represents part I of a publication series entitled "Smart Surfaces of Biobased Materials".

Literature Cited

1. Simmons, W.C. In *Smart Materials Technologies*. Simmons, W.C., Aksay, I.A., Huston, D.R., eds., Proceedings SPIE, SPIE: Bellingham, WA, 1997, Vol. 3040, pp 2-7
2. Lyman, D.J., Knutson K. In *Polymeric materials and pharmaceuticals for biomedical use*. Goldberg, E.P., Nakajima, A., eds., Academic Press, NY, 1980
3. Perez-Luna, V.H., Horbett, T.A., Ratner B.D. *J. Biomed. Mat. Res.*, 1994, 28, 1111
4. *Biomaterials and interfacial approach*. Hench, L.L.,. Ethridge, E.C., eds.. Biophysics and Bioengineering Series. Academic Press, NY, 1982, Vol. 4
5. Harrison J.H. *Am. Jt. Surg*, 1958, 95, 3
6. Lyman, D.J., Muir, W.M., Lee, I.J. *Trans. Am. Soc. Artif. Intern. Organs*. 1965, 11, 301,
7. Poncin-Epaillard, F., Legeay, G., Brosse, J. *J. Appl. Polym. Sci.*, 1992, 44, 1513,
8. Pan, D.H.,. Prest, W.M. Jr. *J. Appl. Phys.* 1985, 58 (8), 2861,
9. Hariharan, A., Kumar, S.K., Russell, T.P. *J. chem. Phys.* 1993, 98 (8), 6516,
10. Hariharan, A., Kumar, S.K., Russell, T.P. *Macromolecules*, 1991, 24, 4909,
11. Hariharan, A., Kumar, S.K., Russell, T.P. *J. chem. Phys.* 1993, 98 (5), 4163,
12. Hiemenz, P.C. *Polymer Chemistry. The basic concepts*. Marcel Dekker, Inc., NY, 1984
13. Cowie, J.M.G. *Polymers: Chemistry and physics of modern materials*. Blackie Academic and Professional, London, 1991
14. Sealey, J.E., Samaranayake, G., Todd, J.G.,. Glasser, W.G. *J. appl. Polym. Sci. B: Polym. Phys.*, 1996, 34, 1613,
15. Mezger, T.,. Cantow, H.-J. *Polym. Photochem.*, 1984, 5, 49,
16. deOliveira, W., Glasser, W.G. *Polymer*, 1994, 35 (9), 1977,
17. deOliveira, W., Glasser, W.G. *Cellulose*, 1994, 1, 77,
18. Wang, J.-H., Claesson, P.M., , Parker, J.L., Yasuda, H. *Langmuir*, 1994, 10, 3887

Chapter 24

Relation Between the Conditions of Modification and the Properties of Cellulose Derivatives: Thermogelation of Methylcellulose

J. Desbrieres, M. Hirrien, and M. Rinaudo

Centre de Recherches sur les Macromolécules Végétales (CERMAV-CNRS), affiliated with Joseph Fourier University, BP 53, 38401 Grenoble Cedex 9, France

The most important natural polysaccharides consist of cellulose, amylose and amylopectin produced by plants, and chitin from crustaceous shells. Because they are rich in hydrogen bonds, they are insoluble or have low solubility in water and have high cohesion. Usually their derivatization is performed in heterogeneous conditions leading to a heterogeneous distribution of the substituents along the macromolecular chain and non-reproducible properties (depending on their origin). In this work conditions for homogeneous chemical modification of cellulose are proposed. A specific substitution on C-2 and C-3 positions is also carried out. A wide spectrum of samples with different degrees of substitution and different distributions of substituted units is obtained. These results lead to a better understanding of the methylcellulose (MC) gelation mechanism and of the role of the chemical structure on the properties of MC solutions (either in dilute or semi-dilute regime). The experimental data presented allow discrimination among the proposed mechanisms. In particular, the gelation phenomenom is initiated by hydrophobic interactions involving zones of trisubstituted units.

Naturally occurring polymers such as cellulose, starch components, or chitin are produced each year on an enormous scale. Each offers advantages, such as being renewable and biodegradable, properties that are more and more valuable for environmental protection. Nevertheless, in their native forms they have a limited range of applications. Therefore, derivatives are needed in which both the nature of the substituent and the degree of substitution can be varied, in order to obtain material adequate for the considered application. For these important polysaccharides, derivatization is able to cover the range of solubilities in organic solvents and water or to produce thermoplastic materials.

Due to their structure and organization in plants or animals, polysaccharides lack solubility and usually result in heterogeneous chemical modifications. Such reactions lead to non uniform substituent distribution on repeating units. Heterogeneity of the substitution gives a poor predictability of the derivatives properties and may explain, for example, the variety and variability of the coefficients relating a physical property to the molecular weight found in the literature. To compete with synthetic polymers, polysaccharides must have a regular distribution of substituents and must produce samples with reproducible characteristics. In order to control the conditions of reaction, the reactions must be carried out in a homogeneous phase. In this paper the homogeneous conditions to prepare methylcellulose (MC) and the relationships between the structure of such derivatives and their physicochemical properties are presented. They are compared with commercial heterogeneously prepared samples, and the mechanism of the thermogelation of aqueous solutions is discussed.

Results and discussion

Reaction conditions. Generally the chemical modification of polysaccharides begins in heterogeneous conditions producing progressively soluble derivatized molecules which separate from the substrate when the synthesis is performed in solubilizing conditions for the derivatives. To perform a homogeneous substitution in all the range of degrees of substitution (DS), it is necessary to first solubilize the polysaccharide. This is, in our opinion, the most important step. Very few direct solvents are available especially for cellulose and chitin. The swelling and/or the dissolution in different experimental conditions depend on the morphology of the native substrate. The accessibility of the reactive groups will depend on this supermolecular structure in addition to the degree of crystallinity (diffusion control of the reactants...) and the intra- or interchain hydrogen bonds giving a modulation of the reactivity of the hydroxyl groups in 2, 3 or 6 positions. Only complete destruction, passing through a solution state, will enable the memory effect of the original state of organization to be avoided. Along the macromolecular chain, statistic substitution will be obtained, giving a well-defined derivative with predictable physical properties, especially in relation to the molecular weight and the degree of substitution.

For cellulose, the solvent DMAc/LiCl proposed from 1979 seems to be the most promising (*1,2*). The advantage of this mixture is to be a direct solvent without derivatization nor depolymerization. Usually the LiCl content adopted is between 5 and 9% wt/wt related to DMAc solvent, and the reactions are often carried out under mild conditions (*3*). Due to its polar aprotic character, a range of organic reactions is possible and in particular the modifications of hydroxyl groups. Many reactions are listed in the papers from McCormick and Dawsey (*1-3*). The role of the solvent is first to disrupt the hydrogen bond network and then to be a solvent of the polyhydroxylated polymers (i.e., to be polar). Another method usually applied for etherification uses alkaline conditions. The advantage of alkaline conditions (depending on the temperature and alkaline concentration) is not only to disrupt the supermolecular structure but also to form the alcoolate (Cell-O⁻) in concentrate conditions (pH>13) (*4*), which is the first step in etherification. These alkaline conditions remain powerful for etherification (carboxymethylcellulose production). Some time ago sodium in liquid ammonia was used to produce CMC with very high DS (up to 3). These conditions were adopted to obtain intercrystalline and crystalline -OH accessibility (*5*).

Industrial preparation of commercial methylcellulose (DS from 1.3 to 1.7) involves the use of a heterogeneous slurry of cellulose in a so-called solvent that swells but does not dissolve the polymer. It leads to a non-uniform substituent distribution on glucose repeat units along the chain and the presence of blocks of tri-, di-, mono, or unsubstituted units, especially for intermediate degrees of substitution due to different accessibilities of cellulose areas to the reagents (*6*). Hence, there is a poor control of the reaction and poor predictability of product properties.

It is possible, at a laboratory scale, to prepare methylcellulose under a homogeneous process allowing a uniform substituent distribution on cellulose units. The conditions adopted are presented in Tables I and II. The solvent used is DMAc/LiCl and dimethylsulfinyl anion as reactive intermediate (*7*). Moreover, specific methylation may be performed by intermediate specific tritylation on C-6 position (*7, 8*).

Table I : Conditions for methylcellulose preparation and consequences on substituent distribution

Industrial process	Laboratory process	
HETEROGENEOUS phase synthesis	HOMOGENEOUS phase synthesis	Specific methylation
Swelling in caustic medium + CH3I or CH3Cl	Cellulose dissolved in DMAc / LiCl 6% + sulfinyl anion + CH3I	Tritylation : selective on C-6 position Methylation with CH3I Detritylation Deacetylation
Limited -OH accessibility (surface of crystalline zones, amorphous zones)	Better -OH accessibility	
Substituent distribution by ZONES	REGULAR distribution of substituents	Methyl groups on C-2 and C-3 positions
Sample A4C	Samples M22, M15, M12, M29, M18	Sample S23

Table II : Experimental conditions for homogeneous synthesis of methylcelluloses

Degree of substitution (DS)	Cellulose / DMAc-LiCl (g/g)	NaH / DMSO (g / mL)	Methyl Iodide (mL)	Reaction Time (h)
1.2	5.8 / 800	15 / 80	20	24
1.7	6 / 740	20 / 100	30	36
2.1	6 / 700	10 / 50	20	7 days

The methylcellulose samples investigated have the substituent distributions given in Table III, where the fraction of the different modified monomeric units are listed.

Solubility properties (role of the substitution pattern). The role of the distribution of the methyl groups on solubility was previously mentioned. Miyamoto and co-workers demonstrated clearly that the degree of substitution necessary for water solubility is lower for homogeneously prepared materials than for similar materials prepared heterogeneously (9). This better solubility is probably caused by the more uniform distribution of the substituents of the homogeneously-derivatized products along the chain. In our work homogeneous water-soluble MC was prepared with DS between 0.9 and 2.2, but commercial samples need a degree of substitution larger than 1.3 to be water soluble. A water-soluble MC sample with a DS value of 0.9 is the minimum substitution that will suppress the intermolecular hydrogen bonds in cellulose. The better regularity of the substituent distribution is clearly demonstrated when we compare two products with the same average degree of substitution (1.7), but prepared under heterogeneous (A4C) or homogeneous (M29) conditions : in M29 there are fewer glucose units but a larger content of monomethyl derivatives with the same average amount of trimethylated units (Table III).

Methylcellulose solution properties were studied either in the dilute (the polymer concentration below the overlap concentration c*) or the semi-dilute regime.

Properties in the dilute regime. In dynamic light scattering experiments, we observed the size of the molecules in aqueous solution, expressed as the hydrodynamic radius R_H, as a function of the temperature (Figure 1). At 20°C and up to 50°C the correlation function exhibits a single relaxation, and the inverse relaxation time is q^2-dependent, indicating a diffusive relaxation. Only one size of molecule is present in solution, with R_H around 20 nm, of the same order of magnitude as the radius of gyration of similar cellulosic derivatives. At 60°C the 20 nm molecules are still present but in smaller quantity, and a second population appears (R_H around 200 nm) corresponding to the formation of aggregates.

The evolution of the refractive index signal of the material eluted in Steric Exclusion Chromatography for different MC samples is examined as a function of temperature (10). For the lowest DS (smaller than 1.5) the chromatograms do not change much when temperature increases; on the contrary, for A4C and higher DS samples new peaks appear for lower elution volumes corresponding to molecules with higher dimensions. Each sample was filtered off before the experiments and maintained one night at the chosen temperature. The quantity of the polymer actually eluted through the columns is indeed a good indication of the formation (or not) of aggregates retained at the front of the column when the temperature increases. The ratio between m_c, the eluted weight calculated from the area of the refractive index peak and the refractive index increment, and m_i, the injected weight of sample, is a good indication of the ratio of polymer which is eluted (Figure 2). At low temperature this ratio is close to 1, indicating all the polymer is eluted and confirming that the methylcellulose solution is a true solution. Two behaviours were obtained according to the DS values. For DS smaller than 1.5, this ratio does not vary significantly while, that for DS larger than 1.5, this ratio decreases greatly above 45°C. For temperatures higher than this value, interactions occur leading to the formation of aggregates.

The presence of interactions has also been observed using fluorescence spectroscopy. Pyrene was used as a probe because the vibrational structure of its fluorescence emission spectrum is sensitive to the polarity of its environment (11). The ratio between I_1 and I_3, respectively, the intensities of the first (at 373 nm) and the third (at 383 nm) peak of the fluorescence emission spectrum, is equal to 0.6 in hexane and

Table III : Characterization of some methylcelluloses prepared.

Sample	A4C	M22	M15	M12	M29	M18	S23
$\overline{DS}$ a	1.7	1.2	1.3	1.5	1.7	2.2	1.3
% Non S	10	9	9	5	4	9	16
% MonoS	29	68	63	51	45	12	47
% DiS	39	19	21	29	32	36	36
% TriS	22	4	7	15	19	43	-

a   determination by 13C n.m.r. in DMSO-d6 (353K)

338

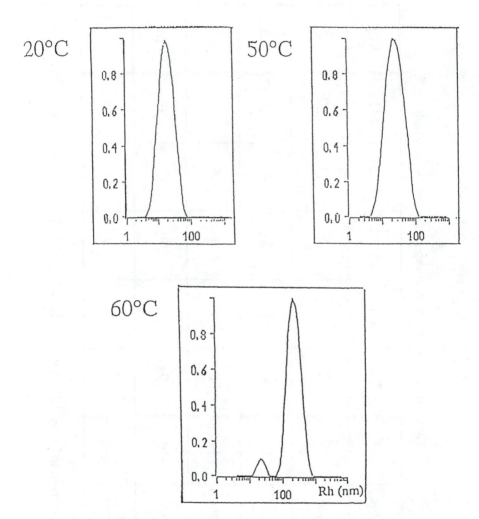

Figure 1. Evolution of the hydrodynamic radius (R_H) of A4C in water with temperature (DS=1.7)

(a) $\overline{DS}$ < 1.5

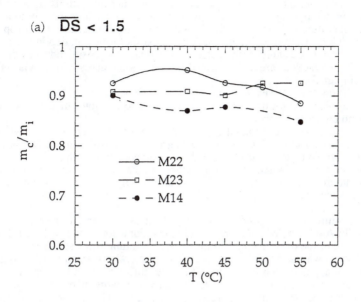

(b) $\overline{DS}$ > 1.5

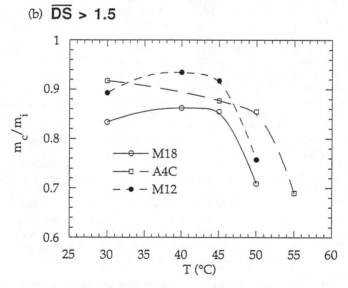

Figure 2. Evolution of the mass loss of the polymer eluted in steric exclusion chromatography of methylcellulose samples in water with temperature (a) DS < 1.5, (b) DS > 1.5

1.9 in water. Experiments with aqueous MC solutions were performed up to a temperature at which turbidity is observed. For A4C in water this ratio is equal to 1.8 at 10°C, signifying that pyrene is in a polar environment, and decreases quickly to a value of 1.5 for temperatures higher than 50°C, indicating the formation of hydrophobic domains in which pyrene stays (Figure 3). A hysteresis is formed on cooling as a proof of interactions established at higher temperature. The relative viscosity indicates the same mechanism; the interactions are demonstrated from the large increase in the viscosity of the solution (Figure 3).

From the dilute regime experiments one can conclude :
* methylcellulose in water forms, at low temperature, a true solution which can be filtered and contains no undissolved matter,
* for samples with DS larger than 1.5, observable aggregates are formed when the temperature is greater than 45°C; for lower degrees of substitution, no formation of aggregates is observed up to 55-60°C within the experimental time scale,
* as interactions occur when the temperature increases, they are essentially of hydrophobic nature, as is confirmed by the role of electrolytes such as sodium chloride which decreases the gelation temperature (*12*).

Properties in semi-dilute regime. Oscillatory shear experiments were carried out; a typical curve for A4C sample is presented in Figure 4. For a concentration larger than 2.5 g/L (in the range of the overlap concentration), the evolution of the storage modulus G' shows two distinct waves. At low temperature the solution is clear, and in the range 30-50°C a weak gel appears; then the solution becomes turbid above a temperature which depends on the concentration. An elastic turbid gel with a large increase of viscosity is observed above 60°C within the experimental time scale. This second wave is more pronounced with the higher polymer concentrations. The concentration 2.5 g/L may be considered for the A4C sample as the critical polymeric concentration needed for gelation. It is surprising that the wave at low temperature is apparently not dependent on the polymer concentration.

The gelation phenomenom is time dependent and reversible (*10*). The influence of the substitution distribution and of the DS values was studied further using the same procedure as for the A4C sample (Figure 5). With the low DS samples (DS < 1.5) only one wave was observed, with a slight increase of G' for temperatures higher than 60°C. For the sample without trisubstituted units (S23), G' remains constant with temperature; G" decreases following the viscosity of the solvent. Hysteresis was also observed in relation to the presence of cooperative interchain interactions, except for the S23 sample (*10*). Furthermore, with low DS samples the temperature for which G' is equal to G" is shifted toward higher temperatures. For the sample with the highest DS value (2.2), a behaviour similar to that of A4C was observed in the presence of two waves in the temperature domain studied, a finding opposite to that of several authors (*13*). With this degree of substitution, the proportion of trisubstituted units is relatively high and the presence of blocks of such units may be suspected. Moreover, a sol-gel transition may be observed even with methylcellulose samples with low DS ($\cong$ 1.3) prepared from an homogeneous process. The evolution of G' and G" with frequency for a solution of a sample with DS of 1.3 at 70°C is typical of a weak gel state rheogram (Figure 6). At low frequency the G' value is nearly constant and higher than G". The obtained gel is not turbid at this temperature.

Calorimetric experiments were carried out on methylcellulose solutions as were performed on other polysaccharide gels (*14,15*). With the A4C sample, during heating an endothermic peak is observed but two exothermic peaks are present during cooling (Figure 7). Only the energy of interaction associated with the phase separation (and the large increase in viscosity) is observed on heating. It is necessary to heat up to 50°C and 65°C to observe peaks at 30°C and 40°C respectively on cooling (*12*). These two peaks seem to correspond to interactions of a different nature : the peak at 30°C seems to be

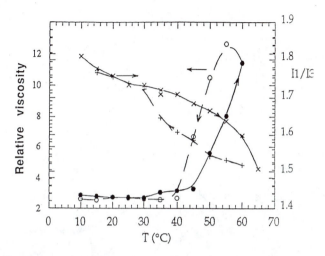

Figure 3. Influence of temperature on fluorescence (I_1/I_3 ratio) and relative viscosity for A4C methylcellulose (DS=1.7, solvent water, c=2g/L)

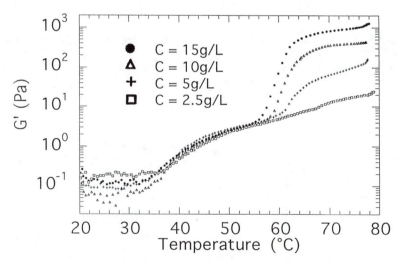

Figure 4. Influence of polymer concentration and temperature on the elastic modulus of A4C solutions (G' (Pa) modulus at 1 Hz, DS=1.7, solvent water), reproduced, by permission, from *Journal of Chimie Physique*, Elsevier, ref. 10

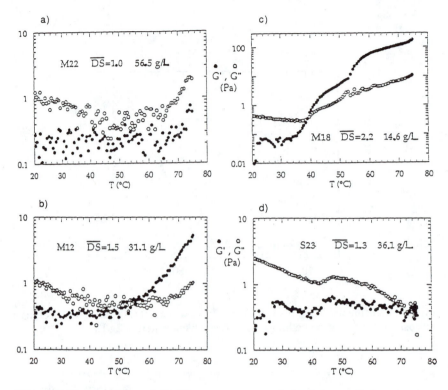

Figure 5. Rheological moduli of homogeneous methylcellulose solutions in water as a function of temperature (G' and G" moduli at 1 Hz) : (a) M22, DS=1.0, c=56.5 g/L, (b) M12, DS=1.5, c=31.1 g/L, (c) M18, DS=2.2, c=14.6 g/L, (d) S23, DS=1.3, c=36.1 g/L

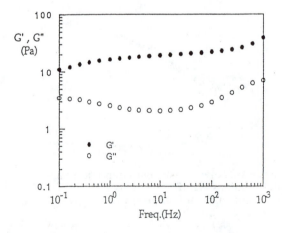

Figure 6. Influence of frequency on G' and G" rheological moduli (M15, DS=1.3, c=19.8 g/L , T=70°C, solvent water), reproduced, by permission, from *Journal of Chimie Physique*, Elsevier, ref. 10

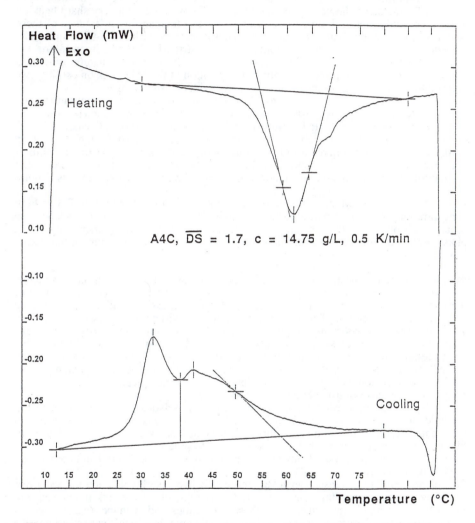

Figure 7. Calorimetric thermograms of aqueous solutions of A4C methylcellulose (DS=1.7, c=14.75 g/L, temperature rate 0.5 deg/min)

related to interactions involving blocks of trisubstituted units, and the 40° peak to di- and monosubstituted units. This assumption is proposed from the observation of thermograms performed with MC samples containing low trisubstituted units (12).

For homogeneously prepared methylcelluloses with a degree of substitution of 2.2 (i.e., for the same rheological behaviour as A4C), we observe a very sharp peak on heating and only one peak on cooling (Figure 8). The temperature difference of these peaks is much smaller than those characterizing hysteresis with the A4C sample. The presence of one peak indicates an homogeneous distribution of substituent along the chain, but with commercial samples the presence of blocks of different natures of units is revealed by the different peaks on cooling. The position of the M18 peak on cooling also indicates a smaller energy of interaction compared with the A4C sample.

From fluorescence spectroscopy M18 and A4C have a similar behaviour (Figure 9). An increase of interactions was observed related to the decrease of the I_1/I_3 ratio at a temperature very close to the temperature for which a modification of behaviour occurs in calorimetry and rheology. But this technique can be carried further. Even on methylcellulose substituted on only 2- and 3- positions, a decrease of the ratio is obtained; it is the only technique that indicates the presence of loose hydrophobic interactions without gelation, as demonstrated from the rheology experiments (Figure 10).

Nevertheless, when solutions of S23 or methylcellulose with low DS at high concentration (60 g/L) are placed in an oven at 90°C for a long time (up to weeks) phase separation is observed. This phenomenon, which is shared by most polymers, means that phase separation related to substitution is promoted by temperature increase but with a low kinetics.

Mechanism of gelation. For many years mechanisms of thermogelation of aqueous methylcellulose solutions were proposed. The major discussions concern the nature of the zones responsible for gelation. Savage et al. (16) express the ability to gel from the presence of zones coming from the original cellulosic structure; they were refuted by Heyman (17) who has studied highly substituted methylcelluloses. Rees (18) speaks about micellar interactions and Sarkar (19) postulates that gelation is due to hydrophobic or micellar interactions. Khomutov et al. (20) proposed the gelation is due to crystallization, Haque and Morris (21) imply crystalline zones within the gelation process while Kato et al. conclude that the "crosslinking loci" of methylcellulose gels consist of crystalline sequences of trimethylglucose units (22).

From all these observations and from comparisons between the different methylcellulose samples (prepared from heterogeneous or homogeneous processes) with different degrees of substitution and different structural characteristics, a mechanism of themogelation of aqueous methylcellulose solutions is proposed :
* we start at low temperature from a true solution which can be filtered and contains no undissolved matter,
* when temperature increases, whatever the sample, hydrophobic interactions occur due to methyl groups. According to the structure (distribution of substituents and presence or not of blocks of highly substituted units) these interactions lead to an increase of viscosity and sometimes to the presence of small hydrophobic domains. In this case these are due to the presence of zones of predominant trisubstituted units leading to a clear gel.
* increasing the temperature more leads to phase separation and turbidity : polymer-concentrated domains connect and a turbid gel is obtained. The temperature at which phase separation or "gelation" occurs depends upon the molecular weight of the macromolecular chain and the polymer concentration of the solution besides the structure.

Whatever the sample, the same steps are observed but with modified characteristics depending on the substitution : the critical temperatures for physical changes are displaced to higher values when the substitution is more regular even if the average substituent content is the same (due to the presence of less highly substituted zones). But in any case the presence of trisubstituted units is compulsory for observable gelation (high viscosity, turbidity, and elastic character).

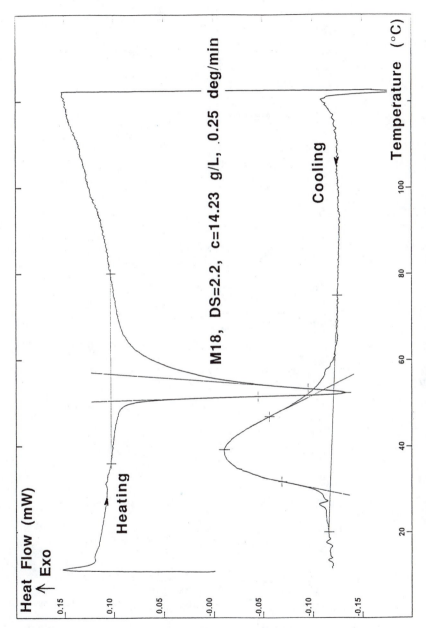

Figure 8. Calorimetric thermograms of M18 sample (DS=2.2, c=14.23 g/L, temperature rate 0.25 deg/min)

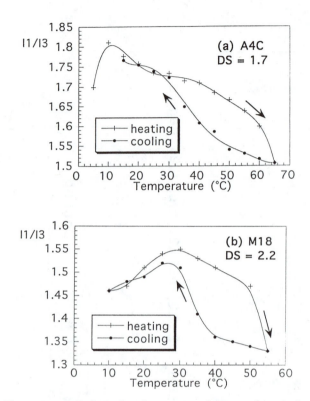

Figure 9. Fluorescence I_1/I_3 ratio of aqueous solutions of methylcellulose : (a) A4C, DS=1.7, c=7.4 g/L, (b) M18, DS=2.2, c=4.9 g/L

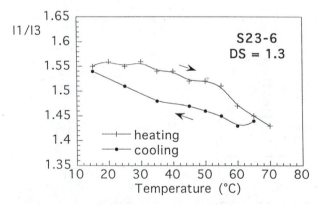

Figure 10. Fluorescence I_1/I_3 ratio of aqueous solutions of S23 methylcellulose (DS=1.3, c=7.4 g/L)

Conclusion

The distribution of the substituents clearly controls the main properties of methylcellulose. This distribution depends on the conditions for chemical substitution, i.e. whether conditions are homogeneous or heterogeneous. The solubility needs less methyl substituent regularly disposed along the chain avoiding the packing of cellulosic chains by cooperative hydrogen bonds. The stability of the junctions, a consequence of hydrophobic interactions, in the clear gel phase is directly related to the existence of highly substituted blocks and mainly exist in the commercial samples obtained by heterogeneous chemical modification. The rheology depends directly not only on the average degree of substitution but on the substituent distribution.

Experimental

The commercial methylcellulose sample (Methocel A4C) was kindly provided by Dow Chemical.

The synthesis of homogeously prepared methylcellulose samples is carried out according to the procedure previously described (7).

^{13}C n.m.r. measurements were realized on a Bruker AC300 spectrometer at 353K using DMSO-d6 as solvent.

Total hydrolysis of methylcellulose was carried out with 2N trifluoroacetic acid within 4 hours at 120°C. After evaporation of the acid in vacuo, water was distilled from the samples ten times to remove traces of the acid before h.p.l. chromatography.

H.p.l.c. analyses were performed with a Waters-Millipore chromatograph equipped with a differential refractometer detector (Waters 410). A reversed-phase column Nucleosil C18 5μ (250*4.6 mm) with water as eluent at ambient temperature was used. This technique gives the content in unsubstituted and substituted monomers.

The aqueous solutions were prepared by dissolution of freeze dried methylcellulose in water at 5°C over 24 hours to assure a complete solubilization.

The quasi-elastic scattering experiments were performed using the ALV apparatus. The scattered light of a vertically polarized $\lambda_0 = 488$ nm argon laser (Spectra Physics model 2020, 3W, operating around 0.3W) was measured at different scattering angles. The autocorrelation function of the scattered intensity was obtained using the ALV-5000 autocorrelator and was analyzed by the constrained regularization method (CONTIN). The results presented are obtained for a scattering angle of 130°.

Steric Exclusion Chromatography (SEC) experiments were carried out using water as eluent and the multidetection equipment described previously (23). The refractive index increment dn/dc for the MC is taken to equal 0.136. The experiments were performed on grafted silica gel columns (Shodex OHpak B-804, B-805).

Fluorescence experiments were performed using a LS50B luminescence spectrometer from Perkin Elmer.

Oscillatory shear measurements were carried out using a CS50 rheometer from Carri-Med in the linear regime at a frequency of 1 Hz (25% deformation) during heating and cooling scans at 0.5 deg/min.

Calorimetric experiments were performed using a micro DSC III calorimeter from Setaram. The temperature rate was 0.5 or 0.25 deg/min.

Literature Cited

1   Dawsey, T.R. In *Cellulosic polymers, blends and composites*; Gilbert, R.D. Ed; Hanser : Munich, 1994, 157
2   Dawsey, T.R. *Rev. Macromol. Chem. Phys.* **1990**, *C30* (34), 405

3   Mc Cormick, C.L.; Callais, P. *Polymer* **1987**, *28*, 2317
4   Elmgren, H; Norrby, S. *Die Makromol. Chem.* **1969**, *123*, 265
5   Hudry Clergeon, G.; Rinaudo, M *J. Chim. Phys.* **1967**, *64*, 1746
6   Nevell, T.P.; Zeronian, S.H. In *Cellulose chemistry and its applications*; Nevell, T.P., Zeronian, S.H., Edts, John Wiley : New York, 1985, 15
7   Hirrien, M.; Desbrieres, J.; Rinaudo, M *Carbohydr. Polym.* **1996**, *31*, 243
8   Kondo, T.; Gray, D.G. *Carbohydr. Res.* **1991**, *220*, 173
9   Takahashi, S.I.; Fujimoto, T.; Miyamoto, T.; Inagaki, H. *J. Polym. Sci. Part A : Polym. Chem* **1987**, *25*, 987
10 Vigouret, M.; Rinaudo, M.; Desbrieres, J. *J. Chim. Phys.* **1996**, *93*, 858
11 Kalyanasundaram, K.; Thomas, J.K. *J. Amer. Chem. Soc.* **1977**, *99*, 2039
12 Hirrien, M.Thesis, Grenoble, 1996
13 Savage, A.B. *Ind. Eng. Chem.* **1957**, *49*, 99
14 Watase, M.; Nishinari, W.; Clark, A.H.; Ross Murphy, S.B. *Macromolecules* **1989**, *22*, 1196
15 Watase, M.; Nishinari, W. *Makromol. Chem.* **1987**, *188*, 1177
16 Savage, A.B.; Young, A.E.; Maasberg in *Cellulose and cellulose derivatives. Part II*; Ott, E., Spurlin, H.M., Grafflin, M.W., Ed; Interscience : New York, 1963; 904
17 Heymann, E. *Trans. Faraday Soc.* **1935**, *31*, 846
18 Rees, D.A. *Chem. Ind. London* **1972**, 630
19 Sarkar, N. *J. Appl. Polym. Sci.* **1979**, *24*, 1073
20 Khomutov, L.I.; Ryskina, I.I.; Panina, N.I.; Dubina, L.G.; Timofeeva, G.N. *Polym. Sci.* **1993**, *35*, 320
21 Haque, A.; Morris, E.R. *Carbohydr. Polym.* **1993**, *22*, 161
22 Kato, T.; Yokoyama, M.; Takahashi, A. *Colloid and Polym. Sci.* **1978**, *256*, 15
23 Tinland, B.; Mazet, J.; Rinaudo, M. *Makromol. Chem., Rapid Comm.* **1988**, *9*, 69

INDEXES

Author Index

Subject Index

Bestsellers from ACS Books

The ACS Style Guide: A Manual for Authors and Editors (2nd Edition)
Edited by Janet S. Dodd
470 pp; clothbound ISBN 0–8412–3461–2; paperback ISBN 0–8412–3462–0

Writing the Laboratory Notebook
By Howard M. Kanare
145 pp; clothbound ISBN 0–8412–0906–5; paperback ISBN 0–8412–0933–2

Career Transitions for Chemists
By Dorothy P. Rodmann, Donald D. Bly, Frederick H. Owens, and Anne-Claire Anderson
240 pp; clothbound ISBN 0–8412–3052–8; paperback ISBN 0–8412–3038–2

Chemical Activities (student and teacher editions)
By Christie L. Borgford and Lee R. Summerlin
330 pp; spiralbound ISBN 0–8412–1417–4; teacher edition, ISBN 0–8412–1416–6

Chemical Demonstrations: A Sourcebook for Teachers, Volumes 1 and 2, Second Edition
Volume 1 by Lee R. Summerlin and James L. Ealy, Jr.
198 pp; spiralbound ISBN 0–8412–1481–6
Volume 2 by Lee R. Summerlin, Christie L. Borgford, and Julie B. Ealy
234 pp; spiralbound ISBN 0–8412–1535–9

The Internet: A Guide for Chemists
Edited by Steven M. Bachrach
360 pp; clothbound ISBN 0–8412–3223–7; paperback ISBN 0–8412–3224–5

Laboratory Waste Management: A Guidebook
ACS Task Force on Laboratory Waste Management
250 pp; clothbound ISBN 0–8412–2735–7; paperback ISBN 0–8412–2849–3

Reagent Chemicals, Eighth Edition
700 pp; clothbound ISBN 0–8412–2502–8

Good Laboratory Practice Standards: Applications for Field and Laboratory Studies
Edited by Willa Y. Garner, Maureen S. Barge, and James P. Ussary
571 pp; clothbound ISBN 0–8412–2192–8

For further information contact:
Order Department
Oxford University Press
2001 Evans Road
Cary, NC 27513
Phone: 1-800-445-9714 or 919-677-0977
Fax: 919-677-1303

Highlights from ACS Books

Desk Reference of Functional Polymers: Syntheses and Applications
Reza Arshady, Editor
832 pages, clothbound, ISBN 0–8412–3469–8

Chemical Engineering for Chemists
Richard G. Griskey
352 pages, clothbound, ISBN 0–8412–2215–0

Controlled Drug Delivery: Challenges and Strategies
Kinam Park, Editor
720 pages, clothbound, ISBN 0–8412–3470–1

Chemistry Today and Tomorrow: The Central, Useful, and Creative Science
Ronald Breslow
144 pages, paperbound, ISBN 0–8412–3460–4

Eilhard Mitscherlich: Prince of Prussian Chemistry
Hans-Werner Schutt
Co-published with the Chemical Heritage Foundation
256 pages, clothbound, ISBN 0–8412–3345–4

Chiral Separations: Applications and Technology
Satinder Ahuja, Editor
368 pages, clothbound, ISBN 0–8412–3407–8

Molecular Diversity and Combinatorial Chemistry: Libraries and Drug Discovery
Irwin M. Chaiken and Kim D. Janda, Editors
336 pages, clothbound, ISBN 0–8412–3450–7

A Lifetime of Synergy with Theory and Experiment
Andrew Streitwieser, Jr.
320 pages, clothbound, ISBN 0–8412–1836–6

Chemical Research Faculties, An International Directory
1,300 pages, clothbound, ISBN 0–8412–3301–2

For further information contact:
Order Department
Oxford University Press
2001 Evans Road
Cary, NC 27513
Phone: 1-800-445-9714 or 919-677-0977
Fax: 919-677-1303